A cluster of intellectual innovations appearing within scholastic natural philosophy in the fourteenth century played a critical role in the future development of scientific thought. Beneath these innovations lay a profound reconceptualization of nature. This book attempts to analyze the components of this reconceptualization and to uncover the pressures and concerns that shaped it. To do so, it looks both within the university and beyond it, to the monetized society that surrounded and supported it. It argues that the transformation of the conceptual model of the natural world within scholastic natural philosophy c. 1260–1380 was directly linked to the social and economic process of monetization that transformed European society over this same period. It illustrates how those perceptual shifts essential to the emergence of modern scientific thought – the shifts toward quantification, geometric representation, multiplication, relativity, probability, mechanistic order, and dynamic equilibrium – were grounded in the experience and comprehension of monetized society.

The book's earlier chapters analyze scholastic writings on economic questions (particularly those found in commentaries on Aristotle's discussion of exchange in *Nicomachean Ethics* v.5), focussing on the six new categories of analysis and description devised by philosophers to make sense of their experience of monetized society. The concluding chapters investigate the transmission of philosophical insights from the comprehension of the monetized marketplace to the comprehension and construction of nature. They reveal how intellectual developments pioneered within each of the six new categories of economic analysis lie at the base of the most forward-looking conceptual advances within scholastic natural philosophy – advances that proved to be crucial to the further development of scientific thought.

D0907314

Cambridge Studies in Medieval Life and Thought: Fourth Series

ECONOMY AND NATURE
IN THE FOURTEENTH CENTURY

Cambridge Studies in Medieval Life and Thought
Fourth series

General Editor:

D. E. LUSCOMBE

Leverhulme Personal Research Professor of Medieval History, University of Sheffield

Advisory Editors:

R. B. DOBSON

Professor of Medieval History, University of Cambridge, and Fellow of Christ's College

ROSAMOND MCKITTERICK

Professor of Early Medieval European History, University of Cambridge, and Fellow of Newnham College

The series Cambridge Studies in Medieval Life and Thought was inaugurated by G. C. Coulton in 1921. Professor D. E. Luscombe now acts as General Editor of the Fourth Series, with Professors R. B. Dobson and Rosamond McKitterick as Advisory Editors. The series brings together outstanding work by medieval scholars over a wide range of human endeavour extending from political economy to the history of ideas.

For a list of titles in the series, see end of book.

GRACE LIBRARY CARLOW COLLEGE
PITTSBURGH, PA 15213

ECONOMY AND NATURE IN THE FOURTEENTH CENTURY

Money, market exchange, and the emergence of scientific thought

JOEL KAYE
Barnard College

HG
220
A2
K39
2000

CAMBRIDGE
UNIVERSITY PRESS

CATALOGUED

PUBLISHED BY THE PRESS SYNDICATE OF THE UNIVERSITY OF CAMBRIDGE
The Pitt Building, Trumpington Street, Cambridge CB2 1RP, United Kingdom

CAMBRIDGE UNIVERSITY PRESS
The Edinburgh Building, Cambridge CB2 2RU, United Kingdom
40 West 20th Street, New York, NY 10011-4211, USA
10 Stamford Road, Oakleigh, Melbourne 3166, Australia

© Joel Kaye 1998

This book is in copyright. Subject to statutory exception
and to the provisions of relevant collective licensing agreements,
no reproduction of any part may take place without
the written permission of Cambridge University Press.

First published 1998
First paperback edition 2000

Printed in the United Kingdom at the University Press, Cambridge

Typeset in Bembo 11/12 pt [VN]

A catalogue record for this book is available from the British Library

Library of Congress cataloguing in publication data
Kaye, Joel, 1946–
Economy and nature in the fourteenth century : money, market exchange, and the emergence
of scientific thought / Joel Kaye.
p.   cm. – (Cambridge studies in medieval life and thought : 4th ser., v. 35)
Includes bibliographical references.
ISBN 0 521 57276 2 (hardbound)
1. Money – History.   2. Exchange – History.   3. Science, Medieval – Philosophy.
I. Title.   II. Series.
HG220.A2K39   1998
332.4'9—dc21   97–10271   CIP

ISBN 0 521 57276 2 hardback
ISBN 0 521 79386 6 paperback

CONTENTS

Contents

ACKNOWLEDGMENTS

I would like first to acknowledge my debt to Edward Peters, Ruth Mazo Karras, and R. Dean Ware, without whose initial support this book could not have been written. I have worked on this project for the last five years while teaching at Barnard College, and I want to thank my colleagues here for the warm and intellectually alive atmosphere they have created. I am only sorry that Bill McNeill, who helped bring me to Barnard and encouraged me in my project, is not alive to see its conclusion.

Having the opportunity to talk with and come to know Caroline Bynum, my colleague across the street at Columbia University, has been my privilege and my great good fortune. I am profoundly grateful to her for her friendship, advice, and encouragement, and for the generosity she has shown me over these past five years. Other colleagues at Barnard and Columbia have also been generous with their time and expertise: I am indebted to Robert Somerville for his many kindnesses, including his careful reading of my chapters touching on Roman and canon law; Wim Smit read and offered valuable comments on the entire manuscript; Christopher Baswell, Serge Gavronsky, Robert Hanning, Martha Howell, John Mundy, Herbert Sloan, Lisa Tiersten, and Lars Trägårdh have either commented on particular chapters or helped me with particular points. My student assistant, Sophie Forrester Bolt, provided valuable editorial advice and was a great help in the production of the manuscript.

Outside the Barnard/Columbia community, I want to express my great debt to Lester Little for reading the manuscript and for offering me his gracious support. Julius Kirshner kindly read the chapters on economic history and provided valuable suggestions. Kate Cooper and Conrad Leyser have helped me with their advice and support at many points. I am indebted to Steven Marrone for the care he put into his detailed and constructive comments on the entire manuscript and, similarly, to David Luscombe, the editor of this series, for the time and concern he has given to this project.

ix

Acknowledgments

Since the origins of this work stretch back to graduate school, I would like to acknowledge my debt to the Mellon Foundation for awarding me two years of Dissertation Fellowship, and to the University of Pennsylvania for the Penfield Grant that allowed a year of study in European libraries. Barnard College's enlightened policy of making leaves available to its assistant professors gave me the year I needed (1995–96) to give final form to the book. During this year I was able to complete my manuscript research at the Vatican Library through the assistance of the American Academy in Rome, where I was in residence as a visiting scholar.

I would like finally to thank my friends, my sisters and their wonderful families, and my parents whose memories I hold dear. Above all, I want to thank my wife, Thea, who has shared this journey with me and has helped me with wonderful grace and humor at every step. I dedicate this book to her with love.

INTRODUCTION

Intellectual innovations within fourteenth-century natural philosophy occupy an important place in the history of scientific thought. Over the course of the century, philosophers subjected elements of the Aristotelian model of the natural world to critical analysis, advancing claims of logic, mathematical consistency, and empirical evidence against Aristotelian authority. The selective critique of Aristotle was informed by a wealth of new questions and innovative speculations. Beneath these speculations lay a profound reconceptualization of nature.[1]

In broad terms, the conceptual landscape that emerged in the fourteenth century resulted from a striking shift in the models derived to represent order and activity in the natural world: from a static world of numbered points and perfections to a dynamic world of ever-changing values conceived as continua in expansion and contraction; from a mathematics of arithmetical addition to a mathematics of geometrical multiplication, newly accepting of the approximate and the probable; from a world of fixed and absolute values to a shifting, relational world in which values were understood to be determined relative to changing perspectives and conditions; and from a philosophy focused on essences and perfections to one dominated by questions of quantification and measurement in respect to motion and change. Each of these new directions proved to be of great importance to the future of scientific thought.[2]

---

[1] Marshall Clagett, "Some Novel Trends in the Science of the Fourteenth Century," in Charles Singleton (ed.), *Art, Science, and History in the Renaissance* (Baltimore, 1967), 275–303, esp. 302 – 03; Anneliese Maier, "La doctrine de Nicolas d'Oresme sur les 'configurationes intensionum,'" in Maier, *Ausgehendes Mittelalter*, 3 vols. (Rome, 1964–77), vol. I, 335–52, esp. 336–37. For additional bibliography on this point, see chapter 6.

[2] For an assessment of the importance of several of these innovations to science – geometricization of space, concepts of relativity, a mathematics of the approximate and of limits – see Alexandre Koyré, *From the Closed World to the Infinite Universe* (Baltimore, 1957), 1–27. Koyré, however, situates these innovations later than the fourteenth century in almost all cases. For an appreciation of the

Introduction

Proto-scientific speculation in the fourteenth century developed within the rigorous intellectual culture of the university, particularly at Paris and Oxford. The abstraction, logical density, and technical complexity characterizing this speculation engage and impress logicians and historians of science to this day. Given the highly refined and formal intellectualism of scholastic natural philosophy, it is understandable that with few exceptions historians of medieval science have hesitated to step outside the sphere of intellectual culture in their search for the factors influencing its development.

This book suggests a broader historical explanation for the new directions taken in fourteenth-century natural philosophy than has so far been offered. It argues that the transformation of the conceptual model of the natural world, accomplished within the technical disciplines of the universities of Oxford and Paris c. 1260–1380, was strongly influenced by the rapid monetization of European society taking place over this same period, beyond the university and outside the culture of the book. It analyzes the impact of the monetized marketplace on the most striking and characteristic concern of natural philosophy in this period: its preoccupation with measurement, gradation, and the quantification of qualities. It investigates the transference of insights from the philosophical comprehension of the monetized marketplace to the philosophical comprehension of nature. It traces how those perceptual shifts essential to the emergence of modern science – the shift toward quantification, geometric representation, multiplication, relativity, probability, mechanistic order, and dynamic equilibrium – were grounded in the experience and comprehension of monetized society.

In the early years of the fourteenth century, English natural philosophers associated with Merton College, Oxford, initiated a vital new approach to the study of motion and qualitative change.[3] These scholars, now known collectively as the "Merton School" or the "Oxford Calculators," constructed a highly technical logic and mathematics of measurement.[4] They applied mathematical rules and quantitative schemata to a

contributions of fourteenth-century thinkers in these areas, see John Murdoch, "From Social into Intellectual Factors: An Aspect of the Unitary Character of Late Medieval Learning," in John Murdoch and Edith Sylla (eds.), *The Cultural Context of Medieval Learning* (Dordrecht and Boston, 1975), 271–348, esp. 287. James Weisheipl characterized fourteenth-century developments in natural philosophy at Oxford as a "revolution in scientific thought," in "The Place of John Dumbleton in the Merton School," *Isis* 50 (1959), 439–54, 439.

[3] Edith Sylla, "Medieval Quantifications of Qualities: The 'Merton School,'" *Archive for History of Exact Sciences* 8 (1971), 7–39; James Weisheipl, "Developments in the Arts Curriculum at Oxford in the Early Fourteenth Century," *Mediaeval Studies* 28 (1966), 151–75; Marshall Clagett, *The Science of Mechanics in the Middle Ages* (Madison, 1959), 206.

[4] John Murdoch, "*Mathesis in philosophiam scholasticam introducta*: The Rise and Development of the Application of Mathematics in Fourteenth-Century Philosophy and Theology," in *Arts libéraux et*

wide range of philosophical questions concerning qualities and motions, including the question of motion in space.[5] In the process of refining their logico-mathematical analysis of qualitative change, the Calculators laid the foundations for a future mathematical physics.

By the second quarter of the fourteenth century, masters at the University of Paris began to adopt the intellectual interests and methods of the English Calculators. As they did so, the passion to measure and quantify that characterized the proto-science of *calculationes* quickly invaded every realm of scholastic thought, including theology. Soon not only entities that had never been measured before, but also those that have never been measured since, were subjected to a kind of quantitative analysis. Theological questions concerning the most subjective and seemingly immeasurable qualities, such as the strength of Christian charity, or the comparison of human love to Christ's love, or the means by which the quality of grace increases in the soul, were routinely treated as problems of quantification, and subjected to analysis according to the latest developments in the logic and mathematics of measurement.[6]

Struck by the application of measuring schemata to the solution of philosophical and theological problems, John Murdoch asked: "How and why did the near frenzy to measure everything imaginable come about in the fourteenth century?"[7] Murdoch's question, first posed more than two decades ago, is central to this present study. How are new conceptual possibilities created? Why are new intellectual problems and approaches suddenly brought to prominence? With characteristic forthrightness, Murdoch admitted that he could not answer with certainty how a measuring "mania" came to dominate speculation in this period, but he suggested a number of possibilities. His first suggestion pointed to a creative dynamic produced solely from within the logic of the intellectual-philosophical debate.[8] His second suggestion broadened the range of influence to include the "catalyst" of theological concerns.[9] Here he

philosophie au moyen âge: Actes du Quatrième Congrès international de philosophie médiévale (Montreal, 1969), 215–46. For additional bibliography on the Calculators, see chapters 6 and 7.

[5] The conceptual linking of motion and qualitative change is central to the natural philosophy of this period and particularly characteristic of the Calculators' work. Weisheipl ("Dumbleton," 447) notes, for example, that the Calculator Thomas Bradwardine treated velocity as if it were a qualitative ratio capable of being intensified or lessened in the same manner as color and heat. On this linkage, see Clagett, *Science of Mechanics*, 199–219.

[6] John Murdoch, "*Subtilitates Anglicanae* in Fourteenth-Century Paris: John of Mirecourt and Peter Ceffons," in Madeleine Pelner Cosman and Bruce Chandler (eds.), *Machaut's World: Science and Art in the Fourteenth Century* (New York, 1978), 51–86; Murdoch, "Unitary Character," esp. 280–303; Murdoch, "*Mathesis*," 238ff. The Calculators' application of mathematical rules to theological problems is discussed further in chapters 6 and 7.   [7] Murdoch, "Unitary Character," 287.

[8] Murdoch (*ibid.*) stresses the influence of the philosophical thought of Duns Scotus and William of Ockham. For the unifying influence on scholastic philosophy of Aristotelian authority, see Murdoch, "Unitary Character," 308.   [9] Murdoch, "Unitary Character," 288–89.

3

Introduction

cited, above all, the many dimensions of measurement involved in the central theological question of the relation of finite humanity to an infinite God.[10]

In explaining the new frenzy to measure qualities primarily as the working out of problems generated from within the philosophical tradition, Murdoch articulates a position on philosophical and scientific innovation prevalent among historians of medieval science. Murdoch goes further, however, when he suggests as a third area of influence the social and intellectual milieux of the university.[11] Edith Sylla has investigated this connection between philosophical speculation and its university setting in depth. She has shown the relationship between the vibrant, sometimes fierce, "disputational context" of the university and the evolving mathematical and logical form of both the questions asked and the answers considered successful in philosophical debate.[12] While acutely conscious of the impact of scholastic society on the shape and direction of philosophical inquiry, Sylla has limited her consideration of social factors to the society of scholars within the schools. With few exceptions, historians of medieval science have hesitated to step outside the university and outside the sphere of a refined intellectual culture in their analysis of fourteenth-century natural philosophy.[13]

[10] Murdoch, "Unitary Character," 289, 298–303. For a general discussion of the application of measurement languages to theological problems, see Murdoch, "Unitary Character," 289–307. For more on the role of theological controversy in the development of philosophical languages of measurement, see Edith Sylla, "Autonomous and Handmaiden Science: St. Thomas Aquinas and William of Ockham on the Physics of the Eucharist," in Murdoch and Sylla, Cultural Context, 349–91; Sylla, "Godfrey of Fontaines on Motion with Respect to Quantity of the Eucharist," in A. Maierù (ed.), Studi sul XIV secolo in memoria di Anneliese Maier (Rome, 1981), 105–41, esp. 107–09.

[11] Murdoch, "Unitary Character," 309. Although Murdoch considered the influence of the social environment of the university, he concluded that, "At best, social factors account for available possibilities; they seldom say anything about which ones were taken up and why" (309). Murdoch also (272 n. 2) briefly mentioned the possible influences on philosophy from the disciplines of medicine and law. This valuable insight is considered further in chapters 6 and 7.

[12] Edith Sylla, "The Oxford Calculators," in Norman Kretzmann, Anthony Kenny, and Jan Pinborg (eds.), The Cambridge History of Later Medieval Philosophy (Cambridge, 1982), 541–63; Sylla, "The Oxford Calculators in Context," Science in Context 1 (1987), 257–79.

[13] There are notable exceptions to this statement. In the specific case of calculationes, Janet Coleman ("Jean de Ripa, OFM, and the Oxford Calculators," Mediaeval Studies 37 [1975], 130–89, 131) suggests a connection between the expansion of commercial calculation in the fourteenth century and the increasing concern for calculation and measurement in university thought of the same period. For a specific argument linking economic thought and philosophical speculation, see William Courtenay, "The King and the Leaden Coin: The Economic Background of 'Sine qua non' Causality," Traditio 28 (1972), 185–210. Courtenay's more recent work continues to explore these connections. For a discussion of medieval scientific development at every point sensitive to its social and economic context, see Brian Stock, "Science, Technology, and Economic Progress in the Early Middle Ages," in David Lindberg (ed.), Science in the Middle Ages (Chicago, 1978), 1–51. The influence of practical measurement in social life on modes of measurement developed within scholastic natural philosophy is recognized by Pierre Souffrin, "La quantification du

4

Introduction

As the reading of any text in natural philosophy from this period indicates, the great majority of positions taken and points made were in response not to external experiences or influences, but to questions and positions defined by the ongoing debate. In this literature, logic dominates and leads. Evidence of direct experience in the world beyond the text is rare. Historians of medieval science have, therefore, directed their attention primarily toward the internal elements of scholastic debate and of university culture.

While the internal analysis of texts and traditions provides the base on which all intellectual history must rest, at certain points its limitations in explaining intellectual innovation become apparent even to those most committed to its practice.[14] These limitations are especially clear in the case of major shifts in perception and direction like those that defined fourteenth-century natural philosophy.[15] It is difficult to see (to take one set of examples) how influences coming solely from within the schools can explain the strength of the intellectual movement of *calculationes*, or the preoccupation with measurement and relation that informed it, or the faith in the potential of quantification that underlay it, or the new perceptual models of the natural world from which it arose. Understanding intellectual innovation of this magnitude requires an approach that focuses on the *interaction* between the culture of the schools and new conceptual models coming from beyond the schools – in this case models that took shape as scholars directly (and sometimes painfully) experienced and sought to comprehend the dynamic of the monetized marketplace.

mouvement chez les scolastiques: la vitesse instantanée chez Nicole Oresme," in Jeannine Quillet (ed.), *Autour de Nicole Oresme: Actes du Colloque Oresme organisé à l'Université de Paris* XII (Paris, 1990), 63–83, esp. 66. For connnections between commercial culture, numeracy, and mathematical innovation, see Warren Van Egmond, *The Commercial Revolution and the Beginnings of Western Mathematics in Renaissance Florence, 1300–1500* (Ph.D. dissertation, Indiana University, 1976); Alexander Murray, *Reason and Society in the Middle Ages* (Oxford, 1978); Richard Hadden, *On the Shoulders of Merchants: Exchange and the Mathematical Conception of Nature in Early Modern Europe* (Albany, N. Y., 1994). Particularly relevant here is the work of Michael Wolff investigating the connection between scholastic economic theory and natural philosophy. See Michael Wolff, *Geschichte der Impetustheorie: Untersuchungen zum Ursprung der klassischen Mechanik* (Frankfurt a/M, 1978); also, Michael Wolff, "Mehrwert und Impetus bei Petrus Johannis Olivi," in Jürgen Miethke and Klaus Schreiner (eds.), *Sozialer Wandel im Mittelalter: Wahrnehmungsformen, Erklärungsmuster, Regelungsmechanismen* (Sigmaringen, 1994), 413–23.

[14] Witness the puzzlement of John Murdoch (cited 3) concerning the "frenzy" to measure.

[15] On this point, another statement of John Murdoch's is revealing: "When one looks for evidence in the sources that might provide a tolerably clear answer to such a question [i.e., why the use of logic as an analytic tool increased so greatly between the thirteenth and fourteenth centuries] (or to many other questions having to do with new trends that occurred in the fourteenth century), one almost feels that, between those fourteenth-century works that 'have it' and those thirteenth-century ones that don't, some evil genius has systematically destroyed the key works that could provide an answer." See "The Involvement of Logic in Late Medieval Natural Philosophy," in Stefano Caroti (ed.), *Studies in Medieval Natural Philosophy* (Florence, 1989), 3–28, 5.

Without the vibrant intellectual culture of the medieval university in place – its disputational context, its characteristic attention to detail and logical rigor, its passion to synthesize – social experience and economic insights would never have been translated into new conceptual models, much less into mathematical and logical languages capable of expressing and refining these models. But at the same time, the rigid formality of scholastic discourse, and the focus of fourteenth-century natural philosophy on the technical intricacies of mathematics and logic, has disguised the rich layer of contact between philosophical speculation and social experience in this period. The biography of virtually every natural philosopher of note from the late thirteenth century reveals that the world of higher thought was not bounded by the walls of the university, whether actual or metaphorical. The new conceptual model of nature's form and activity arose in the minds of men who were deeply involved in the life of their society and highly conscious of the transformative process of monetization taking place within it.

The students and masters at Oxford and Paris lived in urban settings where the effects of monetization and commercialization were everywhere to be seen and experienced.[16] Were the student to venture into High Street in Oxford, or to cross the Grand pont in Paris, he would likely be caught up in crowded markets as he measured the price of a coveted pen or book or tankard against the coins in his purse. If he were a foreigner to the city, as was most likely, he would be brought into frequent contact with moneychangers and the complex equations that converted his currency into locally accepted coin. He would be required to calculate and husband his resources in a society that provided numerous opportunities for spending. It is hardly surprising that the earliest surviving letters from university students witness their preoccupation with monetary shortages and record their pleas for financial aid.[17]

The involvement of the student in the world of the market was not limited to periodic encounters, nor did it end with his inception as master. If anything, it seems to have increased in the case of those scholars most responsible for the new proto-scientific direction in natural philosophy. The one constant in the biographies of these scholars is that they were repeatedly given responsibilities in practical affairs drawing on thinking far removed from their philosophical and theological training. Indeed, the most innovative and influential scholars of the century seem

[16] For a remarkable fourteenth-century description of Paris as a giant marketplace written by a contemporary scholar and natural philosopher, see Jean de Jandun, *Tractatus de laudibus Parisius*, in Le Roux de Lincy and L. M. Tisserand (eds.), *Paris et ses historiens aux XIVe et XVe siècles* (Paris, 1867), 32–79.

[17] Charles Homer Haskins, "Letters of Mediaeval Students," in *Studies in Mediaeval Culture* (New York, 1929), 1–35, esp. 7–21, 27–28.

almost to have been on an informal bureaucratic track, with their experience as administrators beginning during their university careers and ending with exalted positions in the bureaucracies of Church or civil government.[18]

Within the universities, every examination taken, every grade passed and degree earned, had a price attached to it. The surviving administrative records from Oxford and Paris attest to the range of fees charged and the amount of conscious effort required of the master-scholars in the assessment and collection of these fees – for the teaching masters at the university were in almost all cases its administrators as well.[19] The minute monetary regulation and gradation of university life was further complicated by the habit of varying each fee levied in proportion to the ability of each student to pay. Again it was ordinarily the teaching masters who were charged with assessing, collecting, accounting for, and recording these fees.[20] The evidence in surviving university records for continued bureaucratic involvement has led one modern scholar to conclude that university masters of the fourteenth century spent as much of their time performing administrative duties as they did writing and lecturing.[21]

The economic writings left by these masters (examined in chapters 3–5) reveal how strongly their administrative experience influenced their perceptions of money and exchange. The same scholars whose economic writings are studied in this book and whose work demonstrates the clearest insight into the structure of economic life – Godfrey of Fontaines, Henry of Ghent, Peter John Olivi, John Duns Scotus, Geraldus Odonis, Walter Burley, Richard Kilvington, Jean Buridan, Nicole Oresme – all made significant and forward looking contributions in the area of natural philosophy. To take but one example, each of these thinkers played a vital role in refining the concept of qualitative intensity,

---

[18] For more detail on the administrative involvement of Merton Calculators (especially Thomas Bradwardine and William Heytesbury) and on the equally influential Parisian natural philosophers Jean Buridan and Nicole Oresme, see chapter 1. Additional biographies illustrating this pattern of administrative activity on the part of influential natural philosophers are given in chapters 5 and 6.

[19] William J. Courtenay, "The Registers of the University of Paris and the Statutes Against the *Scientia Occamica*," *Vivarium* 29 (1991), 13–49, esp. 18. Evidence from the surviving records of the university for the Parisian masters' continual involvement in university administration is considered in chapter 1.

[20] For more on the minute, bureaucratic, and monetary grading of student activity, see Jacques Verger, "Le coût des grades: droits et frais d'examen dans les universités du Midi de la France au moyen âge," in Astrik L. Gabriel (ed.), *The Economic and Material Frame of the Mediaeval University* (Notre Dame, Ind., 1977), 19–36. For the involvement of Merton Scholars in these college administrative tasks, see chapter 1.

[21] William Courtenay, "The London *Studia* in the Fourteenth Century," *Medievalia et humanistica* 13 (1985), 127–41, esp. 131–32.

a concept central to the intellectual movement of *calculationes* and to proto-scientific thought in the fourteenth century.[22]

Despite ample evidence for the involvement of fourteenth-century natural philosophers in the economic life of their time, not one directly acknowledges the impact of social and economic experience on his philosophical speculation. While in their economic writings they continually state that "money measures all things" (*inventum est nummisma ut sit medium et mensura omnium commutabilium*), and while they investigate in great depth how money performs its function of measuring, relating, and equalizing, they never directly acknowledge its influence on their philosophical preoccupation with these same questions of measurement, relation, and equalization. Although they often remark on money's extraordinary success as an instrument of gradation and commensuration, they never acknowledge it as a model for the conceptual instruments they themselves devised to perform similar functions within philosophical discourse. At a number of points in the following pages I discuss this lack of acknowledgment (or recognition) and what I believe are its probable causes, but some general observations can be made here.[23]

In contrast to modern scientific attitudes, scholastic thinkers expressed strong doubts that scientific truths could be based upon personal and particular experiences of an ever-changing object world. They believed that observations drawn from personal experience lacked the necessity, universality, and truth-value required by science. There was, consequently, a concerted effort to cleanse philosophical discourse from the taint of its contact with contingent experience. Insights drawn from the experience of nature were quickly denatured – translated into propositions and logical terms deemed to be the proper subjects of scholastic debate. In a paradox that has often been remarked upon, many of the most important works in natural philosophy of this period contain not a single reference to personal observations of nature.

Experiences drawn from economic life carried with them a double weight of negative connotations since, even when scholastic thinkers recognized the importance of economic activity, they remained suspicious of it. When the influential natural philosopher, Jean Buridan, writes on economic questions, he shows conclusively that he understands the multi-faceted role of money as an instrument of measurement in ex-

---

[22] For an essay that considers the contribution of each of these thinkers to the question of qualitative intensity, see Anneliese Maier's influential study, "Das Problem der intensiven Grösse in der Scholastik," in *Zwei Grundprobleme der scholastischen Naturphilosophie: Studien zur Naturphilosophie der Spätscholastik* (Rome, 1951), 1–79. See also chapter 6 in this book.

[23] See chapter 6, 196–98, and chapter 7, 209–10.

8

change – particularly money's capacity to bring the most diverse goods into a common system of measurement and relation.[24] Faced with a parallel question of measurement in his commentary on Aristotle's *Physics*, Buridan asks whether money as a common measure of diverse goods can serve as a "proper" model for the philosophical measurement of diverse species. Here, however, in the context of natural philosophy, he concludes that it cannot. Speaking within a tradition of medieval economic commentary, he notes that money can measure only relative and ever-changing economic values, not the essential qualities and natures that are of concern to philosophy. Even more damaging in his eyes was the association of monetary measurement with fraud. The economic value that money measures as selling price is, Buridan writes, often distorted through a bargaining process in which deception is intended by both buyer and seller. Thus, although Buridan is acutely aware of money's function as the common measure of all goods in his society, he cannot accept its influence as a measure in philosophy.[25] A similar disjunction is found in the work of every philosopher considered in this book. Given the strict requirements for truth, universality, and necessity in the highly formal discourse of scholastic natural philosophy, medieval thinkers never explicitly acknowledge the influence of any model drawn from the tainted sphere of the marketplace on their philosophical speculation.

In the absence of this conscious recognition (the "smoking gun," so to speak), I have relied throughout on the method of isolating and comparing elements of the scholastic model of the monetized marketplace (including the model of money as measure) with the defining elements of the proto-scientific model of nature that emerged in the fourteenth century. After isolating the elements that I believe define both intellectual spheres, I detail the verbal and formal similarities existing between them and the many levels of connection joining them. To make it easier for the reader to test the strength of these connections, I use the same category headings in my chapters on scientific thought as I do in my chapters on economic thought.

In order to explore the relationship between economic insights into the monetized marketplace and philosophical insights into the workings of nature, it is necessary to link areas of historical investigation rarely

[24] Buridan's rich insights into money and exchange are found at many points in chapter 5; his contributions to natural philosophy are discussed in chapters 6 and 7.

[25] Jean Buridan, *Quaestiones super octo phisicorum libros Aristotelis diligenter recognite et revise a magistro Johanne Dullaert de Gandavo* (Paris, 1509; reprint, Frankfurt a/M, 1964), VII.6, 106va–107rb. Buridan's determination on this point is considered in greater detail in chapter 6. The failure on the part of scholastic thinkers to consciously link money as economic measure to money as philosophical measure is considered at many points in chapters 5–7.

considered together: economic history, the history of economic thought, and the history of science. The goal of integrating the findings in these areas determines the structure of this work. In chapter 1, I provide definitions of monetization, market development, and monetary consciousness that are followed throughout the work. I then consider the progress of monetization and the level that market organization had achieved in England and France by the middle of the fourteenth century. The chapter concludes with details from the biographies of leading fourteenth-century natural philosophers, illustrating their involvement in a range of administrative duties that brought them into close contact with the economic life of their society.

Scholastic philosophers inhabited an intellectual universe as well as a social universe. Authoritative texts and contemporary philosophical debates framed and mediated their social experience. Knowledge of the rich tradition of scholastic writing on economic subjects is essential to understanding how scholastic thinkers "experienced" the new dynamic of the monetized marketplace. In chapters 2 through 5 I investigate the scholastic literature on money and market exchange, considering both the authoritative texts inherited on these subjects and the important additions and corrections made to these texts in the thirteenth and fourteenth centuries.

Chapter 2 considers Aristotle's detailed discussion of money and economic exchange in Book v of the *Nicomachean Ethics*. Aristotle's conception of economic exchange as a dynamic process of equalization, one that could be represented both as a mathematical equation and as a geometric figure, had great influence on scholastic economic thinking. His markedly mathematical, geometrical, and relativist treatment of exchange provided an important textual ground for the later linking of scholastic economic thought with proto-scientific speculation.

In chapter 3 I consider the two earliest and most influential commentaries on Aristotle's *Ethics*: those of Albertus Magnus and Thomas Aquinas. These commentaries reveal the understanding, acceptance, and even expansion of Aristotle's sophisticated analysis of money and exchange by Christian thinkers of the middle of the thirteenth century. They provide as well the point against which we can measure the rapid development of money and market consciousness from the late thirteenth century.

Chapter 4 is divided into two parts, both of which are concerned with the central question of equality and equalization in exchange. The first part considers the definition of equality in writings on usury and just price theory from the earliest church councils through the thirteenth century. The second part follows the history of the changing philosophical conception of equality and equalization through the late thirteenth century:

from a knowable and numerically definable point of equality to an estimated, geometrically defined range (or "latitude") of approximation. I illustrate the conflict between older arithmetical and newer geometrical conceptions of economic equality through a heated philosophical-economic debate that occurred at the end of the thirteenth century between two Parisian masters, Henry of Ghent and Godfrey of Fontaines.

Chapter 5, the last of the economic chapters, investigates developments in economic thinking that occurred over the course of the fourteenth century. The focus here is on the economic writings of those scholars who at the same time made pivotal contributions to fourteenth-century natural philosophy: Peter John Olivi, John Duns Scotus, Walter Burley, Richard Kilvington, Geraldus Odonis, Albert of Saxony, Jean Buridan, and Nicole Oresme. Although scholastic economic thinking has often been characterized as abstract, unrealistic, and bounded by authority, a close reading of fourteenth-century texts reveals the willingness of philosophers and theologians to reformulate economic conceptions in order to comprehend and accurately describe the changing realities of the monetized marketplace.

In the speculation considered in chapter 5, we come most forcefully to those economic insights that had the greatest influence on scientific speculation: the recognition of money as a common measuring continuum for all commodities in exchange; the social geometry of a world perceived to be tied together by economic exchange and the *medium* of money; the comprehension of the relativist determination of all values in the marketplace; the understanding of the necessity of a mathematics of the approximate in the exchange equation; and, perhaps most important, the emerging sense of the marketplace as a dynamic, supra-personal, self-ordering system of equalization.

I have come to recognize that these insights can be organized into six conceptual categories, in each of which developments in scholastic economic thought connect to proto-scientific speculation: (1) mathematics and the geometry of exchange; (2) equality, the mean, and equalization in exchange; (3) money as *medium* and as measure; (4) relation and the relativity of value in exchange; (5) common valuation in exchange; and (6) the social geometry of monetized society. The chapters on scholastic economic thought are divided into headings determined by these categories.

The concluding chapters, 6 and 7, are focused on new directions in the comprehension and representation of nature pursued within fourteenth-century natural philosophy. Under each of the six category headings, defining elements of the new scholastic understanding of monetized exchange are related to specific innovations within fourteenth-century natural philosophy. It is, I believe, within this fertile area of connection between perceptions of a dynamic, monetized society and perceptions of a dynamic nature that philosophers created the conceptual landscape within which Western science later developed.

In discussing the complex relationship between economic and proto-scientific thought in the fourteenth century, a question arises concerning the direction in which the insights flowed. Were they insights originally derived to make sense of pressing new economic and social realities that were then applied to the understanding of the natural world, or were they insights developed within the philosophical and intellectual tradition of the schools that were then applied to the understanding of economic problems? This question cannot be answered unconditionally. Clearly there was influence in both directions.[26] The very act of creating conceptual models rested on the existence of a sophisticated university culture, proficient in the exercise of logic and criticism, and confident in its powers of creative speculation. But, while allowing for the strength of scholastic culture, my study of particular intersections between economic and proto-scientific thought has led me to conclude that the creative impulse behind the fertile new models of measurement, relation, and equalization emerging within the schools came largely from the experience and comprehension of unsettling social and economic developments that were transforming the society beyond the schools.[27]

The interchange between economic and scientific perceptions is considered at many points in the following chapters. Taken together they provide an outline of a mechanism of transference between the scholar's conception of the social world and his conception of the natural world, between his insights into the working of a monetized society and his insights into the working of a newly quantifiable and measurable nature. To scholastic thinkers influenced by Roman and canon law and Aristotle's analysis of exchange in *Ethics* v, the monetized marketplace came to be perceived as a dynamic system of equalization, governed by its own proper mathematics and logic. Take exchangers with unequal needs exchanging unequal products at crossed purposes (each wanting to buy cheap and to sell dearly); add money as a line instrument of commensuration and relation; allow a relativist price to fluctuate in relation to need

---

[26] Jean Buridan's use of a newly invented mathematical function to measure the increase and decrease of commodity value is a clear case of a model of measurement flowing from mathematical to economic thought. This is discussed in chapter 5, 145–46.

[27] For works that focus on the transference of economic models and insights into literary form and content, see Marc Shell, *The Economy of Literature* (Baltimore, 1978); Eugene Vance, "Chrétien's *Yvain* and the Ideologies of Change and Exchange," *Yale French Studies* 70 (1981), 42–62; R. A. Shoaf, *Dante, Chaucer, and the Currency of the Word: Money, Images, and Reference in Late Medieval Poetry* (Norman, Okla., 1983).

and calculated benefit; expand the system of equilibrium from the individual exchange to the community of exchangers; and the result was, somehow, market order and equality. What is more, from the middle of the thirteenth century, the order established by the self-equalizing marketplace of free exchangers was coming to be recognized as resistant to, and in some ways superior to, an order imposed by decree, even when motivated by the best social intentions.[28]

In the religious and philosophical universe of the thirteenth century, where the existence of order necessarily, as Aquinas said, implied the existence of an active ordering intelligence,[29] where in the Aristotelian universe even the movement of the celestial spheres required the constant intervention of active intelligences, there was no other system that conformed to the scholastic conception of the self-equalizing system of market exchange.[30] Serious ethical and theological questions were raised by an order in which equality and justice resulted from an impersonal, almost mechanical, common process, rather than from a conscious ethical decision on the part of individuals.[31] And the model of market equalization was worse than merely impersonal. As the fourteenth century progressed, scholastic thinkers realized that market equality was the geometric product of *willed* inequalities − *crossed* diagonals − each exchanger seeking to benefit more than the other from the exchange.[32] Such production of order and equality out of willed inequality violated the essence of the traditional metaphysical and physical understanding of the *ordo rerum*.

As the power and weight of the marketplace within society grew over

[28] See chapter 1 for the difficulties experienced by French and English kings when they attempted to control market prices in the early fourteenth century, and the sharp recognition of these difficulties by contemporary chroniclers. The theoretical recognition of the superiority of market equalization in literature ranging from civic ordinance to scholastic *quaestio* is considered in chapters 3, 4, and 5.

[29] "Intellectus solius est ordinare." See de Wulf's comment on this general position of Aquinas in *Philosophy and Civilization in the Middle Ages* (Princeton, 1922), 193. Aquinas' association of order with an active intelligence is considered further in chapter 4.

[30] For the association of intelligence and order in the conception of the physical world in the middle of the thirteenth century, see Albertus Magnus, *De mineralibus* (*Book of Minerals*), Dorothy Wyckoff (tr.) (Oxford, 1967), I.I, 8 (30): "Aristotle said that every work of Nature is a work of intelligence." Albert's overstatement of Aristotle's position (e.g., *Physics* II.8 [199a3]) is revealing here.

[31] This question is considered in chapter 4. See for example the theological insistence, voiced by Aquinas and Henry of Ghent, that the individual must remain responsible for instituting and maintaining equality in the exchange process, rather than relying on the impersonal market process of equalization.

[32] See chapter 4, 90–93, for the recognition of this condition in civil law in the thirteenth century. Its recognition by scholastic philosophers writing in the fourteenth century is considered in chapter 5, 132, 159. For the reflection of this insight in fourteenth-century natural philosophy, see chapter 7, 220–23.

the course of the twelfth and thirteenth centuries, opposition between the economic order and traditional models of natural order led to continuing attempts (seen most clearly within scholastic usury theory) to force economic definitions to conform to traditional definitions of "natural" equality.[33] The distinction between natural order and market order created great tension within an intellectual culture whose habit was to unify and synthesize. The tension grew as the power and position of the market in society grew, until, by the late thirteenth century, as a result of this continued opposition, it was the conception of the *natural* order that began to give way.[34] Scholastic natural philosophers began to create a new model of nature, one that could comprehend the order and logic of the marketplace – dynamic, self-equalizing, relativistic, probabilistic, and geometrical – a nature constructed and bound together by lines in constant expansion and contraction. It was within this new model of nature that science emerged.

[33] See chapter 4, 80–87.
[34] This transformation is discussed at the conclusion to chapter 4, in the context of a debate between Henry of Ghent and Godfrey of Fontaines over the nature of economic equality.

Chapter 1

THE ECONOMIC BACKGROUND: MONETIZATION AND MONETARY CONSCIOUSNESS IN THE THIRTEENTH AND FOURTEENTH CENTURIES

When historians and economists speak of the monetization of European society in the twelfth and thirteenth centuries, they are speaking in relative rather than absolute terms of a multi-faceted social process.[1] Increases in the volume of coinage in circulation and the frequency of monetary transactions are only two of many factors involved.[2] The process of monetization was inextricably tied to what has been called "the commercial revolution of the thirteenth century": the rapid growth of trade, markets, and towns; the acceleration of agricultural and craft production; the evolution of specialized commercial enterprises and techniques; and the penetration of monetary and commercial values into all areas of social life.[3] In this sense, the process of monetization is first

---

[1] Michael M. Postan, "The Rise of a Money Economy," *Economic History Review* 14 (1944), 123–34, esp. 127.

[2] Nicholas Mayhew estimates the volume of coinage in circulation in England in 1086 at c. 37,500 pounds, and in the decades following 1300 at c. 1,100,000 pounds. See Mayhew, "Modelling Medieval Monetisation," in R. H. Britnell and B. M. S. Campbell (eds.), *A Commercialising Economy: England 1086–1300* (Manchester, 1995), 55–77, esp. 62–65; Mayhew, "Population, Money Supply, and the Velocity of Circulation in England 1300–1700," *Economic History Review* 48 (1995), 238–58. Annual output of the English Mint has been calculated to have grown from one million pennies in 1170 to fifteen million by 1250. On this, see R. H. Britnell, *The Commercialisation of English Society 1000–1500* (Cambridge, 1993), 102; C. E. Challis (ed.), *A New History of the Royal Mint* (Cambridge, 1992). For the expansion of minting on the Continent, see the figures provided in Carlo M. Cipolla, *Before the Industrial Revolution: European Society and Economy* (New York, 1976), 192–93.

[3] Peter Spufford, "Le rôle de la monnaie dans la révolution commerciale du XIIIe siècle," in John Day (ed.), *Etudes d'histoire monétaire* (Lille, 1984), 355–95. An expanded version of this paper appears as chapter 6 in Spufford, *Money and Its Use in Medieval Europe* (Cambridge, 1988), 245–63. See also R. H. Britnell, "Commercialisation and Economic Development in England 1000–1300," in Britnell and Campbell, *A Commercialising Economy*, 7–26; Britnell, "The Proliferation of Markets in England 1200–1349," *Economic History Review* 34 (1981), 209–21; Christopher Dyer, "The Consumer and the Market in the Later Middle Ages," *Economic History Review* 42 (1989), 305–27. For the penetration of the money economy into the countryside, see Georges Duby, *Rural Economy and Country Life in the Medieval West*, Cynthia Postan (tr.) (Columbia, S. C., 1968), 139, 151,

apparent in the Italian cities of the late eleventh and twelfth centuries.[4] For England and France, historians identify the "long" thirteenth century – that is, between approximately 1180 and 1320 – as the period of most rapid monetization.[5]

The accelerated use of money had ever-expanding social, economic, and intellectual consequences. As the process of monetization gathered speed, habits of thought and perception initially restricted to those actively engaged in commerce came to be adopted by members of all segments of society. Among the most characteristic of these new habits were: the focus on monetary profit and loss in a wide range of decision making; the recognition of the importance of detailed written records for this calculation; the resulting broad development of literacy and numeracy; and the translation of qualitative values into quantitative, often monetary, terms as a way to simplify the process of calculation.[6] The gradual spread of these habits had important historical consequences, especially when they were adopted by administrators of the incipient royal and papal bureaucracies.[7] Administrators discovered that they could greatly simplify the process of assessing and collecting the dues their

232–59; Spufford, *Money and Its Use*, 239–43, 382. For the growth of market-oriented agriculture in England in the thirteenth century, see Michael Postan, *The Medieval Economy and Society: An Economic History of Britain 1100–1500* (Berkeley, 1972), 100–01; Bruce Campbell, "Measuring the Commercialisation of Seigneurial Agriculture c. 1300," in Britnell and Campbell, *A Commercialising Economy*, 132–93.

[4] David Herlihy, "Treasure Hoards in the Italian Economy 960–1139," *Economic History Review* 10 (1957), 1–15. For a discussion of the effects of monetization on the social, intellectual, and religious life of Italy from the late eleventh century, see Lester K. Little, *Religious Poverty and the Profit Economy in Medieval Europe* (Ithaca, N. Y., 1978); also, Little, "Evangelical Poverty, the New Money Economy, and Violence," in David Flood (ed.), *Poverty in the Middle Ages* (Werl, 1975), esp. 11–15. Marvin B. Becker discusses the "exponential increase" in credit and the circulation of money over the course of the twelfth century in Italy in *Medieval Italy: Constraints and Creativity* (Bloomington, Ind., 1981), 19. For the interaction between monetary and cultural developments in Italy during this period, see Brian Stock, *The Implications of Literacy: Written Language and Models of Interpretation in the Eleventh and Twelfth Centuries* (Princeton, 1983).

[5] Spufford, *Money and Its Use*, 245–63; Britnell, "Proliferation of Markets"; Postan, "Rise of a Money Economy," 128. It has been estimated that the volume of coinage in England increased six to eight times in the century following 1180. See D. M. Metcalf, "The Volume of English Currency," in N. J. Mayhew (ed.), *Edwardian Monetary Affairs* (Oxford, 1977).

[6] For a discussion of the influence of money on these developments, see Alexander Murray, *Reason and Society in the Middle Ages* (Oxford, 1978), pts. I and II; Michael Clanchy, *From Memory to Written Record: England 1066–1307* (Oxford, 1993), 44–73; Nancy F. Partner, *Serious Entertainments: The Writing of History in Twelfth-Century England* (Chicago, 1977), chapter 6, "The Temporal World," esp. 152–59; Fernand Braudel, *Capitalism and Material Life 1400–1800*, Miriam Kochan (tr.) (New York, 1973), 325–26; Wesley C. Mitchell, "The Role of Money in Economic History," in Frederick C. Lane and Jelle C. Riemersma (eds.), *Enterprise and Secular Change* (Homewood, Ill., 1953), 200–04.

[7] Clanchy, *From Memory to Written Record*, 62–68; Ellen Kittell, *From Ad Hoc to Routine: A Case Study in Medieval Bureaucracy* (Philadelphia, 1991); John Baldwin, *The Government of Philip Augustus: Foundations of French Royal Power in the Middle Ages* (Berkeley, 1986), 55–58, 405–23; Joseph Strayer, *The Reign of Philip the Fair* (Princeton, 1980), 148–56.

institutions required by replacing older, less quantifiable determinants of wealth and status with the easily gradable, numerable, and standardized determinant of monetary income from landed and moveable property.[8] Monetization and bureaucratization thus grew up together. Side by side they facilitated measurement and calculation by translating loosely defined grades and qualities into numbered, quantitative terms.

The growth of the consciousness of money's place and function in society paralleled the rise of the merchant *estat* from a lowly position to one of great social and political power over the course of the long thirteenth century. The merchants' evident social success presented a newly rationalized model of behavior and perception, leading individuals far removed from commerce to imitate their modes of thought and activity.[9] The commercial habit of measuring and calculating in monetary terms was extended to all manner of things. Where previously, for example, forests had been primarily valued by lords for the status and pleasure they afforded, they came in the thirteenth century to be seen as income opportunities, as resources to be rationally exploited toward the end of profit. Peter Spufford has characterized this broad translation of older values into monetary terms as a "revolution in attitudes toward money."[10] It can equally be seen as the adoption by non-commercial classes of commercial attitudes toward the exploitation of opportunities for profit.[11] Despite the fervent warnings of moralists, money was becoming the measure of all things.[12]

In the monetized society of the thirteenth century, even in informal discourse, social rank came to be graded and defined according to

[8] For the early monetary rationalization of taxes in England, see G. L. Harris, *King, Parliament, and Public Finance in Medieval England* (Oxford, 1975), 33ff.; Britnell, *Commercialisation*, 40–43. For an outline of French taxation to 1356, see John Bell Henneman, *Royal Taxation in Fourteenth-Century France: The Development of War Financing 1322–1356* (Princeton, 1971), esp. 27–39, 303–29; Raymond Cazelles, *Société politique, noblesse, et couronne sous Jean le Bon et Charles V* (Geneva, 1982), 11–16.

[9] Raymond Cazelles, *Nouvelle histoire de Paris de la fin du règne de Philippe Auguste à la mort de Charles V 1223–1380* (Paris, 1972), 85–118, discusses the profound impact of the Parisian bourgeoisie on thirteenth- and fourteenth-century French political, social, and economic life. For an interesting case study of the power of wealth to determine status in the fourteenth century, see the story of Simon de Mirabello, pawnbroker, moneylender, and regent of Flanders, in Raymond De Roover, *Money, Banking, and Credit in Mediaeval Bruges* (Cambridge, Mass., 1948), 153–55. The social and political power of English merchants is discussed in Sylvia Thrupp, *The Merchant Class of Medieval London 1200–1500* (Chicago, 1948), esp. 238ff.

[10] Spufford, *Money and Its Use*, 245.

[11] Campbell, "Measuring Commercialisation," 186, 189–92. Campbell notes, however (193), that the form in which manorial accounts were kept better served to detect fraud than to calculate opportunities for profit. For a discussion of the extremes of monetized rationality in the activity of the merchants of this period, see Yves Renouard, "Le rôle des hommes d'affaires italiens dans la Méditerrané au moyen âge," in *Etudes d'histoire médiévale* (Paris, 1968), 405–18.

[12] For the acknowledgment of money as universal measure in the economic thought of the fourteenth century (*Nummisma est mensura omnium commutabilium*), see chapter 5, 138–39.

numbered income.[13] Many observers were distressed by this development, and examples abound of complaints that aristocratic qualities were
being swamped by the multiplying powers of money.[14] A thirteenth-
century tale clearly expresses this perception. Crossing the Grand pont in
Paris (or the Pont au change as it was already called in that day because it
was the official location of the Paris moneychangers and was covered
over its length by their busy, open booths), a man meditates on the great
reputation of money:

Why shouldn't silver be well thought of? After all, with it one buys ermine
clothes, abbacies and church benefices, cities and castles, great lands and pretty
women. It is silver that disinherits the orphan, absolves the excommunicate,
renders justice to the scoundrel, and pardons injuries more effectively than pretty
words . . . It turns a peasant into a gentleman, makes a happy man of a
melancholic, a wise man out of a fool . . . It is silver that ends wars, leads armies,
makes ignoble families illustrious. In short, it commands the whole world.[15]

Medieval attitudes toward money cannot be grasped without understanding the intense dualism surrounding the judgment of its existence
and action in the world. It was seen, often by the same person, as both a
remarkably successful instrument of economic order, balance, and gradation, and at the same time as the great corrosive solvent, the overturner,
the perverter of balance and order.[16] This dualistic attitude of admiration

[13] For the common practice of measuring a man by the size of his purse, see Jacques Rossiaud, "The
City Dweller," in Jacques Le Goff (ed.), Lydia G. Cochrane (tr.), *Medieval Callings* (Chicago,
1990), 139–79, esp. 149; Becker, *Medieval Italy*, 2. Murray (*Reason and Society*, 186) notes that,
whereas in pre-monetized society status determined numbered *wergeld*, in monetized society,
numbered income came to determine status.

[14] See for example Matthew Paris' accusation (*Matthew Paris' English History*, J. A. Giles [ed. and tr.],
3 vols. [London, 1852–54], vol. III, 15) against the earl of Gloucester that he had fallen from the
nobility of his ancestors to the status of a usurer or common trader, for marrying his legitimate heir
to the daughter of a wealthy but non-noble family for 5,000 crowns. The swamping of aristocratic
values by money is a constant theme of the French *fabliaux* in this period. On this, see Charles
Muscatine, *The Old French Fabliaux* (New Haven, 1986), esp. 82–84. For the effect of monetization on the composition of the nobility of France, see Edouard Perroy, "Social Mobility Among
the French Noblesse in the Later Middle Ages," *Past and Present* (1962), 25–38.

[15] *Fabliaux ou contes, fables, et romans du XIIe et XIIIe siècles*, Legrand d'Aussy (ed.), 5 vols. (Paris, 1829),
vol. III, 216, extract of "De dom argent": "car enfin, à quoi dom Argent n'est-il pas bon? C'est
avec lui qu'on achète péliçons et manteaux d'hermine, abbayes et benefices, cités et châteaux, les
grandes terres et les jolies femmes. C'est lui qui fait déshériter un orphelin, absoudre un
excommunié, rendre justice à un villain, et pardonner les injures plus efficacement que de beaux
sermons . . . Argent fait d'un villain un homme courtois, d'un mélancolique un homme gai, d'un
sot un homme d'esprit . . . Enfin il termine les guerres, conduit les armées, illustre les familles
ignobles . . . et commande à toute la terre." The subject of medieval monetary satire is treated
brilliantly in John A. Yunck, *The Lineage of Lady Meed: The Development of Mediaeval Venality Satire*
(Notre Dame, Ind., 1963). For Yunck's judgment of the primary place of money as a subject of
medieval satire, see 185. For the privileged place of Rome in this satire, see especially 71–108. See
also Duby, *Rural Economy*, 259; Little, *Religious Poverty*, 19ff., and Little, "Evangelical Poverty,"
15; Stock, *Literacy*, 86.   [16] Little, *Religious Poverty*, 33.

and fear, which gained intellectual authority from its place within Aristotle's thought, underlay even the most enthusiastic scholastic appreciations of money's capacity to facilitate measurement in administration and exchange.[17]

The best place to examine the impact of monetary and market consciousness on university thought is at Paris – one of the busiest commercial centers of medieval Europe and the home of its greatest university. Scholars studying at Paris in the fourteenth century would surely have crossed the Pont au change (as our poet did a century earlier) and witnessed the busy moneychangers calling out the current rates of exchange, surrounded by their scales and their piles of coin.[18] As foreigners to the city, they would have been particularly dependent on the services of these changers to convert the money received from home into coin acceptable to local merchants. Stepping out of their doors, they would have experienced the feverish activity of a city in which the power of money and the fruits of commerce were everywhere apparent.[19]

The surviving royal laws or *Ordonnances* of France provide an excellent source for examining the growth of monetary and market awareness in Paris from the late thirteenth century.[20] The great majority of the hundreds of extant *Ordonnances* from this period were directly or indirectly concerned with monetary questions.[21] They reveal the extreme complexity of the monetary situation in France due to the expanding economic needs of the monarchy, wildly fluctuating values in the coinage, and the exigencies of a century of war, famine, social dislocation, and plague.[22] Often the complex *Ordonnances* dealing with royal monetary

---

[17] For Aristotle's position, see chapter 2, esp. 52–54. For the scholastic understanding and appreciation of money as an instrument of balance and order, see chapters 3, 4, and 5. For the complex Christian heritage in regard to money and wealth, see especially chapter 4.

[18] A short history of the Pont au change is given in Hercule Géraud, *Paris sous Philippe le Bel: le rôle de la taille imposé sur les habitants de Paris en 1292* (Paris, 1837), 376–79. See also Cazelles, *Histoire de Paris*, 104.

[19] The hectic commercial activity of Paris and the Grand pont is beautifully illustrated in miniatures from a fourteenth-century manuscript reproduced in Virginia Wylie Egbert, *On the Bridges of Medieval Paris* (Princeton, 1984).

[20] *Ordonnances des roys de France de la troisième race*, vol. I, *1254–1327* (Paris, 1723), vol. II, *1327–1355* (Paris, 1729), E. de Laurière (ed.); vol. III, *1355–1364* (Paris, 1732), Denis François Secousse (ed.).

[21] For a description of the criteria for inclusion into this collection by the first two editors, see François Olivier-Martin, "Ordonnonces des rois de France de la troisième race," in *Les travaux de l'Academie des inscriptions et belles-lettres: histoire et inventaire des publications* (Paris, 1947), 37–44. For a later, more complete survey of the *Ordonnances* from this period, see J. M. Pardessus, *Table chronologique des Ordonnances des rois de France de la troisième race* (Paris, 1847).

[22] One of the most striking effects of monetization was on the economics of warfare. Many of the economic and administrative strategies pursued by governments after the middle of the thirteenth century were driven by the enormous new costs of war in this period. See Joseph R. Strayer, "The Costs and Profits of War," in Harry Miskimin, David Herlihy, and A. L. Udovitch (eds.), *The Medieval City* (New Haven, 1977), 269–91; W. M. Omrod, "The Crown and the English Economy 1290–1348," in Bruce Campbell (ed.), *Before the Black Death: Studies in the "Crisis" of the*

policy are prefaced by a statement detailing both the economic changes in society that necessitated the new law, and the economic assumptions upon which the law is designed. Through these one can chart the escalating levels of monetary awareness and the growing sophistication of governmental economic thought over the course of the fourteenth century.

The *Ordonnances* provide clear evidence of a rapid increase in concern with monetary affairs from the reign of Philip IV (1285–1314).[23] This concern was in no small measure connected to the debasement of French coinage that began in Philip's reign and inaugurated a period of extreme monetary instability that plagued France for the next three-quarters of a century.[24] Modern historians give two primary explanations for the policy of debasement. The first is rather simple and was the one most often recognized by contemporaries: the prince debased his coinage, continually calling in old coins and reminting new coins with lower silver content, because he gained considerable sums for the royal treasury (called *seigneurage* or *moneyage*) with each new minting. Raymond Cazelles has estimated that, in certain periods of rapid debasement, the king's profits from *moneyage* constituted more than one-half the receipts of the royal treasury.[25]

Recent historical studies of debasement have tended to downplay (but not deny) the role of the king's exploitation, emphasizing instead the rising value of silver due to its short supply, and the resultant need to decrease the silver content of the coinage in order to bring it into line with the changing market value of silver.[26] One of the positive results of

Early Fourteenth Century (Manchester, 1991), 149–83; Harry Miskimin, *The Economy of Early Renaissance Europe 1300–1460* (Englewood Cliffs, N. J., 1969), 92–98; Miskimin, "L'or, l'argent, la guerre dans la France médiévale," *Annales ESC* 40 (1985), 171–84; Spufford, "Le rôle de la monnaie," 368–72.

[23] This impression is undoubtedly heightened by the relative scarcity of surviving documents from the reigns of Philip's predecessors, which is in turn connected to advances in bureaucratic efficiency. See Strayer, *Reign of Philip the Fair*, 100–05, 180–86, 219–20; Henneman, *Royal Taxation*, 27–31. For related bureaucratic advances in England, see Clanchy, *From Memory to Written Record*, 62–78; and for Flanders at the end of the thirteenth century, see Kittell, *From Ad Hoc to Routine*, 81–86.

[24] The four periods of most extreme monetary instability were 1295–1305, 1326–29, 1337–43, and 1346–60. After 1360 France enjoyed twenty-five years of relative stability. See Raymond Cazelles, "La stabilization de la monnaie par la création du franc (décembre 1360) – blocage d'une société," *Traditio* 32 (1976), 293–311, 294. For an overview of the economic condition of France in this period, see Edouard Perroy, "A l'origine d'une economie contractée: les crises du XIVe siècle," in *Annales ESC* 4 (1949), 167–82. For the earlier history of monetary mutation in both France and Spain, see Thomas N. Bisson, *Conservation of Coinage: Monetary Exploitation and Its Restraint in France, Catalonia, and Aragon (c. AD 1000–c. 1225)* (Oxford, 1979), esp. 8–14.

[25] Cazelles, "La stabilization," 296.

[26] Etienne Fournial has argued forcefully for the importance of market factors (particularly the supply and value of silver) in the explanation of debasement. See his *Les villes et l'économie d'échange en Forez aux XIIIe et XIVe siècles* (Paris, 1967), esp. 276–86; also, Fournial, *Histoire monétaire de*

this newer approach has been to focus attention on the existence, from the middle of the thirteenth century, of an entrenched international market in precious metals of sufficient size and power to determine the value of silver and gold for large areas of Europe.[27] Strict prohibitions of precious-metal exports, and escalating threats of punishment against those *malicieuses genz* who dealt in silver or engaged in export activities, recur regularly in the *Ordonnances*.[28] The illegal market, however, survived every governmental attempt at control and essentially came to dictate the price of gold and silver to the royal mints.[29]

At each issuance of a newly debased coin or series of coins, the king would proclaim a disproportionate reduction in the value of the coins already in circulation and the notice of a certain date after which the old coins would cease to have any legal value. All who held the old coins were forced to take them to government-licensed moneychangers to be exchanged at the legal rate for the new coins.[30] In periods of rapid debasement, such wholesale replacement could occur several times in the space of a year. In the most unstable period, from 1337 to 1360, there were eighty-five separate legally proclaimed mutations and recalls of the coinage.[31] Monetary changes of this magnitude seriously disrupted the economic life of Paris. The affairs of its merchant *estat*, whose economic, social, and political power had grown immensely over the century, were thrown into disarray, exacerbating social tensions.[32]

The rise to power of Etienne Marcel, provost of the Parisian merchants, clearly indicates the degree of social dislocation in the city in this period. In February 1358 Etienne led his townsmen in an attack on the

---

l'occident médiéval (Paris, 1970), 88–95. Fournial (*Histoire monétaire*, 115) also recognized that the king at times changed the money in order to procure needed financial resources for the crown.

[27] See Michael Prestwich, "Early Fourteenth-Century Exchange Rates," *Economic History Review* 32 (1979), 470–82; Nicholas Mayhew, *Coinage in France from the Dark Ages to Napoleon* (London, 1988), 79; De Roover, *Money, Banking, and Credit*, 77; Mavis Mate, "High Prices in Early Fourteenth-Century England: Causes and Consequences," *Economic History Review* 28 (1975), 1–16, esp. 9–10; Spufford, *Money and Its Use*, 285; Raymond Cazelles, "Quelques réflexions à propos des mutations de la monnaie royale française (1295–1360)," *Le moyen âge* 72 (1966), 83–105, 251–78, esp. 92.

[28] Prestwich, "Exchange Rates," 480. For examples of *Ordonnances* dealing with the prohibition of silver exportation and the drain on silver caused by exchange dealers, see: vol. I, 372 (1303), 423 (1304), 770 (1322), 792 (1326); vol. II, 255 (1346). In addition to the problem of market pressures on exportation, the rising price of silver also increased the incidence of thesaurization which was, as well, a constant theme in the *Ordonnances*. See, for example, the *Ordonnances* in vol. I, 324 (1294), 347 (1302), 532 (1313); vol. II, 86 (1332).   [29] See Mayhew, *Coinage in France*, 80–85.

[30] A typical *Ordonnance* from 1343 (vol. II, 184–85) states that anyone who has not turned in his old coins after a certain date and is found exchanging them will have all his money seized and his body put at the king's disposal.   [31] Fournial, *Histoire monétaire*, 98.

[32] *Ordonnances*, vol. II, 255 (1346): "dont il advient chascun jour que quant li bon marchanz vendent leurs denrées à certain priz, selon la valüe de la monnoie qui court au jour de la vente, et iceux marchanz donnent aucun terme de leur payement, le priz desdictes monoies est si creu par les voies dessusdites, avant ledit payement, que lesdiz marchanz perdent une grande partie de leur debte."

palace of the dauphin, where they assassinated a number of noble minis-
ters before the dauphin's eyes and forced the future Charles V to wear the
colors of popular insurrection.[33] Scenes such as these, often violent,
profoundly unsettling, and turning on the growing power of money and
commercial interests in society, surrounded the students and masters in
Paris from the beginning of the century. The bold innovations in
measurement and quantification found in the writings of these masters
evolved in the context of a society being transformed by the social reality
of monetization.

Each of the more than one hundred mutations over the first half of the
fourteenth century required a publicly posted *Ordonnance* listing the
proportional equivalences between old and new coins (sometimes more
than a dozen circulating coins were involved), so that rents, debts, and
contracts could be paid or repaid in proportional values. The complexity
of such constant, legally mandated reproportioning was enormous and
touched every level of society.[34] Not surprisingly, in order to make the
best out of a bad situation, people began to counter the negative effects of
debasement by paying the closest attention to shifts in the monetary
market and by trying to anticipate the changes they knew would come.[35]
Hoarding and the melting down of coins with relatively high silver
content were practiced on a large scale, at every level of society.[36] Still,
the complications and loss of income associated with monetary debasement were real and painful, especially to the propertied classes who saw
their rental income progressively diminished as the coinage weakened.[37]
Echoes of the complaints of both rich and poor concerning monetary
instability began to appear in the *Ordonnances* by 1303. In that year the
king declared that he was well aware of the "grief, damage, and loss" his
subjects had suffered and continued to suffer due to the mutation of the
coinage.[38] He promised on behalf of the "common profit" that, when the

[33] Michel Mollat and Philippe Wolff, *The Popular Revolutions of the Late Middle Ages*, A. L.
Lytton-Sells (tr.) (London, 1973), 120ff. Raymond Cazelles ("La stabilization," 298) notes that
popular fury directed against monetary changes prepared the way for Etienne Marcel's revolt. In
the background were the financial strains placed on the French crown by the ransom of King Jean.

[34] Spufford, *Money and Its Use*, 290.   [35] Cazelles, *Société politique*, 24ff.

[36] Fournial, *Histoire monétaire*, 88.

[37] Cazelles, "La stabilization," 294–96. Cazelles (*Société politique*, 21–24) speculates that the Valois
princes and their economic advisors were well aware that debasement helped modify economic
inequalities between renter and *rentier*, and that, in fact, periodic debasements were used as a means
of easing the economic burden on the poorer sector of the population.

[38] *Ordonnances*, vol. I, 384–85 (1303): "Nous qui regardons les griés, les domages, et les pertes que il
ont longuement eu, souffert et soustenu, ont et souffrent de jour en jour, par les müemens de noz
Monoies, à la requeste et à la priere de noz diz Prelaz et Barons presens, octroions et promettons
[within a year] . . . faire bonne monoie . . . de la valüe de quoi estoient ceux qui couroient au temps
le Saint Roy Loys nostre ayeul."

economic demands of the current war had abated, he would reform the money and restore it to the stable value it had held in the time of St. Louis.[39]

When Philip IV, temporarily relieved from the economic burdens of war, made good his promise to revalue the currency in 1306, he and his advisors were shocked by the immediate and furious reaction on the part of the Parisian lower classes or *menu peuple*.[40] Infuriated by the king's policy, which led to the effective tripling of their rents, townspeople rushed to the home in which the king was staying and trapped him inside, refusing for a day to let him leave. They then turned to the home of Philip's chief economic councilor, Etienne Barbette, and in their fury destroyed his house and his garden. The unprecedented actions of the crowd against their king served to awaken even insulated observers to the reality of the new power of money and monetary interests within their society.

French chroniclers of the period not only reported this incident but attempted to explain its economic causes, particularly the hardships on renters produced by the revaluation.[41] Such economic understanding and detail was new to French chronicles in the fourteenth century.[42] Before the 1290s the chronicles had remained almost innocent of economic

---

[39] *Ordonnances*, vol. I, 389 (1303). The silver content of the *denier* or *gros* of St. Louis remained the standard to which successive French kings sought to return the debased coinage in those rare periods when they found themselves financially able to strengthen the currency. The strong currency of St. Louis ceased to be held as an ideal only after 1343 when the king's economic advisors had come to recognize that there were simply insufficient supplies of silver for such a revaluation. On this, see Cazelles, "Mutations de la monnaie," 92–96.

[40] There were anticipations of problems, particularly in the repayment of debts and the payment of rents in the newly strong currency. The solution was to order that all debts be paid back in strong money proportional to the value of the weak money at the time of the loan. See *Ordonnances*, vol. I, 443–47 (1306). Note the legal requirement for the citizen to reason and act proportionally – proportionality defined here as numbered ratios of old and new coin values.

[41] *Chronique Latine de Guillaume de Nangis de 1113 à 1300, avec les continuations de cette Chronique de 1300 à 1368*, Hercule Géraud (ed.), 2 vols. (Paris, 1843), vol. I, 355: "Occasione mutationis monetae debilis in fortem, damnosa seditio, praecipue propter locationes domorum, Parisius exorta est. Cum enim cives Parisius locare domos, et earum locationis pretium in forti moneta, juxta regale statutum, recipere niterentur; (quod tamen communis populis multitudini grave nimium propter triplicationem consueti pretii videbatur) tandem aliqui ex popularibus ipsis tam contra regem quam contra cives conjuncti. " See a similar account in the *Chronicon Girardi de Fracheto et anonyma ejusdem operis continuatio*, vol. XXI (Paris, 1855), in *Recueil des historiens des Gaules et de la France*, and in French in the *Chronique Parisienne anonyme de 1316 à 1339*, vol. XI in A. Hellot (ed.), *Mémoires de la Société de l'histoire de Paris et de L'Ile-de-France* (Paris, 1885), 18–20.

[42] For earlier notices in French chronicles on economic matters, see Guillaume de Nangis' record of the revolt of the townsmen of Rouen against the king's tax collectors in 1292 (*Chronique Latine de Guillaume de Nangis*, vol. I, 282). See also the notice by the monks of St. Denis of the economic battle between Philip IV and Boniface VIII over clerical taxation in 1296, and their analysis of the economic dimension of the war with Flanders in 1302 in *Les Grandes Chroniques de France*, Jules Viard (ed.), vol. VIII (Paris, 1934), 167, 201.

matters, remarkably so in comparison to the financial interest and knowledge displayed by thirteenth-century English chroniclers such as Matthew Paris.[43] By the early years of the fourteenth century, however, the undeniable evidence of monetization had opened even the highly formalized genre of the French chronicle to the cognizance of economic and monetary factors as historical forces.[44]

In the aftermath of the Parisian renters' riot, twenty-eight of the ringleaders were hanged. But if Philip was able to maintain his control over monetary policy at this time, he had considerably less success in his attempt to regulate and similarly control the crucial markets in grain and bread. In theory, medieval princes maintained the right to intervene in the market, especially during times of scarcity and steep price rises. In the fourteenth century, however, the kings of both France and England found that they were unable to do so – that the market resisted their attempts at control.

A systematic conception of the marketplace, in which it is understood that prices rise and fall according to the availability of goods in relation to common need (*indigentia communis*), is almost a requirement for that segment of the population whose livelihood centers on trade. An examination of the journals of fourteenth-century merchants reveals how sophisticated their understanding of market prices had become, and how central this understanding was to every aspect of their activity.[45] Noteworthy in our period was the penetration of the concept of the market as a dynamic, self-regulating system into the consciousness of people not directly engaged in commerce – most importantly for our purposes, into the consciousness (and economic writings) of those schoolmen whose speculation transformed the model of nature in the fourteenth century.[46]

The comprehension of the market as a kind of mechanism was rein-

---

[43] For Matthew's "rather mercenary outlook on life," see Richard Vaughan, *Matthew Paris* (Cambridge, 1958), 6, 145–46. Such economic precocity on the part of Matthew can be explained partly by his background and personality, and partly by the advance in English royal bureaucracy, tax collection, monetization, and market development over that of France. As Matthew himself said, everyone in England, from baron to prelate to merchant to agricultural worker, was laboring under the "pestilential infection" of royal and papal taxation: *Matthew Paris' English History*, vol. II, 510. For the precocious development of English monetary and market consciousness as witnessed by Matthew's accounts, see his inclusion of year-end observations on the shifting market prices for grain, bread, and wine in vol. II, 42, 287; vol. III, 60, 110, 155, 265, 291.

[44] The extension and refinement of this consciousness over the first half of the fourteenth century can be witnessed in the work of the last Dionysian redactor of the *Grandes Chroniques* from 1337 to 1347. See *Les Grandes Chroniques de France*, vol. IX (Paris, 1934), esp. 164, 235, 245, 285.

[45] See for example Yves Renouard, "Les hommes d'affaires italiens et l'avenement de la Renaissance," in *Etudes d'histoire médiévale*, 419–37; Iris Origo, *The Merchant of Prato: Francesco di Marco Datini* (London, 1957), esp. 13–20, 73–75.

[46] For the early development of the concept of market order within scholastic economic thought, see chapter 4. For later, fourteenth-century refinements, see chapter 5, 120–33 and 159–61.

forced by the growing observation of its capacity to define values not only in the absence of, but also in *spite* of, attempts by governing agencies to regulate and to control it. Both the *Ordonnances* and French Chronicles offer a clear view into the growth of this market consciousness. The earliest case recorded in either source of the king's failure to regulate the market occurred in 1304–05. A scarcity of grain in that year, combined with monetary instability, had sent prices soaring, and a series of *Ordonnances* were issued to counter this rise. The order went out for royal officers to inspect all the towns and villages surrounding Paris to discover how much grain the farmers were hoarding and how much each family needed for food and seed. Everything not necessary for survival was to be taken to market, but as the law ordered, *petit à petit*, so that the market would not be dislocated further.[47] When this failed, and prices continued to rise, a *Mandemant* set maximum grain prices at one-third their current price, and again the farmers were ordered to bring their grain to market.[48] The law had precisely the opposite of its desired effect. Its attempted enforcement proved disastrous. Grain promptly disappeared throughout Paris, to the extent that bakers were forced to close their shops.

The continuator of Guillaume de Nangis' *Chronicle* covered this economic incident in hitherto unmatched detail. He noted the specific price of wheat at the time of the famine, the maximum price set by the edict of King Philip, the failure of the market to conform in any way to the king's edict, the worsening of the situation due to the attempt at governmental control, and the resulting violence of the crowds surrounding the closed shops of the bakers.[49] Recognizing his failure after a month of continuing unrest, King Philip, who was not in the habit of reversing himself on any subject, was forced to revoke his maximum price edict and to substitute an edict requiring simply that grain not be hoarded or sold to middlemen, but be sold openly and publicly at whatever its market price.[50]

A similar English attempt to regulate the market through maximum price edicts on common foodstuffs during the famine year of 1315 met

47 *Ordonnances*, vol. I, 424–25.
48 *Ordonnances*, vol. I, 426. See Harry A. Miskimin, *Money, Prices, and Foreign Exchange in Fourteenth-Century France* (New Haven, 1963), 21–22, for a similar example from England in the same year.
49 *Chronique Latine de Guillaume de Nangis*, vol. I, 346: "Porro cum regio proclamatum fuisset publice edicto ne ultra quadriginta solidos venderetur, necdum tamen propter hoc cessavit caristia, sed adeo magis invaluit, ut Parisius panifici, qui panem venalem ad sufficientiam habere non poterant, claudere compellerentur fenestras et ostia, ne eis a pressura communis populi per violentiam auferrentur."
50 *Ordonnances*, vol. I, 426: "Mandemant au Bailli de Vermandois, contenant reglement pour le prix des Grains" (1304). Farmers are given the order: "vendre et donner pour tel pris comme il en pourra avoir, et que l'acheteur en voudra donner atrempement toutesvoies, ou au mains faire savoir, et crier communelment par les marchiez, que il a grain à vendre, en disant la quantité et que volentiers le vendra qui achater le voudra."

with similar failure and was also promptly withdrawn.[51] It too drew the attention of a contemporary chronicler:

The regulations formerly made about food were completely abolished . . . For as a result of that statute little or nothing was exposed for sale in the markets, whereas formerly there had been an abundant market in goods, though they seemed dear to travelers. But it is better to buy dear than to find in the case of need that there is nothing to be had. For although scarcity of corn raises the price, subsequent plenty will improve the situation.[52]

Throughout this period there remained certain customary and legislated constraints on medieval prices and wages. Both princes and city councils continued to regulate (often in minute detail) the quality, quantity, and, at times, the prices of goods in exchange. Manufactured articles in particular were subject to price regulations. Nevertheless, in the all-important area of foodstuffs, a market sensitive to the interplay of supply and common need and operating independently of external control both existed and, by the fourteenth century, was commonly understood to exist.[53]

A final demonstration of the organization of markets beyond the king's control in this period, or beyond any conscious control for that matter, was the evolution in France of a floating market price for money itself in response to constant monetary debasement. The unofficial market in money functioned independently of the king's legally established *cours* or price, and was maintained in direct contravention of his commands and threats.[54] Once again the king blamed malicious changers, exchange

---

[51] Miskimin, *Money and Prices*, 22. See also the revocation in 1315 of Philip IV's attempt to control the traffic on the Seine through duties and tolls: *Ordonnances*, vol. I, 598.

[52] N. Denholm-Young (ed. and tr.), *Vita Edwardi Secundi: The Life of Edward the Second* (London, 1957), 69: "Ordinationes super uictualibus prius facte penitus dissoluuntur . . . Nam ex quo processit illud statutum, nichil vel modicum in foro reperiebatur expositum, cum tamen prius habundaret forum venalibus, licet cara viderentur transeuntibus. Porro melius est emere care quam nichil emendum ad opus inuenire. Nam licet raritas annonam facit cariorem, habundantia subsequens reddet meliorem."

[53] In the judgment of Jacques Rossiaud ("The City Dweller," 150): "all city-dwellers willy-nilly were sensitive to the stability of their money, to movements of the money market, and to events that affected supply and demand, wholesale and retail." See also Britnell, *Commercialisation*, 93–94; Postan, *Medieval Economy*, 228; Miskimin, *Money and Prices*, 21; Philippe Wolff, *Commerces et marchands de Toulouse (vers 1350–vers 1450)* (Paris, 1954), 560.

[54] *Ordonnances*, vol. II, 254 (January, 1346): "Et pour ce que il Nous avoit esté rapporté que in nostredict Royaume, tout communement se mettoient et prenoient toutes monnoies d'or et d'argent de quelques coings que il fussent, tant du nostre, comme d'autrui, et mettoit chascun sur lesdites monoies, tant d'or comme d'argent, tel pris comme il li plaisoit, et à la volenté. " For a similar complaint, see *Ordonnances*, vol. II, 390 (March, 1350). Fournial (*Histoire monétaire*, 27) noted that, although all coins received an official *cours*, they were, long before 1346, commonly fixed in value by what he called a "collective consensus." The king's exasperation here reflected the fact that the legal *cours* had become at this point completely ignored and unenforceable.

dealers, and silver exporters for causing the price of coins to change from day to day without regard to the legal *cours*.[55] Here, even in an area in which the king's right of control was recognized by all, shifting market prices had established themselves in place of mandated price.

In both England and France the economic situation was even more tumultuous after the Black Death than before it. The dislocation of the market in land, labor, and agricultural products was so severe that, beginning in 1349, governments of both countries repeatedly passed laws to hold wages and prices within acceptable bounds. The general failure of these laws, often admitted in the prologues of their subsequent re-issues, further reinforced the recognition of the marketplace as a system functioning independently of external direction and control. Nevertheless, the severe economic dislocations resulting from the plague, primarily to the disadvantage of the landowning class, led administrators and lawyers to redouble their efforts to establish legal limits for prices and wages, covering an extraordinarily wide range of products and services.[56] It led them to institute, as well, a complicated system of graded monetary penalties for those who ignored the limits imposed. In the comprehensive *Ordonnance* of February 1350, for example, twelve paragraphs were devoted to setting limits for bread prices (with twenty-six separate breads and breadweights listed), twenty paragraphs to regulating tavern keepers, twelve to butchers, and almost fifty to fisherman. Laborers of all kinds were forbidden to charge more than a third more for their labor than they had been accustomed to receive before the great mortality.[57] Administrators even entered the slippery area of commercial profits, limiting merchants to 10 percent on unimproved goods (two solidi out of twenty) and promising one-fifth of the collected fines to those who brought the offenders to the attention of the authorities.[58] In the decade following the Black Death unprecedented administrative interference in minute details of economic life affected all levels of French and English society, as

---

[55] *Ordonnances*, vol. II, 254–55: "plusieurs malicieuses genz et cauteleus, en venant presomptueuse-ment contre nostredit cry et deffense, et pour decevoir et defrauder les bons marchanz et les autres bonnes genz, qui ladite fraude pas ne cognoissent, prennent encore et mettent toutes monnoyes d'or et d'argent, en leur donnant tel prix comme il leur plaist."

[56] A large number of documents illustrating legal and governmental attempts at market control in England after the plague have been collected and translated in Rosemary Horrox (ed.), *The Black Death* (Manchester University Press, 1994), esp. 276–338.

[57] *Ordonnances*, vol. II, 369 (February, 1350): "ne pourront prendre et avoir pour leurs labeurs et journées que le tiers plus outre ce qu'on en soulout donner avant la mortalité, tant en tasche comme en journée, et non plus." Both the English *Ordinance of Labourers* (1349) and the *Statute of Labourers* (1351) set post-plague wages and prices at the level customarily obtaining in the year before the plague. [58] *Ordonnances*, vol. II, 377–78.

agricultural laborers, crafts workers, merchants, mercenary soldiers, and even royal bureaucrats were required, on pain of punishment, to follow administrative guidelines on wages and prices.[59] And all this occurred in a period in which the French currency was reaching its stage of maximum instability, with fifty-one separate mutations of the coinage between 1355 and 1360.[60]

Every major *Ordonnance touchant les monnoies* from 1313 ended with a solemn command to the bailiffs and royal officers of France to publish and publicly announce the details of the law in all the towns and "notable places" within their jurisdiction. This must be done, according to the *Ordonnance* of 1313 and those that followed, so that no one might be able to excuse himself from punishment by reason of ignorance of the law's content.[61] In short, from the early years of the fourteenth century, no one within the sphere of the king's law, from noble to peasant, could have afforded to remain ignorant of the frequent detailed changes in the coinage. No one could have avoided the complex and tedious reproportioning of new coins to old prices and debts that each new issue demanded. The penalties for ignorance were too stiff – economic loss, confiscation, imprisonment, and (in the most desperate of the *Ordonnances*) bodily harm.[62]

It is difficult to imagine that anyone living in Paris, including the students and masters from all Europe who spent their years at the university in its center, could have remained untouched by the social and economic upheavals of this half-century. Fortunately, we do not have to rely on supposition to assert the strong involvement of Parisian scholars in the economic and social life of their time. From what is known of the financial and administrative complexities surrounding the Nations and colleges at the University of Paris, and the rotating involvement of the teaching masters in the routine performance of administrative duties, an assumption of scholarly isolation from economic life is impossible to

---

[59] The full range of administrative involvement in economic life appears in an *Ordonnance* from November 1354, vol. II, 564–602. Note that, even while insisting on setting administrative limits to prices and wages, the authors recognize that market values must be taken into consideration: "C'est assavoir que il ordonnerons commant et pour quel pris les denrées de vivres, et toutes autres denrées vendables seront vendues, et aussi taxeront justement les pris des journées de tous Ouvriers et Laboureurs de certaines saisons, et temps à autres, et les loyers et salaire de tous Servans et Servantes, eue consideration au marché et a la cherté des choses estans en leur pays."
[60] Fournial, *Histoire monétaire*, 98.
[61] *Ordonnances*, vol. I, 524: "Et pour ce que nul ne se puisse escuser d'ignorance, Nous mandons et commandons a tous seneschaux, Baillis et Prevosts que il hastivement et sans nul delay en leurs assises, et en toutes les bonnes villes . . . fassent solemnement publier, crier et preconiser toutes les Ordonnances et deffenses dessusdites."
[62] Bodily harm is threatened in *Ordonnances*, vol. II, 255 (January, 1346), and 391 (March, 1350).

maintain.[63] Unlike the bureaucracies of pope and king, the university had no professional administrative staff, no scribes whose primary employment was the preparation of university records. Whatever records were kept, including financial records, were the responsibility of those masters who filled the rapidly rotating positions of procuror and rector.[64] The continual distribution of the weight and care of administration among numerous masters was assured by the practice followed by each of the Nations at Paris of holding a new election for procuror *monthly*.[65] The participation of masters in the administration of their university is clearly illustrated in the biographies of two of the most illustrious scholars of the fourteenth century, Jean Buridan and Nicole Oresme.

Of all the fourteenth-century thinkers who wrote on economic questions, Buridan and Oresme had the deepest insights into the workings of the monetized marketplace.[66] One of the factors most responsible for this economic sophistication was their active involvement in university administration. As was often the case in the fourteenth century, the intellectual lives of Buridan and Oresme combined an acute perception of economic life with the most penetrating speculations on the natural world.

The life of Nicole Oresme, student, teacher, philosopher, bishop, advisor at the court of King Jean II and later at the court of King Charles V, provides a model of that interaction of influences and duties – religious, philosophical, administrative, and financial – so common to the scholar's experience in this period. It was within the university that Oresme was first raised to positions of administrative power and responsibility. He began his years at Paris as a poor student on scholarship. After receiving his Master of Arts degree in 1348, he turned to the study of theology while in residence at the College of Navarre, the largest of Paris'

---

[63] The best surviving records of the practical administration of the University of Paris in this period are the *Libri procuratorum* of the various Nations. Financial transactions and concerns dominated these administrative records and, consequently, the administrative energies of those masters who were elected to serve as *procuror*. Unfortunately, for the period of the early fourteenth century, only the records from the English (or German) Nation survive: *Liber procuratorum Nationis Anglicanae (Alemanniae) in Universitate Parisiensi*, H. Denifle and E. Chatelain (eds.), in *Auctarium chartularii Universitatis parisiensis*, vol. I, *1333–1406*, 2nd edn. (Paris, 1937).

[64] This point is made forcefully by William Courtenay in "Registers of the University of Paris," 18.

[65] For information on these offices see *Liber procuratorum*, vol. I, xxi–xxiv; Lowrie J. Daly, *The Medieval University 1200–1400* (New York, 1961), 51–58. For the monetary measurement of virtually every stage of a student's academic life at the university, see Verger, "Le coût des grades." For the complex financial transactions involved in the foundation and maintenance of even a small college at the University of Paris in the early fourteenth century, see Astrik L. Gabriel, *Student Life in Ave Maria College, Mediaeval Paris* (Notre Dame, Ind., 1955), esp. 61–81.

[66] The important contributions of Oresme and Buridan to both economic and scientific thought in the fourteenth century are discussed at many points in chapters 5, 6, and 7.

colleges in the fourteenth century. In 1355 or 1356 he incepted as master of theology and was elected grand master of his college.[67] Such a position carried with it great administrative and financial responsibilities along with great honor, as did all university offices at this time. Part of Oresme's duties as grand master involved the direct receiving and paying out of monies for the college's expenses. In order to do this, along with his office he was presented with one of three keys to the college treasure box. Furthermore, he had the responsibility at the annual audit to verify the accounts of the college's other financial officers for the entire year.[68]

Oresme undertook these administrative duties at the same time that French coinage was in its most unstable period. Between 1355 and 1360 there were more than fifty separate mutations of the coinage, mostly devaluations, necessitating the almost continual reproportioning of prices, fees, and expenditures.[69] One can imagine the problems involved in keeping accurate financial accounts during this tumultuous period. We do not have to surmise that Oresme was conscious of, and greatly troubled by, the monetary anarchy in the society around him. We have the record of his knowledge of monetary affairs and his sharp complaints about the damage to society caused by monetary mutations in his remarkable treatise on money and minting, the *Tractatus de origine et natura, iure, et mutacionibus monetarum*, written sometime between 1355 and 1360.[70] This work – in effect Europe's first practical treatise on money, minting, and monetary policy – remained the most influential treatise on these subjects into the sixteenth century.[71] In it, when Oresme complains that

[67] Good, short biographies of Oresme can be found in the introduction to *Maistre Nicole Oresme: Le livre de éthiques d'Aristote*, Albert Douglas Menut (ed.) (New York, 1940), and the introduction to *Nicole Oresme: "De proportionibus proportionum" and "Ad pauca respicientes,"* Edward Grant (ed. and tr.) (Madison, 1966). For a study of Oresme's clerical career, see François Neveux, "Nicole Oresme et le clergé normand du XIVe siècle," in Quillet, *Autour de Nicole Oresme: Actes du Colloque Oresme*, 9–36.

[68] The description of Oresme's duties as grand master is from Francis Meunier, *Essai sur la vie et les ouvrages de Nicole Oresme* (Paris, 1857). See also Ernest Borchert, *Die Lehrer von der Bewegung bei Nicholas Oresme* (Münster, 1934), 4–7. An officer with the stature of grand master of Navarre might be asked to take on additional administrative tasks from outside the college. See, for example, Gabriel, *Ave Maria College*, 124, where the smaller Ave Maria College asked the grand master of Navarre to act as one of the college's administrative and financial executors.

[69] Fournial, *Histoire monétaire*, 98.

[70] This work has been edited with facing-page translation by Charles Johnson, *The "De moneta" of Nicholas Oresme and English Mint Documents* (London, 1956). For the earlier date of 1355, see Emile Bridrey, *La théorie de la monnaie au XIVe siècle: Nicole Oresme* (Paris, 1906), 52. More common is the date given by Menut of 1356. A recent study of the work is more cautious in allowing the period 1355–60. On this, see Denis Menjot, "La politique monétaire de Nicolas Oresme," in P. Souffrin and A. Ph. Segonds (eds.), *Nicolas Oresme: tradition et innovation chez un intellectuel du XIVe siècle* (Paris, 1988), 179–93.

[71] For information on the manuscripts, translations, and printed editions of this work, see Johnson in Oresme, *De moneta*, xii–xviii; Lucien Gillard, "Nicole Oresme, sujet théorique, objet historique," in Quillet, *Autour de Nicole Oresme: Actes du Colloque Oresme*, 195–233. Oresme's work on minting

as a result of the king's constant monetary debasements, "innumerable perplexities, obscurities, errors, and insuperable difficulties occur in accounts of expenditure and receipts," we can hear the voice of a harried university administrator.[72] If the current dating of the *De moneta* to between 1355 and 1360 is correct, then Oresme gained the economic experience to write this influential work during his university days, either while a student pursuing his doctorate in theology, or at the latest, in his first years as grand master of the College of Navarre.[73]

Oresme's place as the preeminent scholar and natural philosopher at the University of Paris in the middle of the fourteenth century was rivaled only by that of his teacher, Jean Buridan. Born in the last years of the thirteenth century, Buridan is not mentioned in the surviving records of the university before the announcement of his election as rector in 1328.[74] The office of rector was the highest titular and administrative office within the university. It is probable that this *vir venerabilis et discretus* (as he was described in the notice of his election) was chosen for such an important administrative position only after having established his reputation as a capable administrator in lesser offices.[75] In addition to being elected rector a second time in 1340, Buridan is recorded as having been appointed by the university to argue for its exemption from the king's salt tax in 1344, and having been part of a committee that redefined the financial and administrative functions of the *receptor* (receiver or treasurer) of his own Picard Nation at the university.[76] Clearly this was a man at home in the world of administration.

did not have a rival until the treatises written by another great scientist, Nicholas Copernicus, almost two centuries later.

[72] *De moneta*, ch. 21, 35: "Preterea, istis durantibus, quasi innumerabiles perplexitates, obscuritates, errores, et inextricabiles difficultates accidunt in compotis, de mis[i]is et receptis."

[73] The probable date of Oresme's doctorate in theology is 1356, the same year he became grand master. This is also the date Menut considers "fairly certain" for the writing of the *De moneta*.

[74] Edm und Faral, "Jean Buridan, maître ès arts de l'université de Paris," in *Histoire littéraire de la France*, vol. xxxviii (Paris, 1949), 462–605, esp. 469. The dating of Buridan's many works remains uncertain, but it is possible that all the surviving works were written after the period of Buridan's first rectorship. The most comprehensive study of Buridan's life and work is presented in Bernd Michael, *Johannes Buridan: Studien zu seinem Leben, seinem Werken, und zur Rezeption seiner Theorien im Europa des späten Mittelalters*, 2 vols. (Berlin, 1985). For the problems in dating Buridan's work, see Michael, *Buridan*, vol. 1, 259–85; Faral, "Buridan," 494–95.

[75] The announcement of Buridan's election as rector and this assessment of his character are found in *Chartularium Universitatis Parisiensis*, H. Denifle and E. Chatelain (eds.), vol. ii (Paris, 1891), 306, no. 870. C. E. Du Boulay, the seventeenth-century historian of the university, reported that in addition to rector, Buridan had earlier several times occupied the post of procuror for the Picard Nation, but this information can no longer be confirmed since the probable basis of Du Boulay's information, the *libri procuratorum* of the Picard Nation, have been lost: C. E. Du Boulay, *Historia Universitatis Parisiensis*, 6 vols. (Paris, 1665–73), vol. iv, 996.

[76] Faral, "Buridan," 469–77; *Chartularium*, vol. ii, 608–10, no. 1146. Buridan's role in arguing for the university's exemption from the salt tax is reported by Du Boulay but cannot be confirmed by existing records.

Buridan was particularly successful in an economic area of concern to most scholars in this period: the earning of personal income through the award and accumulation of church benefices.[77] He held five rich benefices at one time, which allowed him a considerable income, including servants and extensive properties that required his careful management.[78] His stature in Parisian society beyond the university is indicated by the legends that grew up around him. In them he is celebrated primarily for his cleverness, often in regard to the accumulation of wealth, and always as an actor very much involved in the society outside the schools. A century after his death French poets, including François Villon, were recounting the story of the philosopher who made love to the queen of France and through guile escaped with his life and an armful of silver.[79]

The biographies of Oresme and Buridan demonstrate that the University of Paris in the fourteenth century was an institution whose setting and bureaucratic structure actively connected the lives of its members to the economic life of the society beyond the school. Can the same be said of the other great center of natural philosophy in the fourteenth century, Oxford University, which preceded Paris as the intellectual center for the logical and mathematical investigation of measurement?[80] While Oxford was a bustling market town in this period, it was not, like Paris, a great center of commercial and administrative activity. Were, then, the scholars that spent most of their careers at Oxford more protected from the economic and social developments related to monetization than their Parisian counterparts?

The connection between the Oxford scholar and the larger world of affairs in the fourteenth century has been illuminated by recent studies on Merton College.[81] Merton College, the birthplace of the proto-science

---

[77] For a discussion of the "market value" of an academic degree in an English university, see William Courtenay, *Schools and Scholars in Fourteenth-Century England* (Princeton, 1987), 118–46. See also R. N. Swanson, "Learning and Livings: University Study and Clerical Careers in Later Medieval England," *History of Universities* 6 (1987), 81–103; Jean Dunbabin, "Meeting the Costs of University Education in Northern France c. 1240–1340," *History of Universities* 10 (1991), 1–27.

[78] Faral ("Buridan," 477ff.) talks not only of Buridan's income, but also of his later reputation far beyond the schools as a man of wealth and property. For the details of Buridan's beneficed wealth, see Michael, *Buridan*, vol. I, 206–34; Astrik L. Gabriel, *The College System in the Fourteenth-Century Universities* (Baltimore, n. d.), 10. For other impressive lists of benefices held by influential natural philosophers of this period, see, for example, Walter Burley's in A. B. Emden, *A Biographical Register of the University of Oxford to* AD 1500, 3 vols. (Oxford, 1957–59), vol. I, 313; Richard Kilvington's, in Emden, *Biographical Register*, 1051; and Thomas Bradwardine's, in Emden, *Biographical Register*, 244.   [79] Michael, *Buridan*, vol. I, 286ff.; Faral, "Buridan," 480–85.

[80] For Merton College as the center of studies in the mathematical and physical sciences and for a bibliography of works produced by Merton scholars, see James Weisheipl, "*Repertorium Mertonense,*" *Mediaeval Studies* 31 (1969), 174–224; Weisheipl, "Ockham and Some Mertonians," *Mediaeval Studies* 30 (1968), 163–213, 164; Weisheipl, "Dumbleton," 441.

[81] J. I. Catto (ed.), *The Early Oxford Schools* (Oxford, 1984), vol. I of *The History of the University of Oxford*; in this context see: J. I. Catto, "Citizens, Scholars, and Masters," 151–92; T. H. Aston and

of *calculationes*, was unique among the early schools at Oxford for a number of reasons: its intellectual preeminence, especially in mathematics and natural philosophy; its relatively large size (varying, but averaging approximately forty-five fellows); and its wealth as measured by the extent of its endowed possessions. It was unique as well in the care it took to preserve its administrative records. These records reveal two principal facts: (1) the range and complexity of the economic transactions involved both in the colleges' daily functioning and its continued institutional survival; and (2) the full participation of the students and masters of the college in this practical, everyday administration.[82]

Most of the bureaucratic functions of the college were performed by its fellows. The three bursars who had collective responsibility for the college's monies, the warden who headed the yearly audit and visited the far-flung manors at harvest time to assess the year's taxation in money or in kind, those who oversaw the books and calculated the varying profits of the college's many properties, the overseers of the daily victualing of the college, the agents who made constant visits to nearby manors to supervise their legal and economic administration, the five who assisted the warden and the bursars in preparing the complex yearly audit that lasted almost three weeks – all were in-house administrators; all were recruited from the fellows of Merton College.[83] This was the state of affairs existing at the same time that the fellows of Merton College were leading Europe in their imaginative new approaches to the study of the natural world. Surviving records permit us to go past the general case to consider the administrative careers of two leading fourteenth-century natural philosophers.

Thomas Bradwardine was among the most influential of the Merton Calculators. His treatise on the mathematics of motion, the *De proportione*

Rosamond Faith, "The Endowments of the University and Colleges to c. 1348," 265–310; T. H. Aston, "The External Administration and Resources of Merton College to c. 1348," 311–68; Jean Dunbabin, "Careers and Vocations," 565–606.

[82] Aston, "Administration of Merton," 311–12: "here we can see a body of fellows, for much of the period the most distinguished intellectual group of the whole university, coping with practical problems, ensuring their own support while at the same time accumulating the monies needed for most grandiose building projects, and (we may think) joining together . . . in a range of administrative tasks that touched most facets of rural life." See also Alan Cobban, *The Medieval English Universities: Oxford and Cambridge to c. 1500* (Aldershot, 1988), who considers the active involvement of teaching masters in administrative affairs one of the "most arresting features" of the medieval English university (96).

[83] For the duties of the bursars, see Aston, "Administration of Merton," 331; the warden, 331–32; the profit Calculators, 337–38; the victualers, 348, 362–64; the traveling agents, 333; the yearly audit, 344. According to Aston (364), the college generally preferred to take the profits from its properties in cash rather than in supplies in kind, and to pay cash for its food needs. He surmises that the warden and the bursars consciously did this to give their bailiffs the opportunity to "play the market" with their produce and so increase the profitability of the farm.

velocitatum in motibus (c. 1328), contained the first workable mathematical function describing the proportional relationships involved in the measurement of motion.[84] Once formulated and introduced into philosophical debate, this mathematical function was avidly applied by successive thinkers to a wide range of philosophical and theological problems.[85] Before writing the *De proportionibus*, Bradwardine twice served in the important administrative post of proctor of the university (1326–27, 1327–28).[86] While still at Oxford he became attached to the intellectual circle organized around the wealthy Richard de Bury – Bishop of Durham and at one time treasurer and chancellor of England. Within this circle, Oxford natural philosophers including Bradwardine, Walter Burley, and Richard Kilvington came into close contact with men from the highest levels of the royal administration.[87] Bradwardine's life, like Oresme's in France, illustrates the connection of the university bureaucracy with the bureaucracies of Church and monarchy. After serving as university proctor, he remained at Merton until 1335, at which point he joined the household of de Bury. In 1337 he was appointed chancellor of St. Paul's (1337) and later served the king as an ambassador to France and as royal chaplain, ending his life as Archbishop of Canterbury (1349).

One of the most strenuous duties assigned to the fellows at Merton College was the yearly journey to oversee and administer the most northern properties – a journey of well over 200 miles, requiring more than eight days' travel each way. Performing this office on behalf of the college in 1337 or 1338 was the logician and natural philosopher William

[84] See Bradwardine, *Thomas of Bradwardine. His "Tractatus de proportionibus": Its Significance for the Development of Mathematical Physics*, H. Lamar Crosby, Jr. (ed. and tr.) (Madison, 1961). Bradwardine's formulation involved a conscious refutation of earlier mathematical descriptions of motion, including those derived from Aristotle. See also Anneliese Maier, "Der Funktionsbegriff in der Physik des 14. Jahrhunderts," in her *Die Vorläufer Galileis im 14. Jahrhundert* (Rome, 1966), 81–110.
[85] Bradwardine, *De proportionibus*, 11–18; Murdoch, "Mathesis," 225–32. The many uses of this function in fourteenth-century natural philosophy are discussed in chapter 7.
[86] Emden, *Biographical Register*, vol. I, 244–46; Courtenay, *Schools and Scholars*, 242–43.
[87] For the place of these Oxford scholars in the social circle of Richard de Bury, treasurer and chancellor of England, see Courtenay, *Schools and Scholars*, 133–37, 244; also, Courtenay, "London *Studia*," where Courtenay includes in de Bury's London circle Robert Holcot, Walter Seagrave, John Maudith, and John Acton. For a brief biography of the *Ethics* commentator and natural philosopher Richard Kilvington, see Emden, *Biographical Register*, vol. II, 1050–51; Kilvington, *The Sophismata of Richard Kilvington: Introduction, Translation, and Commentary*, Norman Kretzmann and Barbara Ensign Kretzmann (trs.) (Cambridge, 1990), xvii–xxxiv. The biography of one of the most important Mertonians, John Dumbleton, is all but unknown except for his mention as taking part in a Merton College administrative inquiry in 1338–39. On this, see James E. Thorold Rogers, *A History of Agriculture and Prices in England*, 7 vols. (Oxford, 1886), vol. II, 670–74. The economic writings of Burley and Kilvington are considered in chapter 5. Their contributions to natural philosophy, particularly in the realm of qualitative measurement, are considered in chapter 6.

Heytesbury.[88] The records of Heytesbury's approximately two-week task of auditing accounts, supervising the harvests, and collecting revenues in cash and in kind for Merton College do not survive, but the accounts Heytesbury left of this journey indicate the practical training involved in such an exercise. For each day he noted the amounts spent for food and lodging, adding at the end of his account the additional costs, to the half-penny, of each miscellaneous item and service he purchased.[89]

At the time of this journey on behalf of Merton College, Heytesbury had just completed his major work, the highly technical *Regulae solvendi sophismata*.[90] The logical analysis of local motion, velocity, and acceleration found in the *Regulae* represented an important step in the development of a new mathematical physics by the Mertonians.[91] The evidence of this administrative task requiring practical service and financial involvement is particularly interesting in the case of Heytesbury, because of all the Oxford Calculators – indeed, of all the fourteenth-century natural philosophers – his logical and mathematical speculation is among the most seemingly "abstract" and "logic-dominated." The examples he used in his *sophismata* are as drained of immediacy and life as possible – the language dry, abstract, and technical to an extreme degree.[92]

On the surface, then, no other thinker seems to have been living in a world more purely logical, more bounded by a refined intellectual universe, more detached from the world of experience. Thanks to the

[88] The most complete study of Heytesbury's work is Curtis Wilson, *William Heytesbury: Medieval Logic and the Rise of Mathematical Physics* (Madison, 1956). The seeming disjunction between the Mertonians as highly technical logicians on the one hand and as practical economic administrators on the other may have led Wilson to underestimate Heytesbury's role as an administrator. He interpreted (7) Heytesbury's title as bursar of Merton as indicating solely that he was a recipient of a scholarship (*bursarius*). For the meaning of *bursarius* at Oxford, see Olga Weijers, *Terminologie des universités au XIIIe siècle* (Rome, 1987), 93–99.

[89] Aston ("Administration of Merton," 338) gives the list of expenses and a map of the journey. Heytesbury did not end his service with this mission but is later mentioned in the Merton records as a college bursar in 1352–54 and later still (perhaps 1370–72) as chancellor of the university: Emden, *Biographical Register*, vol. II, 927–28; Courtenay, *Schools and Scholars*, 245–47; Weisheipl, "Ockham and Some Mertonians," 195–99.

[90] For partial editions of this work, see: Heytesbury, *William Heytesbury. On "Insoluble" Sentences: Chapter 1 of His Rules for Solving Sophisms*, Paul Vincent Spade (ed. and tr.) (Toronto, 1979); Heytesbury, *William Heytesbury. On Maxima and Minima: Chapter 5 of Rules for Solving Sophismata, with an Anonymous Fourteenth-Century Discussion*, John Longeway (ed. and tr.) (Dordrecht, 1984).

[91] For the continuing influence of this work, see Wilson, *Heytesbury*, 25–28.

[92] John Murdoch discusses this aspect of Heytesbury's work in "The Analytic Character of Late Medieval Learning: Natural Philosophy Without Nature," in L. D. Roberts (ed.), *Approaches to Nature in the Middle Ages: Papers of the Tenth Annual Conference of the Center for Medieval and Early Renaissance Studies* (Binghamton, N. Y., 1982), 171–213, esp. 191–96. Note Murdoch's explanation for Heytesbury's rigidly logical approach (197–98) in terms of the philosophical position that logic possessed a certainty greater than that attainable through experience. See chapter 6 for a further discussion on this point.

great care to preserve the administrative records of Merton College, we know that it would be incorrect to assume that Heytesbury lived a life of scholarly isolation from the tumult of the monetized marketplace and, most probably, similarly incorrect to assume this of any fellow at Merton in the fourteenth century.[93] Rather than protecting the student from the social and economic life of his society, the college often provided him with his earliest and most direct experience of economic realities and economic choices. Far from being an isolating environment, the university of the fourteenth century was a vitally connected and connecting one.

As inhabitants of bustling cities and market towns; as account keepers, tax assessors, fee collectors, and treasurers within the university; as victualers to their fellow students; as benefice holders and office seekers within religious and civic bureaucracies – scholars of the fourteenth century were required to experience, comprehend, and often accommodate their thinking to the insistent realities of money and market exchange. But administrative service did more than simply connect the scholar to the economic life of his society. It brought with it a highly structured way of conceptualizing problems; it encouraged the habit of translating qualitative grades into quantitative terms; it educated scholars in the techniques of standardization, calculation, measurement, gradation, and relation; and it taught them the clear benefit of these techniques in intellectually ordering the world. In the first half of the fourteenth century one sees these same techniques coming to dominate the philosophical, proto-scientific approach to the study of nature within the schools. Bureaucratic techniques developed to impose order on social life, and generally recognized as being successful in doing so, became part of the intellectual arsenal of the scholar seeking to find order in nature.

---

[93] Heytesbury is the only one of the well-known Merton Calculators who is mentioned by name to have taken part in this administrative journey to the north. That does not mean that the other Calculators avoided these tasks. Most administrative officers remained anonymous. It is Aston's belief ("Administration of Merton," 334) that the considerable administrative tasks of the college were widely shared and that some degree of participation for fellows in the administration of the college "was altogether normal." For the reflection of this bureaucratic activity in the questions the Calculators asked and in the solutions they derived, see chapter 6.

Chapter 2

THE ARISTOTELIAN MODEL OF MONEY AND ECONOMIC EXCHANGE

When economics and science are viewed in modern terms as separate and self-contained subjects of inquiry, it is difficult to envision how easily medieval thinkers exchanged approaches, models, and insights between these spheres. Philosophers of the thirteenth and fourteenth centuries show no hesitation in linking their economic thought (at that time considered under the larger heading of *scientia moralis*) with their speculations on nature. In the Aristotelian system of classification, "practical science" governing the investigation of ethical questions, including the question of justice and just exchange, utilized the same methodology as "theoretical science" governing physical thought; they differed in degree of certainty but not in approach.[1]

The previous chapter outlined social and material developments in the thirteenth century that led to the intensification of monetary consciousness and encouraged the transference of perceptions and techniques from the economic sphere to other spheres of knowledge. From the middle of the thirteenth century this transference was encouraged from within the intellectual culture of the university as well, as the writings of Aristotle came to occupy an ever greater position within the curriculum. Given the great prestige of Aristotle within the university, his mathematical, geometrical, and markedly "scientific" treatment of economic exchange in Book v of the *Nicomachean Ethics* provided an influential, even authoritative, textual ground for the linking of scholastic economic thought and scientific speculation.

Aristotle's two most concentrated discussions of economic questions

---

[1] W. F. R. Hardie, *Aristotle's Ethical Theory* (Oxford, 1968), 28–32; John Herman Randall, Jr., *Aristotle* (New York, 1960), 243; René-Antoine Gauthier, *La morale d'Aristote* (Paris, 1958), 9. For a critique of Aristotle's analysis of justice precisely because of its "quantification of moral value," and its "quasi mathematical-geometrical" form, see Hans Kelsen, "Aristotle's Doctrine of Justice," in James J. Walsh and Henry L. Shapiro (eds.), *Aristotle's Ethics: Issues and Interpretations* (Belmont, Calif., 1967), 102–19. The transference of method and insight between moral and physical sciences continued in scholastic thought. This is discussed in the introduction to chapter 3.

are found in Book I of the *Politics*, chapters 8–11, and Book V of the *Nicomachean Ethics*, chapters 3–5. Of the two, Book V of the *Ethics* provides his most systematic analysis.[2] Although many classical texts touched on economic questions, Aristotle's discussion in the *Ethics* was the only one in which the form of exchange and the function of money were considered as specific intellectual problems with their own proper modes of description and analysis.[3] From the late thirteenth century, the study of the *Ethics* was a required part of the curriculum in the universities of Oxford and Paris. Within the *Ethics*, Book V containing the analysis of money and exchange was given particularly close scrutiny, since it was also the site of Aristotle's most detailed discussion of justice, a subject of the greatest concern to medieval thinkers. Aristotle's decision to place his discussion of economic exchange within his most detailed analysis of *iustitia* assured a central, almost inescapable place to his economic thought within the medieval university.[4]

It is not the purpose of this chapter to analyze Aristotle's economic thinking nor to assess his role in the development of economic theory. A large literature already exists and new and important work continues to be done on these subjects.[5] Virtually all modern commentators agree in characterizing the text of Book V of the *Ethics* as disorganized and disjointed, opening it to a range of economic interpretations.[6] Despite

---

[2] For information on the Latin text of the *Ethics* used in this and succeeding chapters, see n. 11.

[3] S. Todd Lowry, "The Greek Heritage in Economic Thought," in S. Todd Lowry (ed.), *Pre-Classical Economic Thought* (Boston, 1987), 7–30; Joseph Schumpeter, *Economic Doctrine and Method*, R. Aris (tr.) (New York, 1967), 11.

[4] For more on the history of the medieval Latin translations of the *Ethics*, see René-Antoine Gauthier and Jean-Yves Jolif, *L'Ethique à Nicomaque: introduction, traduction, et commentaire*, 2 vols. (Louvain, 1970), vol. I, 115–30; F. M. Powicke, "Robert Grosseteste and the Nicomachean Ethics," *Proceedings of the British Academy* (1930), 85–104; Auguste Pelzer, "Les versions latines des ouvrages de morale conservés sous le nom d'Aristote en usage au XIIIe siècle," *Revue néo-scolastique* 23 (1921), 316–41, 378–412; D. A. Callus, "The Date of Grosseteste's Translations and Commentaries on the Pseudo-Dionysius and the Nicomachean Ethics," *Recherches de théologie ancienne et médiévale* 14 (1947), 186–210, 208. See also chapter 3, nn. 1 and 2.

[5] D. G. Ritchie, "Aristotle's Subdivisions of 'Particular Justice,'" *Classical Review* 8 (1894), 185–92; Joseph Schumpeter, *History of Economic Analysis* (New York, 1954); Joseph Soudek, "Aristotle's Theory of Exchange: An Inquiry into the Origin of Economic Analysis," *Proceedings of the American Philosophical Society* 96 (1952), 45–75; John W. Baldwin, *The Medieval Theories of the Just Price: Romanists, Canonists, and Theologians in the Twelfth and Thirteenth Centuries* (Philadelphia, 1959); Barry J. Gordon, "Aristotle and the Development of Value Theory," *Quarterly Journal of Economics* 78 (1964), 115–28; S. Todd Lowry, "Aristotle's Mathematical Analysis of Exchange," *History of Political Economy* 1 (1969), 44–66; Odd Langholm, *Price and Value in the Aristotelian Tradition* (Bergen, 1979); M. I. Finley, "Aristotle and Economic Analysis," in Jonathan Barnes et al. (ed.), *Articles on Aristotle* II: Ethics and Politics (London, 1977), 140–58; Amleto Spicciani, *La mercatura e la formazione del prezzo nella riflessione teologica medioevale* (Rome, 1977); Joseph J. Spengler, *Origins of Economic Thought and Justice* (London, 1980); Odd Langholm, *Wealth and Money in the Aristotelian Tradition* (Bergen, 1983).

[6] Langholm's description of the chapter (*Wealth and Money*, 47–48) as appearing as if someone has shuffled and haphazardly reconstructed it is most apt.

this disorganization, Aristotle's discussion of exchange here is so compact (chapters 3 through 5 of Book v occupy roughly 200 lines in the Berlin edition of 1886 from which the standard location notation is derived) that a better understanding of his argument can often be derived by reading the text in its entirety than by receiving it at second hand, after it has been pulled apart, reordered, and put back together question by question. My concern here is to write a history of the interconnection of ideas – to investigate only those conceptions and categories found in Aristotle's model of money and exchange that are also found in the most forward-looking scholastic speculations on the natural world. The areas of inter-connection between proto-economic and proto-scientific thought are so extensive in the *Ethics* and in subsequent medieval discussions of econ-omics that there are valuable insights to be gained into both the history of economic thought and the history of science through a study restricted to this interface.

I have isolated six categories of connection between medieval econ-omic and proto-scientific speculation, five of which have their intellec-tual origin in Aristotle's discussion in the *Ethics*: (1) mathematics and the geometry of exchange; (2) equality, the mean, and equalization in ex-change; (3) money as *medium* and as measure; (4) relation and the relativity of value in exchange; and (5) the social geometry of monetized society. This chapter and those that follow it are structured around these categories. Their purpose here is heuristic.[7] For the sake of clarity of exposition and demonstration, I treat them as separable elements, al-though they are bound together and mutually dependent in Aristotle's conceptual scheme and in subsequent scholastic discussions. In the final two chapters of this work, I connect conceptual developments within each of these economic categories to conceptual advances made within fourteenth-century natural philosophy.

The most common theme in the medieval attitude toward money was its role as a solvent. It was seen as a disturbing and distorting element, an overturner of the social order, an instrument of chaos. Again and again, the subject thought to suffer most grievously from money's corrupting effect was the delicate balance of *iustitia*.[8] Aristotle's decision to consider money and exchange in the center of his discussion of justice, and his direct identification of economic transactions with forms of particular justice, served to counteract the negative popular image of money and to

---

[7] The sixth category, "common valuation in exchange," was not present in Aristotle's treatment and therefore is not considered in this chapter. It is considered in chapter 3 as a scholastic addition to the Aristotelian model.

[8] Yunck, *Lady Meed*, 63–69. See chapter 1 for a discussion of negative popular attitudes toward money.

turn the mind of the reader toward an appreciation of the lawfulness and mathematical regularities underlying the system of exchange.[9]

By the fourteenth century, those whose lives were daily affected by monetary and market developments had, out of necessity, developed a sophisticated understanding of market exchange. They recognized it as an interconnected system with its own inherent principles of organization and equilibrium. In order to avoid being caught on the short side of rising prices or left holding worthless coins, it was necessary for everyone, not simply merchants and moneychangers, to pay close attention to the workings of the marketplace. University scholars guarding their resources in the bustling towns and cities would have acquired an understanding of market regularities in the absence of Aristotle's discussion. Clearly, however, Aristotle's unambiguous provision of equilibrium and order as the context of economic exchange served as a textual bridge between the everyday experience of the economic world and intellectual models that could be constructed to comprehend and represent it.[10] The text of the *Ethics* thus contributed to the transformation of the scholastic conception of money and the marketplace: from forces of dislocation and imbalance to instruments of order and equilibrium found at the center of social organization.

MATHEMATICS AND THE GEOMETRY OF EXCHANGE

The central place of mathematics and mathematical models in *Ethics* v.3–5 provides an important point of contact between medieval economic and scientific thought. Aristotle's conceptualization of justice itself was highly mathematical. He writes: "The just, then, is a species of the proportionate (proportion being not a property only of the kind of number which consists of abstract units, but of number in general)."[11]

---

[9] For justice as the context of medieval discussions of economics, see Spicciani, *La mercatura*, 133; Baldwin, *Just Price*, 72.

[10] Langholm (*Wealth and Money*, 37) discusses the influence of the Aristotelian text on the way scholastic thinkers framed their economic perceptions.

[11] [1131a30–32]. Unless otherwise noted, the English translation of the *Ethica Nicomachea* used here is by W. D. Ross, in Richard McKeon (ed.), *The Basic Works of Aristotle* (New York, 1941). *Ethica Nicomachea: translatio Roberti Grosseteste Lincolniensis, recensio recognita*, R.-A. Gauthier (ed.), Aristoteles Latinus 26, 1–3, Fasc. 4 (Leiden, 1973), 458: "Est ergo iustum proporcionale quidem; proporcionale enim non solum est monadici numeri proprium, set totaliter numeri." Since the purpose of this chapter is to analyze Aristotle's economic thought only insofar as it influenced the direction of scholastic economic theory, the Latin text of the *Ethics* cited in the notes is taken from the revised version of Robert Grosseteste's translation from the middle of the thirteenth century. Hereafter, this edition will be cited as Arist. Lat. 26 (1973). This revised version (c. 1260) remained the most widely used Latin text through the fifteenth century. Grosseteste's original translation of the *Ethics* has been dated to 1246–47. On this, see Callus, "Date of Grosseteste's Translations," 200–09.

Aristotle's assertion here that mathematical operations can be performed not only with numbers as abstract units, but also with numbers as they apply to things in the world, had important implications for the future of scientific thought. It encouraged the philosophical use of number as a practical aid in the description, analysis, and measurement of real phenomena. Justice, Aristotle maintains, is achieved through manipulating the relations between subjects and objects according to mathematical principles. "This, then, is what the just is – the proportional."[12]

Aristotle separated justice into two particular forms, distributive (*iustitia distributiva*), and rectificatory, sometimes called corrective or directive (*iustitia directiva*), each form determined by the mathematical process involved in reaching the just mean. Distributive justice involves the distribution of common goods by a central authority in proportion as the recipient had proved himself worthy of reward through service. Since the quality of service and virtue is inherently unequal in men, distributive reward must also be unequal. The determination of distributive justice therefore requires the establishment of a "geometrical" rather than an "arithmetical" equivalence, in which greater service receives proportionally greater reward. (An arithmetical equivalence would entail all receiving a numerically equal reward.) Aristotle calls the form of proportionality governing distributive justice "geometric" because that is the term mathematicians give to ratios in which the whole is to the whole as either part is to the corresponding part.[13] Although this form of justice was primarily designed to cover the distribution of honors by the city to its citizens, Aristotle extended it to economic practices current in his society. Partnerships were specifically included under *iustitia distributiva*, so that funds drawn from the partnership "will be according to the same ratio which the funds put into the business by the partners bear to one another."[14]

Throughout Aristotle's discussion, the definition of distributive justice is framed almost exclusively in mathematical terms of geometric proportion and ratio. The computation of the just involves a ratio containing at least four terms (two people and two rewards or punishments). Aristotle had sufficient confidence in the mathematics of his

---

[12] [1131b16]. Arist. Lat. 26 (1973), 459: "Iustum quidem igitur hoc proporcionale. Iniustum autem quod preter proporcionale." Soudek, "Aristotle's Theory," 45–48, and Lowry, "Mathematical Analysis," concentrate on the mathematical nature of Aristotelian justice, including justice in exchange.

[13] [1131b12–14]. Arist. Lat. 26 (1973), 459: "Vocant autem talem proporcionalitatem geometricam mathematici. In geometria enim accidit et totum comparari ad totum quod quidem pars alterum ad alterum."

[14] [1131b29–30]. Arist. Lat. 26 (1973), 459: "et enim a pecuniis communibus si fiat distribucio, erit secundum proporcionem eandem quam habent ad invicem illatam."

analysis to represent both participants and circumstances as lettered terms in an equation:[15]

The conjunction, then, of the term A with G and of B with D is what is just in distribution, and this species of the just is intermediate, and the unjust is what violates the proportion; for the proportional is intermediate, and the just is proportional. (Mathematicians call this kind of proportion geometrical.)[16]

The second, or what Aristotle called the "remaining," form of justice is called directive or rectificatory justice (*iustitia directiva*).[17] Where distributive justice establishes a proportion between assumed unequals through the geometric cross-multiplication of ratios, directive justice applies to cases in which the participants are assumed to be equals, deserving of equal awards. For this reason *iustitia directiva* is governed by arithmetical rather than geometrical proportion and equality. The intermediate point between gain and loss is found through the arithmetical process of addition or subtraction to an imaginary line of gain and loss. Directive justice is also called "rectificatory" and "corrective" because it applies to those cases in which balance is restored by a judge or intervening orderer. Again we have a striking mathematization of a practical problem; the judge performs mathematical operations on objects and subjects in the world. So clearly defined is the judge by mathematical operations that Aristotle believed his title was derived from the act of bisection.[18] "Now the judge restores equality; it is as though there were a line divided into unequal parts, and he took away that by which the greater segment exceeds the half, and added it to the smaller segment."[19] After saying this Aristotle adds yet more mathematical emphasis with an involved, purely technical discussion on the subject of equalizing unequal lines.[20] Every detail of his discussion of justice reveals that he was conceptualizing and treating the world as if real, existent problems or qualities (justice, service, virtue, gain, loss) could be represented by lines or numbers and solved mathematically. Humans were capable of measuring and manipulating the world as geometers manipulated lines

---

[15] [1131b4–6]. Arist. Lat. 26 (1973), 459: "Est autem et iustum in quatuor minimis, et proporcio eadem; divisa enim sunt similiter et quibus et que; erit ergo ut A terminus ad B, ita G ad D, et permutatim ergo ut A ad G, B ad D; quare et totum ad totum."

[16] [1131b9–14]. Arist. Lat. 26 (1973), 459: "Ergo A termini cum G et B cum D coniunccio in distribucione iustum est, et medium iustum; hoc est preter proporcionale. Proporcionale enim medium, iustum autem proporcionale. Vocant autem talem proporcionalitatem geometricam mathematici. In geometria enim accidit et totum comparari ad totum quod quidem pars alterum ad alterum." [17] [1131b25]. Arist. Lat. 26 (1973), 459: "Reliqua autem una directivum."

[18] [1132a32]. Arist. Lat. 26 (1973), 461: "Propter quod et nominatur *dikayon*, quando dika est, quemadmodum si quis dicat dikayon, et dikastes dicastes."

[19] [1132a24–26]. Arist. Lat. 26 (1973), 460: "Iudex autem adequat, et quemadmodum ligna inaequalia secta, quo meior seccio medietatem superexcedit, hic abstulit et minori seccioni apposuit." [20] [1132b6–10]. Arist. Lat. 26 (1973), 461.

and ratios – all toward the end of equality.[21] Nowhere else in his writings did Aristotle reduce human activity to a mathematically definable process as thoroughly as he did here in Book v of the *Ethics* in relation to justice and economic exchange.

After describing the mathematical forms of distributive and directive justice, Aristotle introduces the concept of reciprocity (*contrapassum*), and it is in this context that he begins his discussion of economic exchange.[22] In the midst of deciding whether *contrapassum* is in itself a particular form of justice as some philosophers maintained, or whether it requires further elaboration and refinement, the question of the mathematical form of economic exchange suddenly appears:

> Now "reciprocity" fits neither distributive nor rectificatory justice. . .But in associations for exchange this sort of justice does hold men together – reciprocity in accordance with a proportion and not on the basis of precisely equal return.[23]

After establishing that the *contrapassum* of proportional reciprocity governs exchange, Aristotle does something quite remarkable. He fits the complex phenomena of economic exchange between any two producers into a geometric form – what he refers to as his "figure of proportionality" (*figura proportionalitatis*):[24]

> Now proportionate return is secured by cross-conjunction. Let A be a builder, B a shoemaker, G a house, D a shoe. The builder, then, must get from the shoemaker the latter's work, and must himself give him in return his own.[25]

---

21 Many manuscripts of the Grosseteste translation had illustrative drawings of bisected and reconstituted lines in the margin at this point. See *Ethica Nicomachea: translatio Roberti Grosseteste Lincolniensis, recensio pura*, R.-A. Gauthier (ed.), Arist. Lat. 26, 1–3, Fasc. 3 (Leiden, 1972), 235 (cited as Arist. Lat. 26 [1972]).

22 There is considerable disagreement among modern commentators as to the status of "reciprocity" in relation to distributive and rectificatory justice. The text is particularly unclear on this point. John Baldwin (*Just Price*, 11) and Joseph Soudek ("Aristotle's Theory," 53–54) believe that Aristotle intended reciprocity to be a third particular form of justice, governed by its own mathematical form. For an early statement of this position, see Ritchie, "Subdivisions," 185–86. S. Todd Lowry in his article on "Mathematical Analysis" attempts to recreate the probable mathematics governing this form. What is important for this study, however, is not what Aristotle might have intended, but how scholastic thinkers understood him. Without exception, the commentaries I have examined took the statement that rectificatory justice was the "remaining" form ("Reliqua autem, una, directivum" [1131b25]) to mean that there were only two particular forms, and that *contrapassum* rather than being a new form was a hybrid of directive and distributive justice, though one governed by a geometrical rather than an arithmetical process of equalization.

23 [1132b24, 31–34]. Arist. Lat. 26 (1973), 462: "Contrapassum autem non congruit, neque in distributivum iustum, neque in directivum . . . Set in concomitacionibus quidem commutativis continet tale iustum contrapassum secundum proportionalitatem et non secundum equalitatem."

24 [1133a34]. Arist. Lat. 26 (1973), 463: "In figuram autem proporcionalitatis."

25 [1133a6–11]. Arist. Lat. 26 (1973), 462: "Facit enim retribucionem eam que secundum proporcionalitatem secundum dyametrum coniugacio. Puta edificator in quo A, coriarius in quo B, domus in quo G, calciamentum in quo D. Oportet autem accipere edificatorem a coriario illius opus, et ipsum illi retribuere quod ipsius."

43

The Aristotelian model of money and exchange

In a large number of Latin manuscripts of the *Ethics* and its commentaries, the proportional figure here described was graphically rendered as an accompanying lettered figure in the margin, indicating the seriousness and concreteness accorded the geometry of exchange by medieval thinkers.[26] The labeled rectangular figure of proportionality with its crossed diagonals formed, for thinkers of the thirteenth and fourteenth centuries, an ideal analogue to the developing conception of the marketplace as a system or mechanism of equalization.

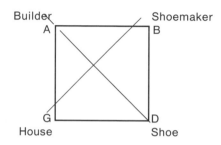

Once he has established it, Aristotle refers to this "figure of proportionality" six more times in the short remainder of his economic discussion in chapter 5. At the conclusion of the chapter, in his beautifully concise definition of the function of money in exchange, he integrates money itself as a geometrical term into the exchange equation. Economic equality is again represented as the product of the cross–conjunction of ratios, but here money serves as the common or continuous third term ("B") making possible the equation of two goods that in themselves are incommensurable ("A" and "C"), so that, A:B = B:C.[27]

All things are measured by money. Let A be a house, B ten minae, C a bed. A is half of B, if the house is worth five minae or equal to them; the bed, C, is a tenth of B; it is plain then, how many beds are equal to a house, viz. five.[28]

There is no escape from the geometry of exchange in Aristotle's *Ethics*. He provides one model only. To consider exchange between producers in Aristotelian terms is to translate the dynamic of the exchange process into a geometric figure, where the direction of exchange

[26] Arist. Lat. 26 (1972), 236, 238, and notes.
[27] Earlier in Book v Aristotle had taken pains to explain the mathematics behind such continuous proportions. See Soudek, "Aristotle's Theory," 69.
[28] [1133b23–26]. Arist. Lat. 26 (1973), 464: "Mensurantur enim omnia nummismate. Domus in quo A, minarum quinque, lectus in quo B, minus dignus; lectus autem quinta pars domus utique erit. Manifestum igitur quanti lecti equale domui, quoniam quinque."

is represented by lines, where exchangers and their goods are represented by mathematical terms, where exchange is represented by an equation of ratios, and where the point of just exchange is suggested by the point of conjunction of the rectangle's diagonals.[29] Money, which gives number to things in the world, and which serves as an expandable, divisible continuum relating diverse commodities in exchange, plays a central role in this geometry of everyday life.

EQUALITY, THE MEAN, AND EQUALIZATION IN EXCHANGE

In contrast to the theories of classical economists and to our modern understanding of economic motivation, in both Aristotelian and scholastic economic theory, not profit and the desire for gain but the establishment of equality is the proper motive and end of exchange. The search for equality and the mean is the central theme of the *Ethics*, and economic exchange is only one of many aspects of human life properly driven by it. The focus on equilibrium and the establishment of equivalences in Aristotle's economic discussion provided a strong basis of connection between economic thought and other areas of philosophical speculation, including speculation in the area of natural philosophy. Since scholastic thinkers were every bit as concerned with the question of equality as Aristotle, the search for underlying principles of equality became one of the great shared themes of medieval economic and proto-scientific thought.

Aristotle's discussion of equality in relation to justice in Book v was exceptional for its dependence on mathematics and geometry. When he applied equality specifically to the process of economic exchange, he added a further *dynamic* component. Equality was not only the proper end of exchange, it also governed the entire process. In his view, commodity equalization was a natural process, set in motion by naturally existing inequalities in the distribution of resources.[30] The possibility of reestablishing equality was the necessary precondition of all exchange. He writes:

If, then, first there is proportionate equality of goods, and then reciprocal action takes place, the result we mention [just exchange] will be effected. If not, the

---

[29] Gauthier and Jolif, *L'Ethique*, 372; Soudek, "Aristotle's Theory," 45.

[30] A full discussion of this origin is found in Book 1 of the *Politics*. For the medieval Latin text of the *Politics*, see *Politica (Libri* I–II), P. Michaud-Quantin (ed.), Arist. Lat. 29, 1 (Bruges, 1961) (cited as Arist. Lat. 29, 1 [1961]). On the origin of trade, see 16 [1257a31–35]: "Magis peregrino enim facto auxilio, per adduci quibus indigent et emittere quibus habundabant, ex necessitate numismatis acquisitus est usus."

bargain is not equal, and does not hold; for there is nothing to prevent the work of the one being better than that of the other; they must therefore be equated.[31]

Nowhere were medieval thinkers more in line with Aristotle than in their insistence that just exchange was a process of equalization, and that money properly used was an instrument devised to facilitate equalization between exchangers and their goods.

Aristotle's fixation with the question of equality involved him with the mathematically and scientifically relevant questions of measurement and commensurability. How, he asked, can equivalences be established between diverse people exchanging diverse goods of diverse quality involving diverse skills of production? For he believed that, without at first the possibility and then the guarantee of equality, even socially necessary exchange would not take place:

> For it is not two doctors that associate for exchange, but a doctor and a farmer, or in general people who are different and unequal; but these must be equated. This is why all things that are exchanged must be somehow comparable.[32]

The recognition that finding a common measure of all goods lay at the heart of exchange led him to focus on the essential role of money as an instrument of relation and commensuration.

MONEY AS *MEDIUM* AND AS MEASURE

In money Aristotle found the solution to the multilayered problem of value commensuration:

> all things that are exchanged must be somehow comparable. It is for this end that money has been introduced, and it becomes in a sense an intermediate [*aliqualiter medium*]; for it measures all things, and therefore the excess and the defect – how many shoes are equal to a house or to a given amount of food.[33]

In medieval Latin translations of the *Ethics*, the term *medium* was used to denote money's complex and dynamic function as a measuring continuum, a connector between exchangers, and an intermediate or third term

---

[31] [1133a11–14]. Arist. Lat. 26 (1973), 462: "Si igitur primum sit secundum proporcionalitatem equale, deinde contrapasssum fiat, erit quod dicitur. Si autem non, non equale, neque commanet. Nichil enim prohibet melius esse alterius opus, quam alterius. Oportet igitur hec utique equari."

[32] [1133a17–20]. Arist. Lat. 26 (1973), 463: "Non enim ex duobus medicis fit communicacio, set ex medico et agricola, et omnino alteris, et non equalibus; *set hos oportet equari*. Propter quod omnia comparata oportet aliqualiter esse, quorum est commutacio" (emphasis added).

[33] Continuing previous citation [1133a20–22]: "Propter quod omnia comparata oportet aliqualiter esse, quorum est commutacio; ad quod nummisma venit, et *fit aliqualiter medium. Omnia enim mensurat, quare et superhabundanciam et defectum.* Quanta quedam utique calciamenta, equale domui vel cibo" (emphasis added).

in the equation of exchange ratios. In medieval works on natural philosophy the term "medium" also frequently appears denoting the *line* as a connector or instrument of measurement. The designation of money in its various functions as a "medium" in Aristotle's text encouraged the later connection between money and the measuring line, a connection that already lay implicit in his rectangular *figura proportionalitatis* governing exchange.[34]

Aristotle wrote: "Money, then, acting as a measure, makes goods commensurate and equates them."[35] It did so by acting as a numbered continuum capable of infinite expansion and contraction, against which all commodities could be commensurated and find relation. He expressed his admiration for the great success of money in accomplishing this multi-faceted task. His immense authority ensured that his appreciation was reflected in medieval thought on the subject. His blanket statement, "All things are measured by money,"[36] had particular resonance within natural philosophy in the fourteenth century because of the growing place within it of problems of measurement and relation.

RELATION AND THE RELATIVITY OF VALUE IN EXCHANGE

But what precisely does money measure? Aristotle's answer here proved most interesting to his medieval commentators. In the course of his discussion there are hints that money measures certain qualities existing within the commodities themselves: the quality of workmanship involved in its production [1133a13–15]; or the value of the variable skills required by the nature of the work (as between a farmer and a doctor) [1133a17–19]; or the different expenditures of labor involved (as between a house and a bed) [1133b23–25]. From the first medieval commentary on the *Ethics* (that of Albertus Magnus) these considerations found a place in scholastic discussions of value.[37] However, judging by the weight Aristotle gave to the concept of inherent value (or "labor value" as it is

---

[34] See chapters 5 and 6 for a discussion of the similarity in profile between the line or latitude as it came to be used by the natural philosophers of the fourteenth century, and the scholastic understanding of money as *medium*. For a discussion of the many meanings of *medium* in scholastic economic theory, and a short history of the understanding and use of this word in medieval commentaries of the *Ethics*, see Langholm, *Wealth and Money*, 69–70.

[35] [1133b16]. Arist. Lat. 26 (1973), 464: "Nummisma utique quemadmodum mensura commensurata faciens, equat."

[36] [1133b23]. Arist. Lat. (1973), 464: "Mensurantur enim omnia nummismate." Walter Burley, for example, an early fourteenth-century contributor to the proto-science of *calculationes*, shows particular fascination with this insight about money as common measure in his commentary on the *Ethics*. See chapter 5, 138–39.

[37] See chapter 3.

sometimes called by modern commentators), it was a decidedly second-ary component in his understanding of price determination.[38]

In contrast, Aristotle states repeatedly and forcefully that money does not measure the essential value of commodities in themselves, but rather the demand or need for those objects experienced by the participants in exchange:

All goods must therefore be measured by some one thing, as we said before. Now this unit is in truth human need which holds all things together . . . but money has become by convention a sort of representation of human need.[39]

Since all things in exchange share the common quality of being needed by someone, money as a measure of need (*indigentia*) creates the possibil-ity of equalization through the establishment of value commensurability. Aristotle's argument has three steps: (1) all goods to be exchanged find their value according to how much they are needed; (2) money measures need as price and represents it as numbered units open to mathematical operations; (3) the value of diverse goods (e.g., a bed and a house) can thus be compared and measured against each other toward the end of equality in exchange. He writes:

Now in truth it is impossible that things differing so much should become commensurate, but with reference to demand they may become so sufficiently. There must, then, be a unit, and that fixed by agreement (for which reason it is called money); for it is this that makes all things commensurate, since all things are measured by money.[40]

---

[38] The weighing of the "objective" value components of labor and expenses against the "subjective" component of human need in Aristotle's analysis of value is perhaps the most controversial question for modern commentators. On this question, see J. J. Spengler, "Aristotle on Economic Imputation and Related Matters," *Southern Economic Journal* 21 (1955), 371–89; Schumpeter, *Economic Analysis*, 60; Finley, "Aristotle and Economic Analysis," 146–48; Baldwin, *Just Price*, 11; Langholm, "Scholastic Economics," in Lowry, *Pre-Classical Economic Thought*, 117–34, 118–23; Langholm, *Price and Value*, 50; Soudek, "Aristotle's Theory," 60, 65–68; John Noonan, *The Scholastic Analysis of Usury* (Cambridge, Mass., 1957), 87. For a balanced presentation of the textual evidence on both sides of this question, see Gordon, "Value Theory." For the continuation of this debate in regard to the earliest commentaries on the *Ethics*, see chapter 3, 68–70.

[39] [1133a26–30]. I have substituted the phrase "human need" (*indigentia*) for the term "demand" that appears in the Ross translation. "Need" more accurately reflects both Aristotle's text and the standard Latin translation of that text available to medieval readers. Arist. Lat. 26 (1973), 463: "Oportet ergo uno aliquo omnia mensurari, quemadmodum dictum est prius. Hoc autem est secundum veritatem quidem continet. Si enim nichil indigerent, vel non similiter, vel non erit communicacio, vel non eadem indigencia que puta propter commutacionem necessitatis num-misma factum est secundum composicionem." For the confusion engendered by variant Latin translations of this passage, see Langholm, "Scholastic Economics," 122–23; see also chapter 3, 68–70.

[40] [1133b18–23]. Arist. Lat. 26 (1973), 464: "Secundum quidem igitur inpossibile tantum differencia, commensurata fieri, convenit sufficienter. Unum utique aliquid oportet esse; hoc autem ex supposicione; propter quod et nummisma vocatur, hoc enim omnia facit commensurata. Mensur-antur enim omnia nummismate." This passage contains many important variants in the medieval Latin translation. See the editor's notes on this point in Arist. Lat. 26 (1973), 464.

Here it is clear that money does not measure the quality or value of an object in itself, but rather a variable quality, accidental and necessarily relative to the object. Economic value continually changes relative to changing need. This conception of price is essentially dynamic and relativistic.

Although Aristotle's works were avidly received by many thinkers of the thirteenth century, there were certain elements of his thought that were difficult to assimilate. One of these was the great freedom with which he employed relativistic thinking – his jumping back and forth between the consideration of a thing in itself and the consideration of the same thing "quoad nos" or "ad aliquid."[41] We have seen this relativist thinking in the concept of geometrical or proportional equivalence in distributive justice, where equivalence in reward is sought relative to unequal talent and service. In discussing the virtue of liberality in Book IV, he wrote that its "magnitude is relative" and must be judged in relation to the agent, the circumstances, and the object offered.[42] Aristotle's habit of framing problems of measurement in terms of the mathematics and logic of the continuum often underlay his relativistic determinations, even in the area of ethics. In Book II of the *Ethics*, for example, he treats virtue as if it could be conceptually reduced to a continuum or line scale of value toward the end of measurement and the determination of relative equivalences.[43] This method of conceptualizing measurement proved attractive to scholastic thinkers. From the late thirteenth century, Aristotle's practice of linking ethical and physical measurement to the logic and mathematics of the continuum became a defining characteristic of both economic and proto-scientific speculation.[44]

Aristotle's conceptual world was shifting and fully relational, with definitions and determinations changing as the point of reference changed. The conceptual world of the thirteenth century was considerably more fixed, static, ordered, and absolute. There is evidence of strong discomfort with and resistance to relativity as a solution to problems of value. This resistance, prevalent in the thirteenth century but easing in the fourteenth, is voiced in writings on economic as well as scientific

---

[41] For a discussion of Aristotle's relativism, see Soudek, "Aristotle's Theory," 51. For the prevalence of "subjective relativism" in Greece in Aristotle's time, see Lowry, "Greek Heritage," 21.

[42] *Ethics* IV.2 [1122a23–26]. Arist. Lat. 26 (1973), 436: "Magnitudo autem ad aliquid."

[43] *Ethics* II.5 [1106a25–32]. Arist. Lat. 26 (1973), 402: "In omnibus itaque continuo et divisibili est accipere hoc quidem plus, hoc autem minus, hoc vero equale, et hec vel secundum ipsam rem, vel ad nos; equale autem medium quid est superhabundancie et defectus. Dico utique rei quidem medium quod equaliter distat ab utroque extremorum, quod est idem et unum omnibus; ad nos autem quod neque habundat neque deficit, hoc autem neque unum neque idem omnibus."

[44] For the central place of the logic of the continuum in scholastic thought (both economic and physical), see chapters 5, 6, and 7 under the heading "Relation and the relativity of value in exchange."

subjects, and was often strong enough to determine positions on these subjects.[45] The textual weight of Aristotle was great, but it required the confluence of the textual and the actual to force a reexamination of this resistance and open the door to relativistic thinking in both medieval economic and scientific thought. In this case, the actual was the everyday observation and experience of the expanding marketplace. The scholar's daily life in the market town of Oxford or the metropolis of Paris would have demonstrated that the same object or commodity, though unchanging in itself, was valued differently from day to day or week to week; that the price of his bread or his wine was constantly redetermined relative to the external factors of time and place, scarcity and need. Such an insight would be greatly expanded if that scholar was involved, as was so often the case, in the administrative tasks of his school, his Church, or his government. The lesson that was being absorbed through experience of the marketplace was this: although the commodities themselves remained fixed, their values changed as the external points of reference (in this case need or *indigentia*) changed. The lesson was simple, even inescapable, but nonetheless difficult to absorb. Such relativity was foreign to the established conceptual scheme of an ontologically ordered universe of graded values, where every object and subject had its fixed place in the order of being from God down to the least of his creations.

Perhaps economic experience alone, over time, would have sufficed to lead scholastic thinkers to an acceptance and understanding of relativistic determinations, especially as one can see such an understanding emerging in the general population in regard to market values. But here in addition to experience was Aristotle's text, available to those scholars who prepared the ground for modern economic theory as they prepared it for modern science, framing, clarifying, and authorizing the recognition of relativity – instituting the continuum as the model for measuring and investigating relative values. The new relativism, derived from economic experience and legitimated by authoritative text, could then be applied to the understanding of activity in the world of nature. By the middle of the fourteenth century, the economic writer and natural philosopher Nicole Oresme was able to recognize that a new relativistic and "geometric" conception of the world had emerged, more complex, more unsettling, but in many ways more "beautiful" than the conception it had replaced.[46] When we look to find the connection between fourteenth-century speculation on nature and sixteenth-century science, we should look less to specific insights and formulae bequeathed by scholastic thinkers, and

---

[45] For an example of the discomfort with the proportional and relative in the determination of economic value, see the position of Henry of Ghent in chapter 4, 107–08.

[46] For Oresme's relativized model of the physical world, see chapter 7. For his recognition of relativity in economic life, see chapter 5, 149–51.

more to their greatly more difficult and important creation of a relativized conceptual world.

THE SOCIAL GEOMETRY OF MONETIZED SOCIETY

The final category to be considered is one frequently overlooked by modern commentators. It concerns the dynamism in Aristotle's conception both of society and of economic relations. His is a vision of society tied together geometrically, with exchange at the center of the geometric figure and money at the center of exchange. One can imagine the difficulty medieval Christian thinkers might have with such a conception. It would seem to violate Christian ideals of social communication. Once again it is the confluence of the textual and the actual – the weight of Aristotelian authority as it combined with and mediated social experience – that permitted such a conception to reemerge in the middle of the thirteenth century.

Aristotle's geometrization of justice and exchange provided the background for his geometry of social forces. He begins chapter 5 of Book v by asserting that economic exchange is governed by proportional reciprocity, and that such reciprocity forms the social basis of the *civitas*.[47] He then introduces the rectangular figure of proportionality governing exchange. In this figure, not only the form of economic exchange is geometricized, but so too is the social *act* of exchange, which is represented by the crossing of the rectangle's diagonals (see p. 44).[48] Society in this scheme is literally held together by the equation of goods and services in the process of exchange. For in Aristotle's view, if the process of exchange did not continually reward a producer's skill and labor with equivalent products from other producers, both production and exchange would cease, destroying the foundation of social communication.[49] In a passage that is often expanded upon by medieval commentators, Aristotle equates the very survival of human industry with the establishment of equivalences in exchange.[50] And yet he gives no plan or

---

[47] [1132b31–34]. Arist. Lat. 26 (1973), 462: "Set in concomitacionibus quidem commutativis continet tale iustum contrapassum secundum proportionalitatem et non secundum equalitatem; per contrafacere enim proporcionale commanet civitas."

[48] [1133a6–7]. Arist. Lat. 26 (1973), 462: "Facit enim retribucionem eam que secundum proporcionalitatem secundum dyametrum coniugacio."

[49] [1133a24–25]. Arist. Lat. 26 (1973), 463: "Si enim non hoc [i.e., equalization], non erit commutacio neque communicacio; hec autem si non equalia sunt, aliqualiter non erit."

[50] [1133a15–17]. Arist. Lat. 26 (1973), 462–63: "Oportet igitur hec utique equari. Est autem hoc et in aliis artibus. Destruentur enim si non fecerit faciens et quantum et quale et paciens hoc et tantum et tale." This statement is treated by certain modern editors as an interpolation. Nevertheless, it appears in both the Grosseteste translation and its revision, and it was often glossed by medieval commentators. On this subject, see Langholm, *Economics in the Medieval Schools: Wealth, Exchange, Value, Money, and Usury According to the Paris Theological Tradition 1200–1350* (Leiden, 1992), 189.

program for ensuring equivalence, no mention of an outside judge to regulate the process, no general rules or criteria of judging. The implication is clear: economic exchange is a self-ordering process, functioning, normally, without the need of an intervening, subjective orderer. Economic equivalence is a dynamic, quasi-mathematical product of the systematic geometry of exchange.[51] And if equality in exchange is required to ensure social communication, and equalization requires commensuration, then money, the line-instrument of commensuration, is the essential connecting element in the geometry of exchange. By binding exchangers and goods through its role as a *medium* or connecting line, money holds the community together:

That demand holds things together as a single unit is shown by the fact that when men do not need one another, i.e., when neither needs the other or one does not need the other, they do not exchange . . . This equation therefore must be established.[52] . . . All goods must have a price set on them; for then there will always be exchange, and, if so, association of man with man.[53]

At the conclusion of his discussion, Aristotle connected the many parts of his argument, providing the clearest statement of his systematic conception of economic exchange:

Money, then, acting as a measure, makes goods commensurate and equates them; for neither would there have been association if there were not exchange, nor exchange if there were not equality, nor equality if there were not commensurability.[54]

In this forceful synthetic statement, Aristotle ties money to the processes of measurement, commensuration, equalization, and exchange, placing it at the center of his social geometry.

NEGATIVE ATTITUDES TOWARD MONEY AND EXCHANGE

Alongside the positive and dynamic aspects of Aristotle's economic thought there existed strongly negative attitudes, motivated in part by his

---

[51] Although Aristotle does not mention the act of free and mutual bargaining in his analysis of exchange, certain of his modern commentators assume that he understood this to be the general rule. See, for example, Lowry, "Mathematical Analysis," 51; also, *Ethics* [1136b7–15]. Finley, on the other hand ("Aristotle and Economic Analysis," 147), strongly denies this.

[52] [1133b6–10]. Arist. Lat. 26 (1973), 463–64: "Quoniam autem indigencia continet quemadmodum quidem ens, ostendit, quoniam cum non in necessitate sunt ad invicem vel utrique vel alter, non communicant, quemadmodum cum quo habet ipse, non indiget quis . . . Oportet ergo hoc equari."

[53] [1133b15–16]. Arist. Lat. 26 (1973), 464: "Propter quod oportet omnia appreciari; sic enim erit semper commutacio, si autem hoc, communicacio."

[54] [1133b16–19]. Arist. Lat. 26 (1973), 464: "Nummisma utique quemadmodum mensura commensurata faciens, equat. Neque enim utique non existente commutacione communicacio erat; neque commutacio, equalitate non existente; neque equalitas, non existente commensuracione."

distrust of economic developments in his own day.[55] An accurate idea of the tension between these elements can be obtained by reading the two concentrated loci of economic comments together, *Ethics* v.3–5, and *Politics* 1.8–11. The most glaring differences between them appear in relation to their treatment of money. Where money in the *Ethics* is viewed as an instrument of social order and integrated into a mathematical scheme of balance and equivalence, in the *Politics* money is treated as a boundless and unnatural element, destructive of social organization. Although Aristotle recognizes in the *Politics* that money when used in the proper way is an essential tool in the socially necessary process of exchange, he focuses instead on how frequently it is misused, how often, in practice, the desire for it is improper and beyond natural bounds.[56] Rather than benefit the *civitas*, it most often serves the immoderate and the excessive.[57] It has no natural value, yet men come to value it above all things. As a measure, instituted by law, it should remain fixed. Instead it multiplies and "breeds," gives birth to itself most unnaturally.[58]

Medieval commentators could not ignore this negative characterization, linked as it was to deep popular fears of money as a disruptive force in their own society. The textual influence of Aristotle's negative vision was further reinforced by Christian contempt for a life spent in pursuit of worldly goods. As a result, Aristotle's distrust of money in the *Politics*, and his condemnation of usury and usurers found there, was often more central to the moral concerns of medieval thinkers than was the positive geometric model of exchange in the *Ethics*. The taint on commerce and profit seeking so clear in the *Politics* remained a potent influence on scholastic moral and economic thought. The legitimacy of intellectual speculation on economic matters remained suspect outside usury theory or commentaries limited to explicating Book v of the *Ethics*. As a consequence, the scope of scholastic discourse on economic questions was considerably diminished.

One of the few propositions acceptable to all modern commentators on *Ethics* v.5 is that while the insights contained in it are remarkable, the text itself is disorganized and often unclear.[59] These difficulties influenced the history of medieval *Ethics* commentaries, since the energies of the commentators went more often simply to clarify and make sense of the text than to expand upon it. Even in our own day, the dislocated text encourages controversies over whether Aristotle accepted inherent or

---

[55] Not only Aristotle but also the leisure-class audience he was addressing held work and business pursuits in contempt. See Schumpeter, *Economic Analysis*, 60.

[56] *Politics* [1257b22–24].   [57] *Politics* [1258a7–8].

[58] *Politics* [1258b1–8]. See Langholm, *The Aristotelian Analysis of Usury* (Bergen, 1984), 58.

[59] For specific passages that may have been corrupted or omitted, see Gauthier and Jolif, *L'Ethique*, vol. I, 351ff., and for the overall sense of the problems of the text, 380.

relativist determinants of market value, or where *contrapassum* fits among his forms of particular justice.[60] There is even disagreement on the basic question of Aristotle's stature as an economic thinker.

Of all modern readers, M. I. Finley is the strongest critic of the idea that Aristotle's discussion in *Ethics* v.5 can be construed in any way as an "economic theory," or that economic "laws" can be found there.[61] In part he is speaking against modern scholars who have tied up the many loose ends in the text, resolved the contradictions, and decided what Aristotle must have meant or would have said if the text had not been so garbled in transmission. Reacting to this situation, Finley's conclusions are too absolute in their denial of cohesion or systematic analysis, but he has clearly recognized the most serious limitation in Aristotle's economic comments: the geometric *figura* of exchange.

With all of its trappings of proportionality and geometry, Aristotle's model was decidedly flawed as an accurate description of exchange in his society in that it was formulated solely in terms of exchanges between two producers. It harks back more to an ancient ideal than it does to a market economy where goods appear in the marketplace at a remove from the producers, having been transported or supplied by third parties. The personality of exchange that Aristotle stresses, where the shoemaker seeks out the carpenter, and where the commodity for sale is marked and can be judged according to the known skill of the producer, simply does not exist in an advanced market, and for the most part no longer existed in the markets of London or Oxford or Paris in the fourteenth century. Although Finley in his study of the ancient Greek economy counteracted the idea that Greece in the fourth century B. C. had anything like an advanced economy in the modern sense,[62] he did not use the argument of Greek economic backwardness to explain Aristotle's backwardness in analysis – just the opposite. For Finley, Aristotle had purposely neglected the sordid reality of trade and exchange as it was practiced in his society in order to return to an economic model more in tune with his social ideal.

Finley believes that Aristotle ignored the role of the merchant or petty trader in exchange because in Athenian culture these tasks were considered "illiberal and irksome," beneath the dignity of Athenian citizens and proper only to the non-citizen or slave.[63] In the *Politics* Aristotle distinguished proper modes of subsistence from subsistence through retail

---

[60] See discussion in nn. 22 and 38 in this chapter.
[61] Finley, "Aristotle and Economic Analysis," 152–58. For a critique of Finley's position, see Lowry, "Greek Heritage," 10, 19.
[62] M. I. Finley, *The Ancient Economy* (Berkeley, 1973), 135–37.
[63] Finley, "Aristotle and Economic Analysis," 149–57. For more on Aristotle's negative attitude toward trade and traders, see Baldwin, *Just Price*, 11; Schumpeter, *Economic Analysis*, 60.

trade[64] and the proper use of an object from its use as an object of trade.[65] If money were natural, he reasoned, then there would be a natural limit to the desire for riches. Since it was clear to him that this natural limit did not exist, he concluded that "retail trade is not a natural part of the art of getting wealth."[66]

A second explanation for the limitations of Aristotle's *figura* of exchange can be tied, paradoxically, to the central place of mathematics and equality in it. How can the profit of the middleman or merchant be squared with a mathematical model governed by strict equality? How can the existence of profit itself be squared with such a model?[67] The geometrical form of the model that so attracted Aristotle and his medieval commentators simply could not do justice to actual economic complexities. Because of its purity and rigidity, Aristotle's mathematical model acted at times to restrict the scholastic perception of new economic realities – above all concerning the role of the merchant and profit in exchange. Medieval thinkers, taken with the model's pleasing symmetries and impressed by its mathematical form, sometimes failed to look beyond it to fashion a more realistic one. Earlier I advanced the idea that the use of this same geometrical model was a crucial element in the creation of new and intellectually productive economic concepts, particularly those centering on the marketplace as a self-ordering mechanism of equalization and on money as a line-instrument of measurement. As we shall see, both positive and negative potentialities of the *figura* were realized in medieval economic writings.

[64] *Politics* [1256b1].   [65] *Politics* [1257a7–16].
[66] *Politics* [1257a17–19]. Arist. Lat. 29, 1 (1961), 15: "Aut est manifestum quod non est secundum naturam crimatistice kapelica; quantum enim sufficiens ipsis, necessarium erat facere commutationem."
[67] Finley, "Aristotle and Economic Analysis," 149; Soudek, "Aristotle's Theory," 75.

Chapter 3

THE EARLIEST LATIN COMMENTARIES ON THE ARISTOTELIAN MODEL OF ECONOMIC EXCHANGE: ALBERTUS MAGNUS AND THOMAS AQUINAS

Shortly after Robert Grosseteste completed the first full translation of the *Nicomachean Ethics* into Latin, Albertus Magnus prepared a detailed commentary on it in the course of a series of lectures given in Cologne c. 1250.[1] More than a decade later, Albert returned to the *Ethics* with a second commentary.[2] Thomas Aquinas was one of Albert's students during the period of the Cologne lectures. In the mid-1260s Aquinas wrote his own commentary on the *Ethics*, based on a revised Latin translation, which showed the continuing influence of Albert's teaching.[3]

---

[1] Albert's first commentary has been edited as vol. xiv of the new Cologne edition of his *Opera omnia*: *Alberti Magni super ethica commentum et quaestiones*, Wilhelm Kübel (ed.) (Monasterii West-falorum, 1968–72). See Jean Dunbabin, "The Two Commentaries of Albertus Magnus on the Nicomachean Ethics," *Recherches de théologie ancienne et médiévale* 30 (1963), 232–50; Auguste Pelzer, "Le cours inédit d'Albert le Grand sur la Morale à Nicomaque, recueilli et rédigé par saint Thomas d'Aquin," *Revue néo-scolastique* 24 (1922), 333–61, 479–520. For the history of the Grosseteste translation, see chapter 2, n. 11. The great importance of the *Nicomachean Ethics* in the medieval schools is indicated by the hundreds of extant manuscripts listed in *Ethica Nicomachea: praefatio*, R.-A. Gauthier (ed.), Arist. Lat. 26, Fasc. 1 (Leiden, 1974), clii–clxvii.

[2] Albert's second commentary is found in *Sententia libri ethicorum*, A. Borgnet (ed.), vol. vii of *Opera omnia* (Paris, 1891). Dunbabin ("Two Commentaries," 245) dates this second commentary to 1267–68. James Weisheipl provides an earlier probable date, 1262–63, in "Albert's Works on Natural Science in Probable Chronological Order," Appendix 1, in Weisheipl (ed.), *Albertus Magnus and the Sciences: Commemorative Essays* (Toronto, 1980), 565–77, 575. The relationship between Albert's two commentaries is complex. Odd Langholm (*Wealth and Money*, 23) believes the earlier commentary edited by Kübel to be superior to the later edited by Borgnet. He considers the latter primarily a truncated version of the former. This characterization does not do justice to the rich insights and details of analysis found in the Borgnet that are not found in the Kübel. See Dunbabin, "Two Commentaries," 245–50. In the following discussion I draw on both commentaries designating each by its modern editor.

[3] *Sancti Thomae de Aquino sententia libri ethicorum*, vol. xlvii of *Opera omnia* (Rome, 1969). Aquinas' commentary has been translated into English in *St. Thomas Aquinas: Commentary on the Nicomachean Ethics*, C. I. Litzinger (tr.), 2 vols. (Chicago, 1964). Baldwin (*Just Price*, 60) dates the revised Grosseteste translation (possibly by William of Moerbeke) to 1260 and the commentary of Aquinas to 1266.

The primary intent of both Albert and Aquinas in their commentaries was straightforward clarification and explication of the text. For the most part they resisted introducing their own or orthodox positions in contrast to Aristotle's formulations, even though a detached presentation of this pre-Christian ethical system presented considerable doctrinal difficulties.[4]

There was a great deal in the *Ethics*, including Aristotle's decidedly mathematical treatment of economic exchange in Book v, that was new and indeed foreign to medieval thinkers. Additional problems of interpreting Aristotle's economic thought were presented to the commentators by the imperfections of the Grosseteste translation and by the many inconsistencies found within the text itself of Book v. The early commentaries of Albert and Aquinas were remarkably successful in clarifying Aristotle's most difficult passages, and nowhere more so than in the chapters dealing with economic exchange.[5] The influence of these first Latin commentaries on the study of the *Ethics* in general and on Aristotle's economic thought in particular proved long-lasting. They remained for centuries the mediators of the Aristotelian text for the young students who struggled with its difficulties, as well as for the scholars who prepared new commentaries in succeeding centuries.[6]

Albert began the prologue to his first commentary on the *Ethics* with a quotation from Ptolemy's *Almagest*.[7] At first sight this seems a strange choice of authorities to introduce the subject of ethics. Albert's selection

---

[4] Dunbabin, "Two Commentaries," 235, 245. For specific doctrinal difficulties introduced by the *Ethics*, see R.-A. Gauthier, "Trois commentaires 'averroistes' sur L'Ethique à Nicomaque," *Archives d'histoire doctrinale et littéraire du moyen âge* 23 (1947–48), 187–336; Gauthier and Jolif, *L'Ethique*, vol. 1, 130–34.

[5] As an aid in the preparation of these first full commentaries, Albert and Aquinas had available to them short Greek commentaries on each of the ten books, translated by Grosseteste at the time of his translation of the *Ethics*. In a number of manuscripts these commentaries were found accompanying the text. See Pelzer, "Versions latines," 411; Powicke, "Grosseteste," 90; Aristotle, *The Greek Commentaries on the Nicomachean Ethics of Aristotle in the Latin Translation of Robert Grosseteste*, H. P. G. Mercken (ed.) (Leiden, 1973). For Book v, Eustratius of Nicaea's commentary was particularly important and continued to be so through the fourteenth century. For an indication of the influence of Eustratius' commentary on later authors, see Langholm, *Wealth and Money*, 24, 85–95; James J. Walsh, "Buridan and Seneca," *Journal of the History of Ideas* 27 (1966), 23–41.

[6] Of the two, Albert's commentaries were the more successful. Gauthier and Jolif (*L'Ethique*, vol. 1, 79) consider them superior to all succeeding medieval commentaries. Noonan (*Usury*, 51–57) is also less impressed with Aquinas' commentary on Book v than with Albert's, believing, as do many, that his most important economic comments are found instead in the *Summa theologiae*. Aquinas' economic thinking outside his *Ethics* commentary will be considered in chapters 4 and 5. Albert's commentaries were also somewhat more influential in their impact on medieval economic theory. For an assessment, see Langholm, *Price and Value*, 61; Langholm, *Wealth and Money*, 22–29; Bridrey, *Théorie de la monnaie*, 413; Baldwin, *Just Price*, 60ff.

[7] Albertus Magnus (Kübel edn.), 1a: "Ptolemaeus in Almagesto: 'Disciplina hominis sui intellectus socius est et apud homines intercessor.'"

of a mathematician and astronomer, however, underscores the connection he assumed between the study of ethics (*scientia moralis*) and the study of mathematics, physics, and metaphysics. For Albert, ethics was the crown of scientific speculation, and he sought to apply to it the same frame he used in his investigations of the natural world.[8] The treatment of ethics as a *scientia practica* continued into the fourteenth century.[9] In Peter Aureole's categorization, for example, *scientia moralis* is considered a "sensual and experimental" science and as such is linked directly with natural philosophy and medicine.[10] The general thesis of this present work that insights were freely exchanged between ethical and proto-scientific thought, between models of economic exchange and models of nature, would have presented little difficulty to Albert or Aquinas in the thirteenth century, or to the proto-scientists and economic theorists of the fourteenth century.[11] Nowhere is Albert's "scientific" approach to ethical questions clearer than in his commentary on Aristotle's discussion of justice and economic exchange in Book v of the *Ethics*.

MATHEMATICS AND THE GEOMETRY OF EXCHANGE

Although Albert was generally cautious in his assessment of the place of mathematics in scientific thought, he fastened on the many openings for mathematical discussion found in Book v.[12] Following Aristotle's point that proportion was a property not only of abstract units but also of things numbered in the world [1131a30–32], Albert introduced a long and involved discussion of the theory of proportions that leaned heavily on the *De arithmetica* of Boethius, and that covers four full columns in the

---

[8] Albertus Magnus (Borgnet edn.), 1a: "Circa moralem scientiam tria hic tanguntur: materia, finis, utilitas." See Benedict M. Ashley, "St. Albert and the Nature of Natural Science," in Weisheipl, *Albertus Magnus and the Sciences*, 73–104, 97–101; Dunbabin, "Two Commentaries," 245.

[9] Arthur Stephen McGrade, "Ethics and Politics as Practical Sciences," in Monika Asztalos, John Murdoch, and Ilkka Niiniluoto (eds.), *Knowledge and the Sciences in Medieval Philosophy*, vol. 1 of *Proceedings of the Eighth International Congress of Medieval Philosophy (SIEPM)* (Helsinki, 1990), 199–219, esp. 200–07.

[10] Peter Aureole, *Commentarium in primum librum sententiarum* (Rome, 1596), 57b: "tres vero sensuales et experimentales, ut Naturalis, Moralis, Medicinalis."

[11] On the new freedom felt by fourteenth-century thinkers to exchange analogies and methods between theology, philosophy, medicine, optics, and mathematics (and I would add economics), see Amos Funkenstein, *Theology and the Scientific Imagination from the Middle Ages to the Seventeenth Century* (Princeton, 1986), 307–09. Jean Buridan's use of a quotation from the *Ethics* to introduce his commentary on Aristotle's *Physics* mirrors Albert's use of Ptolemy to introduce the *Ethics*. Buridan, *Physics*, 2ra: "ut habetur primo Ethicorum quanto est multis communius tanto est melius et divinius."

[12] James A. Weisheipl, "Albertus Magnus and the Oxford Platonists," *Proceedings of the American Catholic Philosophical Association* 32 (1958), 124–39; A. G. Molland, "Mathematics in the Thought of Albertus Magnus," in Weisheipl, *Albertus Magnus and the Sciences*, 463–78, 465–69.

Borgnet edition.[13] It is interesting that Albert interjected what amounted to a primer on proportions here. He was preparing the reader for Aristotle's idea that distributive justice required the establishment of a geometrically proportional mean. Albert's insertion of a technical, mathematical discussion demonstrates that he did not take lightly Aristotle's application of mathematics and proportionality to the questions of justice and just economic exchange. On the contrary, he appears to have been intent on sharpening and refining the use of mathematics in these areas. Aquinas appears similarly concerned to expand Aristotle's application of mathematics to the ethics of exchange.[14] Aquinas strengthened the theoretical case for a mathematical treatment of justice by showing how the use of proportionality in Book v was consistent with Aristotle's fuller definition of proportionality and measurement found in Book x of the *Metaphysics*.[15]

Aristotle, as we have seen, linked the equality achieved by distributive justice (*iustitia distributiva*) to the mathematics of geometric proportionality. One way he illustrated this was through the economic example of a partnership in which the amount each partner received was determined proportional to his investment.[16] Aquinas repeated the case of partnership in his commentary, but he added a new economic example to clarify the mathematics involved. In contrast to Aristotle, Aquinas used specific (if simple) numbers to illustrate the computation of just proportion. In order to link numbers to the abstract idea of proportional reward, Aquinas chose, not surprisingly, an example involving money. How better to illustrate a geometrically proportional equality than to say that, if Socrates works two days to Plato's one, he should receive two pounds (*librae*) in payment to Plato's one. Aquinas further underscored the mathematics of distributive justice by framing his example as an equation involving four terms:

let A be one term, for example, two pounds, and B one pound. But let G be one person, for example, Socrates who has worked two days, and D, Plato, who has worked one day. Therefore, as A is to B so G is to D, because a double

---

[13] Albertus Magnus, Borgnet 345a–347b. Boethius' discussion of proportionality is found in *De arithmetica*, bk. II, 40–53; J.-P. Migne, *Patrologiae cursus completus*, vol. 63, cols. 1145–1166. The discussion of proportions is more fully covered in the Borgnet commentary than in the Kübel (333a–334a). In the Kübel, however, Albert clarifies certain points by stating that proportion is to be found in anything that can be numbered, such as the number of men or horses or stones (333a), and that proportion can be applied to all things "quae habent aliquam quantitatem" (334a).

[14] Baldwin, *Just Price*, 60–62.

[15] Aquinas (1969), 280a: "proportio autem nihil est aliud quam habitudo unius quantitatis ad aliam; quantitas autem habet rationem mensurae: quae primo quidem invenitur in unitate numerali et ex inde derivatur ad omne genus quantitatis, ut patet in x Metaphysicae; et ideo numerus primo quidem invenitur in numero unitatum et ex inde derivatur ad omne aliud quantitatis genus quod secundum rationem numeri mensuratur."   [16] [1131b29–30]. See chapter 2.

proportion is found on the one side and the other. Hence by alternation, as A is to G, so B is to D.[17]

Money, being essentially numbered measure extended into the world, is, as Aquinas recognized, extremely well suited to represent the workings of proportion in everyday life.[18]

Aristotle's second form of particular justice, governed by arithmetical rather than geometrical proportion, gave Albert and Aquinas further opportunities for mathematical digression. Following Grosseteste's translation, Albert called this second form of justice *iustitia directiva*, that is, "directive justice" or "corrective justice." He often added the qualifier, *iustitia directiva in commutationibus et contractibus*, because according to Aristotle this form of justice was particularly applicable to the action of a third-party judge in restoring equality to already completed unjust economic exchanges. It was the judge's task in these cases to equalize the profit and loss between the parties involved through the arithmetical process of addition and subtraction.[19] For Albert as for Aristotle, this second particular form of justice, *iustitia directiva*, was limited in economic transactions to a corrective process undertaken by the judge only after the fact of an unequal exchange. In contrast, the active process of economic exchange itself was governed by *contrapassum*, or "reciprocity."[20]

While both Aristotle and his commentators asserted that equality was the end of all economic exchange, they understood that inequality was its precondition. The skill, effort, and expenses involved in the production of goods for exchange (e.g., between a shoemaker and a carpenter) were necessarily unequal. Furthermore, the needs of each exchanger for the goods of the other were inherently unequal. Therefore, the *contrapassum*

[17] Aquinas (1969), 280b: "Sit ergo A unus terminus, puta duae librae, B autem sit una libra, G autem sit una persona, puta Sortes qui duobus diebus laboravit, D autem sit Plato qui uno die laboravit; sicut ergo se habet A ad B, ita se habet G ad D, quia utrobique invenitur dupla proportio; ergo et permutatim, sicut A se habet ad G, ita se habet B ad D, quaecumque enim sunt ad invicem proportionalia, etiam permutatim proportionalia sunt" (Litzinger trans., vol. I, 406).

[18] Another example of money being used to clarify the idea of proportion is found in Albert's addition of number (and numbered coins) to Aristotle's discussion of the proportional mean [1132a15–16], Kübel 340b–341a: "Et ista aequalitas fit medium, ut si unus habeat sex *marcas* et alius sex; si autem auferatur ab uno una *marca* et detur alteri, habebit ille septem et alter quinque et sic superabit eum in duabus marcis." For a discussion of the relationship between money, numeracy, and mathematics, see Murray, *Reason and Society*, esp. 191–93; Hadden, *On the Shoulders of Merchants*, 85–93.

[19] Note Albert's considerable expansion of the mathematics of such equalization, Borgnet, 352a–353a, and compare to Arist. Lat. 26 (1973), 461 [1132b1–10].

[20] Neither Albert nor Aquinas conceived of this *contrapassum* as a distinct third form of particular justice. Rather they saw it as a hybrid of *iustitia directiva and iustitia distributiva*, but one governed by geometric equalization. See Baldwin, *Just Price*, 62. For conflicting opinions on the relationship of reciprocity to distributive and rectificatory justice, see chapter 2, n. 22.

that governed economic exchange was understood as a process of *geometrical* rather than *arithmetical* equalization. In Aristotle's terms, only through geometric proportionality can equality be established between unequals.[21] It was here in the elucidation of the geometric process of economic equalization that Aristotle made his strongest and most direct use of a mathematical model. He located a geometric "figure of proportionality" (*figura proportionalitatis*) at the heart of economic exchange.[22] It is remarkable that both Albert and Aquinas followed Aristotle's striking exposition of the geometry of exchange without difficulty and, seemingly, without hesitation.

In Aristotle's "figure of proportionality" the relation of the Builder (A) and his House (G) to the Shoemaker (B) and his Shoes (D) is conceived of as a rectangular figure. Although Aristotle does not make the point explicitly, the implications of the figure are that exchange takes place along the diagonals of the rectangle. The point where the diagonals cross would then represent the point of proportional equivalence – the proper end of exchange. While the exchange rectangle is pure Aristotle, Aquinas, and particularly Albert, allowed the figure considerably greater geometrical reality and physicality than the text had provided. Both were aided in this by the fact that many manuscripts of the Grosseteste translation contained one or more figures of exchange (a square with crossed diagonals) actually drawn and labeled in the margins or included in the body of the text itself.[23] The geometric drawings illustrating the square of exchange in Book v were often the only figures accompanying the text of the *Ethics*. As such they would have been likely to attract the attention of students and scholars. Albert passed on this tradition of the drawn figure in his commentary, as did many succeeding commentators.[24]

---

[21] Aquinas (1969), 291a: "Dicit ergo primo quod in communicationibus commutativis verum est quod tale iustum continet in se contrapassum, non quidem secundum aequalitatem, sed secundum proportionalitatem. Videtur autem hoc esse contra id quod supra dictum est, quod scilicet in commutativa iustitia medium accipitur non quidem secundum geometricam proportionalitatem, quae consistit in aequalitate proportionis, sed secundum arismeticam, quae consistit in aequalitate quantitatis. Dicendum est autem quod circa iustitiam commutativam semper quidem oportet esse aequalitatem rei ad rem, non tamen actionis et passionis, quod importat contrapassum, sed in hoc oportet adhiberi proportionalitatem ad hoc quod fiat aequalitas rerum, eo quod actio unius artificis maior est quam actio alterius, sicut aedificatio quam fabricatio cultelli, unde si aedificator commutaret actionem suam pro actione fabri, non esset aequalitas rei datae et acceptae, puta domus et cultelli." See Baldwin, *Just Price*, 62, 73; Schumpeter, *Economic Analysis*, 94.

[22] See Arist. Lat. 26 (1973), 463 [1133a34], for the phrase *figura proportionalitatis*, and chapter 2, 44.

[23] See chapter 2, n. 21. Langholm (*Economics in the Schools*, 182) suggests that the drawn figure first appears in the work of Michael Ephesinus, an eleventh-century Byzantine commentator.

[24] Albert (Kübel edn.), 343a–b: "sicut vides in figura exterius."

61

Where Aristotle had talked of the four terms (the exchangers and their products) as points of the square,[25] Albert recognized that each line or side of the square (*linea lateralis, latera quadrati*) represents a dynamic relationship in the exchange; so that, for example, the relative need or demand (*indigentia*) of the Builder and the Shoemaker is represented by line AB, and the relative value of the House to the Sandals is represented by line GD.[26] In doing so he allowed the geometric line great representational power in the description of the exchange process, more so than had Aristotle. Aquinas followed Albert's lead here. Where Aristotle states that exchange proportions are determined by the conjunction of diameters (*secundum dyametrum coniugacio*), Aquinas' commentary adds physical and active qualities to the geometric *figura*: "Proportional return in exchange is made *through* the conjunction of the diagonals [*facta per diametralem coniunctionem*]."[27]

It is striking how seriously the early commentators took the geometry of the exchange process.[28] As Aristotle had before them, they applied the same cognitive approach to the economic world as they would to the study of nature. The treatment here of exchange as a mathematical equation representable by a geometric figure provided a model for the further extension of mathematics and geometry to other problems concerning measurement in the world. Neither Albert nor Aquinas at any time questioned the appropriateness or accuracy of this geometrical model; rather, they extended it. Albert did, however, at one point digress from the Aristotelian text and observe that economic exchange at times followed patterns unrecognized by Aristotle – patterns that did not conform to his geometric model because they were not the product of purely voluntary exchanges between producers. Demonstrating a clear perception of the state of the marketplace of the middle of the thirteenth century, Albert notes that, during periods of great shortages, the prince may interfere in the market and set maximum prices so that the city as a whole does not suffer. He notes as well that the normal dynamic equival-

---

[25] Arist. Lat. 26 (1972), 236: "Puta edificator in quo A, coriarius in quo B, domus in quo G, calciamentum in quo D."

[26] Albert (Borgnet edn.), 356a–b: "Puta aedificator domus significetur per A, coriarius autem calceum faciens per B, mutata indigentia aedificatoris ad coriarium, et e converso, designentur per lineam AB . . . dependentia domus ad aedificatorem designetur per lineam AG . . . dependentia calceamenti ad coriarium significetur per lineam BD: commutabilitas domus in calceamentum et e converso, signetur per lineam GD, transitus autem domus ad coriarium signetur per diametrum GB, transitus vero calceamenti ad aedificatorem signetur per diametrum AD." See also Kübel, 343a–b.

[27] Arist. Lat. 26 (1973), 462 [1133a5]: "Facit enim retribucionem eam que secundum proporcionalitatem secundum dyametrum coniugacio." Aquinas (1969), 292a: "deinde fiat contrapassum . . . scilicet retributio secundum proportionalitatem facta per diametralem coniunctionem."

[28] See Baldwin, *Just Price*, 60–62, for Aquinas' acceptance of Aristotle's mathematics of justice.

ence established between buyer and seller in exchange can be upset by a pre-existent agreement or pact setting the selling price.[29] In recognizing these exceptions, he demonstrates his association of the geometrical process of equalization (in contrast to the arithmetical process which required an intervening judge) with the free, direct, and dynamic interchange of goods and people in exchange.[30]

Both Aristotle and his early commentators understood that the geometry of exchange rested on the existence of money.[31] At the conclusion of his economic discussion in *Ethics* v, Aristotle created a separate geometric figure to model and represent the function of money in exchange. This figure was critical to the scholastic understanding of the model of money as measure. In the geometrical equation, A was a house worth five minae, B ten minae in coinage, and C a bed worth one mina. In his first commentary Albert was content to restate Aristotle's point that, in an exchange of goods of uneven value, instead of the goods themselves having to be multiplied or divided to attain equality (i.e., five beds exchanged for one house, or one-fifth of a house for one bed), money as the middle-term became the divisible factor, so that the builder could sell his house for five minae and then buy the bed for one of them, instead of having to take all five beds in a direct barter exchange.

In his later commentary Albert brought out further implications of this model not present in Aristotle's text. He used the fact of money's divisibility to underscore the fact that money measured relative rather than essential value. The one-fifth of a house being exchanged for a bed had no meaning in terms of essential value, only in relation to money as measure.[32] His most important contribution consisted of his making explicit what was only implicit in Aristotle's "figure," that is, that money performs the function of the third or intermediate or common term in the continuous proportion A/B = B/C.[33] In Aristotelian terms, the mid-term in such a "continuous proportion" always functions as a

---

[29] Albert (Borgnet edn.), 355b: "Forum enim aliquando disponit princeps pro communi utilitate decernens quod quanti vendendum sit: quia aliter pro necessitate rei aliquando nimis damnificarentur cives: sicut tempore caristiae decernitur, quod tanto et non pluri vendatur modius frumenti. Aliquando tales commutationes fiunt pacto interveniente sine omnis deceptionis fraude: et tunc pro re non datur aequivalens, sed id quod pactum est. Aliquando diriguntur tales commutationes ad rei paritatem, ut aequivalens habeat qui vendit, ad hoc quod alius emit. In his ergo considerationem habere oportet ut contrapassum justum non semper ad idem determinetur, sed aliquando ad par, aliquando ad pactum, aliquando vero ad judicatum repatiatur." For a discussion of this passage, see Langholm, *Usury*, 33–34.

[30] For a strong statement of this position by a fourteenth-century economic thinker, see the position of Geraldus Odonis in chapter 5.

[31] Albert (Borgnet edn.), 358a: "propter commutationem necessitatis quae commutatio fit secundum oppositum proportionis quando componitur opus ad opus, numisma inventum est."

[32] Albert (Borgnet edn.), 360b: "Lectus igitur quinta pars domus erit non secundum essentiam, sed secundum valorem numismatis."   [33] [1133b23–28].

63

continuum or line capable of relating the shifting values of the end-terms. Following Aristotle, Albert envisioned money as the mediating term in all exchange equations.[34] Money functioned as a *line* or *continuum of value*, measuring and numbering the quality of need attached to all goods, making immediately apparent their relative values, and thus permitting their equation. Money is the connecting continuum that permits diverse goods to be "reduced to equality" (*numisma sit medium quo reducuntur ad aequalitatem*). As the line is to geometry, so money is to the geometry of exchange in Albert's model.[35]

EQUALITY, THE MEAN, AND EQUALIZATION IN EXCHANGE

As the search for the mean in human affairs was the central theme of Aristotle's *Ethics*, so it was the central theme in his economic speculation. Virtually every statement Aristotle and his commentators made on the subject of justice and just exchange can be placed under this category heading. In the geometry of exchange, equality had a dual function. It was both the desired end and the engine of exchange. The geometric form was conceived as dynamic: any combination of producer and product could be plugged into the rectangular figure and proportional equality would be the result. In essence, the *figura proportionalitatis* functioned as a primitive *mechanism of equalization*.

Given that there exists a builder who needs to exchange his house for a shoemaker's shoes, how can one assume that the resulting exchange will embody a proportional equivalence – and not only a vague equivalence, but an equivalence sufficiently precise to be represented by a geometric figure and given the name "just." Aristotle had no positive answer to this question, but he did provide a negative one. If, he said, there was no equality, men would cease to exchange. If this were the case, not only would individuals suffer, but the very basis of production in society would be destroyed.[36] Given how fundamental exchange equality is to the functioning of society, it is surprising how few clues Aristotle provides for determining it. Both Albert and Aquinas added economic detail here. According to both, the productive arts would be destroyed if each producer was not at least recompensed for the labor and expenses he had

---

[34] Albert (Borgnet edn.), 360b: "Haec autem figura sic describitur, quod sit inter tres terminos continuae proportionis, scilicet quod valor domus sit A, valor lecti G, B autem numisma sit medium quo reducuntur ad aequalitatem."

[35] Albert (Borgnet edn.), 346b: "Conjuncta autem dicitur, quae medio uno tertio bis utitur, sicut punctum continuans in linea, bis idem est in duplici ratione sumptum, scilicet finis et principii."

[36] [1133a14–16]. Modern editors often consider this passage an interpolation, but it appeared in the Grosseteste translation, and Albert thought it important enough to paraphrase and comment upon it twice in each of his two commentaries. On this, see Langholm, *Economics in the Schools*, 189.

invested in his product.[37] Just *contrapassum* requires in turn that the labor and expenses of each exchanger be equated proportionally.[38]

Aristotle never specifically brought up the subject of "just price" in his economic writings. Albert introduced this fundamental scholastic concept into his commentary in relation to the question of equality in exchange.[39] He conceived of the just price as both the pre-condition and the guarantee of equality. Without just price there would be no proportional equality. Without proportional equality, producers would not be recompensed for their labor and expenses. Without just recompense, the *civitas* would disintegrate.[40] We are left at the end of Albert's analysis not only with a geometrical mechanism of equalization in exchange, but also one on which the life of the *civitas* depended.

MONEY AS *MEDIUM* AND AS MEASURE

The complexities involved in the simplest exchange equation are enormous, as both Aristotle and his commentators clearly recognized:

He [Aristotle] says first, in order that the products of the different workmen be equated and thus become possible to exchange, it is necessary that all things capable of exchange should be comparable in some way with one another so that it can be known which of them has greater value and which less. It was for this purpose that money or currency was invented, to measure the price of such things. In this way currency becomes a *medium* inasmuch as it measures everything, both excess and defect.[41]

---

[37] Albert (Kübel edn.), 343b: "quia artes destruerentur, secundum quod nullus vellet uti eis, si non faciens, idest dans, quantum et quale, scilicet quantum ad expensas et quantum ad laborem." See also Baldwin, *Just Price*, 63.

[38] Albert (Kübel edn.), 345b: "Dicit ergo primo, quod tunc erit contrapassum, in quo est iustitia communicationis, quando aequata sunt operata, et talis aequatio fit per proportionalitatem. Ergo oportet, quod sicut se habet agricola ad coriarium in expensis et labore, ita se habet opus coriarii ad opus agricolae. Et sic oportet in figuram proportionalitatis ducere opera et artifices, ut prius dictum est." For varying assessments of the role of *labores et expensae* in Albert's and Aquinas' thought, see Selma Hagenauer, *Das "justum pretium" bei Thomas von Aquino: ein Beitrag zur Geschichte der objecktiven Werttheorie* (Stuttgart, 1931); Langholm, *Price and Value*, 29ff.; Baldwin, *Just Price*, 74–77; Noonan, *Usury*, 81–99; Jean Ibanès, *La doctrine de l'église et les réalités économiques au XIIIe siècle* (Paris, 1967), 36–41. See also chapter 2, n. 38.

[39] Baldwin, *Just Price*, 68–77.

[40] Albert (Borgnet edn.), 357a: "Primum autem dicimus sicut emptionem justo pretio factam: et quod deinde fit, dicimus sicut pretii solutionem. Si autem non sic fiat in commutationibus talibus, non salvatur aequalitas proportionis, qua non salvata, civitas non commanet: quia non retribuitur in laboribus et expensis." See Baldwin, *Just Price*, 70.

[41] Aquinas, Litzinger trans., vol. I, 425. Aquinas (1969), 294a: "Dicit ergo primo quod, ad hoc quod opera diversorum artificum adaequentur et sic commutari possint, oportet quod omnia illa quorum potest esse commutatio sint aliqualiter ad invicem comparabilia, ut scilicet sciatur quid eorum plus valeat et quid minus. Et ad hoc inventum est nummisma, id est denarius, per quem mensurantur pretia talium rerum, et sic denarius fit quodam modo medium, in quantum scilicet omnia mensurat et superabundantiam et defectum."

Here Aquinas, following Aristotle, defines money as a "medium" because of its two primary functions: (1) "it measures everything [*omnia mensurat*]," and (2) by doing so it brings all things into relation and, potentially, equation. Since questions of measurement and relation were central to late thirteenth- and fourteenth-century natural philosophy, and lay at the very heart of the science of *calculationes*, it is important to gauge just how seriously Aristotle's commentators took the potent idea that money measures and relates all things. From these earliest commentaries, it is clear that the answer is very seriously.

Albert and Aquinas were not content to simply state that money could measure all things. They sought to explain how it performed this remarkable task. Here they clarified Aristotle's argument. Money as a single scale cannot possibly measure the multiplicity of things directly, in themselves (*secundum veritatem*). Diverse commodities must first be reduced to one commonly possessed quality before they can be measured by a single measure. In economic exchange that single quality is *indigentia* or human need.[42] Since all things that are exchanged are needed by the exchangers, the varying degree of need provides a "natural" measurement of all goods. Money is an invented, artificial measure. It measures things second-hand, by measuring the common accidental quality of *indigentia* or need that is attached to all things in exchange.[43] Just *contrapassum* requires the establishment of proportional equality. Money greatly facilitates this equation not only by serving as a *medium* of measurement and relation, but also by introducing number and a numbered scale into the problem of value determination. Thus, all things find their numbered values in exchange through the *medium* of money, the measure of need.[44]

The outline of this elegant argument existed in Aristotle's text, but Aquinas clarified and refined the argument, especially in his emphasis on the dual system of measurement, natural (need) and artificial (money), that permitted commensuration and adequation.[45] True to his scientific

---

[42] Aquinas (1969), 296b: "ostendit per quem modum denarii mensurent. Et dicit quod res tam differentes impossibile est commensurari secundum veritatem, idest secundum proprietatem ipsarum rerum, sed per comparationem ad indigentiam hominum sufficienter possunt contineri sub una mensura." In later economic thought "need" will be replaced by the more easily quantifiable and less value-laden term "demand."

[43] Aquinas (1969), 295a: "Dicit ergo primo quod ex quo omnia mensurantur per indigentiam naturaliter et per denarium secundum condictum hominum, tunc iuste fiet contrapassum quando omnia secundum praedictum modum adaequabuntur." For Albert and Aquinas on money as measure, see Langholm, *Wealth and Money*, 71–73; Langholm, *Price and Value*, 92.

[44] Albert (Borgnet edn.), 357b: "numisma inventum est quod numero per additionem et minutionem mensura uniuscujusque est: et ideo numisma aliqualiter medium fit quo omnia alia proportionantur."

[45] Aquinas (1969), 295a: "omnia mensurantur per indigentiam naturaliter et per denarium secundum condictum hominum." See Langholm, *Price and Value*, 92.

orientation, Albert was even more concerned than Aquinas to analyze money as measure, and more rigorous in his analysis. After introducing Aristotle's statement that money measures all goods toward the end of relation and equalization, he asked, in good scholastic fashion, whether this conception could be reconciled with Aristotle's more developed position on measurement found in Book x of the *Metaphysics* and Book vii of the *Physics*.[46] Here Aristotle states that the measure is always a certain minimum unit of the thing measured, and therefore is always of the same species as the thing measured. All number, for example, is measured by the number one, and all distance is measured by the foot or some other unit of length.[47] But money, after all, shares neither material nor form with the diverse things it measures.[48] Furthermore, the measures used for different species are different; that for oil is different from that for wine. It is impossible, therefore, following *Metaphysics* x, for all things to be measured by a single measure.[49]

Albert then responds to the objections he has raised, and he does so by employing the same dual definition of measure that we have seen in Aquinas. Things can be measured in two ways: according to their nature, or according to an accident that they possess. Aristotle's definition of measurement in the *Metaphysics* pertained only to the measure of essential nature, but measurement is commonly of accidents. The example he gives is very instructive. When we say that cloth (*pannus*) is measured by elbows or ells (*ulna*), we are using a measure that is clearly not of the same species as the thing measured, and so we cannot be measuring the cloth in its essence, but only the "continuous quantity" that it possesses as accident.[50] In such a way, he continues, money does not and cannot measure the nature or essence of things but only the common quality all goods possess of being useful and so being needed by the community (*secundum*

---

[46] Albert (Kübel edn.), 344a: "unumquodque mensuratur sui generis minimo; sed numisma est extraneum a genere artificiatorum aliorum; ergo etc."

[47] *Metaphysics* x.1 [1052b20–1053b8].

[48] Albert (Kübel edn.), 344a: "numisma, cum non habeat in aliquo communitatem cum quibusdam aliis artificiatis, quia neque in materia neque in forma, videtur, quod non possit esse mensura ipsorum."

[49] Albert (Kübel edn.), 344a: "Praeterea, diversorum secundum speciem sunt diversae propriae mensurae, sicut patet in x METAPHYSICAE, quia alio mensuratur vinum et alio oleum; sed non omnia artificiata sunt unius mensurae; ergo non possunt habere unam mensuram."

[50] Albert (Kübel edn.), 344a–b: "Dicendum ad primum, quod aliquid potest mensurari dupliciter: aut quantum ad participationem essentiae vel naturae, et sic unumquodque mensuratur sui generis minimo . . . Potest autem aliquid mensurari quantum ad aliquod accidens, et sic non oportet, quod mensuretur aliquo minimo sui generis, sed aliquo in quo inveniatur illud accidens, sicut pannus mensuratur ulna, quae est minimum potentiae secundum quantitatem continuam in mensuris."

quod veniunt in usum et utilitatem communitatis).[51] The conclusion suggested by Aristotle but made explicit in the commentaries of Albert and Aquinas is that money acts as a *medium* and a measure by quantifying the qualities of use, utility, and need attached to all goods in exchange.[52]

RELATION AND THE RELATIVITY OF VALUE IN EXCHANGE

In the original Greek text, the quality common to all goods that money measured was designated by the word *chreia*, which is properly translated as "need." In the Grosseteste translation that Albert used *chreia* was rendered by the Latin word *opus*, and in the revised translation that Aquinas used was sometimes rendered as *opus* and sometimes as *indigentia*.[53] The confusion as to what determined value – whether it was the labor and expenses (one meaning of *opus*) involved in the production of a good, or, alternatively, the need (a second meaning of *opus*) for that good – can be traced, in part, back to confusions introduced in the early Latin translations of Aristotle's text.[54] Confusion in the text, however, rather than being a hindrance, often proved to be a spur for fruitful speculation on the part of the commentators.

In his first commentary, Albert most often identified *labor et expensae* as the measurable factors common to products in exchange. "However-much the builder exceeds the shoemaker in the labor and expenses he has put into his product, by so much does the value of the house exceed the

---

[51] Albert (Kübel edn.), 344b: "Sic ergo dicimus, quod si fiat mensuratio artificialium secundum esse suae speciei, non mensurantur omnia numismate, sed domus domo et sic de aliis. Si autem mensurantur quantum ad hoc accidens ipsorum, quod est appretiabile esse, secundum quod veniunt in usum et utilitatem communitatis, sic possunt habere omnia mensuram, quae sit certissimi pretii inter alia, quia hoc est dispositio mensurae." Albert later (Kübel, 358) identifies the qualities of use and utility with need: "et opus diximus esse usum vel utilitatem vel indigentiam."

[52] Note that in this statement Albert has substituted the idea of common use and utility "in usum et utilitatem communitatis" for Aristotle's notion of individual *indigentia*. Common valuation has replaced individual valuation in the determination of economic value. This important addition and correction will be considered in greater depth, 73–76.

[53] Arist. Lat. 26 (1972), 237: "Oportet ergo uno aliquo omnia mensurari, quemadmodum dictum est prius. Hoc autem est secundum veritatem quidem, opus quod omnia continet." Compare to the revised Grosseteste translation, Arist. Lat. 26 (1973), 463: "Oportet ergo uno aliquo omnia mensurari, quemadmodum dictum est prius. Hoc autem est secundum veritatem *quidem* [variants: *indigencia quod omnia, opus quod omnia*] continet. Si enim nichil indigerent, vel non similiter, vel non erit communicacio, vel non eadem indigencia que puta propter commutacionem necessitatis nummisma factum est secundum composicionem." See Langholm, "Scholastic Economics," 122; Langholm, *Usury*, 51.

[54] Another example of the shift in meaning attached to the shift in words: Grosseteste, Arist. Lat. 26 (1972), 238: "Quoniam autem opus continet quemadmodum unum quid ens."
   Revised Grosseteste, Arist. Lat. 26 (1973), 463: "Quoniam autem indigencia continet quemadmodum quidem ens."

value of the shoe.''[55] In this early commentary, Albert's treatment of need or demand was limited to the one phrase in the Grosseteste translation in which the concept of *opus* was linked to the concept of *indigentia*.[56] Here need was not yet conceived of as a component of value but simply as the underlying cause of exchange. Men exchanged because they each needed what the other possessed.[57] In the years between the writing of the first and the second commentary, the complex role of *indigentia* in exchange came to be more greatly appreciated by Albert. In the new schema, because *indigentia* was at the basis of human exchange, the varying amounts of it attached to each object actually determined the proportional value of the goods in exchange.[58] Money, then, was invented to give number to human need and thereby to facilitate the relation and equalization of goods in exchange.[59]

While gaining in emphasis, the relational and external determinants of value represented by *indigentia* never entirely supplanted the internal determinants of labor and expenses. Albert's final argument contained contrary elements. He showed that there would be no exchange if there were no equivalence of labor and expenses in production, and no exchange without an equivalence of need. Aristotle had indicated these two possibilities in his confused discussion. Albert, thus, made sense out of the various possibilities presented to him by simply conflating them: he specifically equated the concept of *opus* with use value, utility value, and need value.[60] Within the confusion concerning the determinants of market value there is one fact that is never questioned: money is an invented and artificial measure, instituted by law.[61] In determining the proportion of goods in exchange, the proportion is made not according to the different natures or essences of the goods, which money is incapable of measuring, but according to the relative quantity of some

---

[55] Albert (Kübel edn.), 343b: "quia quantum aedificator superat coriarium in labore et expensis, quas ponit in suo opere, tantum domus superat calceum."

[56] Arist. Lat. 26 (1972), 237: "Hoc autem est secundum veritatem quidem, opus quod omnia continet. Si enim nichil indigerent, vel non similiter, vel non erit commutacio, vel non eadem."

[57] Albert (Kübel edn.), 346a.

[58] Albert (Borgnet edn.), 355b: "Talis enim commutatio non fit per aequalitatem rerum commutatarum, sed potius secundum proportionem valoris rei unius ad valorem rei alterius, proportione habita ad indigentiam quae causa commutationis est."

[59] Albert (Borgnet edn.), 357b–58a: "Quia igitur ista secundum comparationem ad valorem secundum usum indigentiae accipiuntur, numisma inventum est quod numero per additionem et minutionem mensura uniuscujusque est: et ideo numisma aliqualiter medium fit quo omnia alia proportionantur."

[60] Albert (Borgnet edn.), 359a: "opus diximus esse usum vel utilitatem vel indigentiam"; 359b: "quod autem opus sive indigentia vel usus sicut unum quidem ens in omnibus omnia commutabilia contineat."

[61] Aquinas (1969), 295a: "denarius non est mensura per naturam, sed nomo, id est lege; est enim in potestate nostra transmutare denarios et reddere eos inutiles."

common quality possessed by these goods. Nothing possesses proportion in itself but only in relation to other things. Therefore, in exchange, all values must be relative values.[62]

Aquinas is in perfect agreement with this position. It was, in fact, his commentary that was the most important for establishing *indigentia* as the true measure of economic value.[63] To illustrate this point further, he used an example first found in the work of St. Augustine, and one destined to be used over and over again in discussions of money as measurement:

The one thing that measures all things, according to the things themselves, is need [*indigentia*], which all exchanged artifacts have common to them, to the extent that all goods exchanged have reference to human need. In exchange, things are not valued according to the dignity of their natures. If that were so, a mouse, which possesses sensitive life, would be priced higher than a pearl, which is inanimate. But the price of things is determined according to how much men need them because of their usefulness.[64]

Confronted with Augustinian authority, the Aristotelian text, and the lessons to be learned from the everyday function of the market itself, Aquinas recognized that within the system of market exchange, essences and hierarchies have no place. They are, in fact, meaningless. The market is, therefore, a system organized on principles in large part contradictory to the principles understood to govern natural law and Christian society.

THE SOCIAL GEOMETRY OF MONETIZED SOCIETY

Given these contradictions, one might suppose that Albert and Aquinas would resist Aristotle's position in Book v, that economic exchange is the

---

[62] Albert (Borgnet edn.), 357b: "Proportionari autem non possunt qui in uno conveniunt: linea enim ad proportionem valorem non habet: et si ita in se accipiantur sicut substantia lecti et substantia domus, numquam proportionari possunt. Oportet igitur quod ista accipiantur non absoluta, sed comparata aliqualiter ad valorem secundum indigentiae usum, et aliter non erit commutatio ipsorum."

[63] Langholm, *Price and Value*, 50, 93. Aquinas (1969), 296b: "ostendit per quem modum denarii mensurent. Et dicit quod res tam differentes impossibile est commensurari secundum veritatem, id est secundum proprietatem ipsarum rerum, sed per comparationem ad indigentiam hominum sufficienter possunt contineri sub una mensura." At times Aquinas linked the idea of need with the idea of "necessity"; (1969) 295b: "primo ostendit quod necessitas sit mensura secundum rei veritatem; secundo quomodo denarius sit mensura secundum legis positionem." At the same time, Aquinas continued to maintain the importance of recompense for *labores et expensae* in exchange, i.e., (1969), 452a, 453b. See Ibanès, *Doctrine*, 39.

[64] Aquinas (1969), 294b–95a: "Hoc autem unum quod omnia mensurat, secundum rei veritatem est indigentia, quae continet omnia commutabilia, in quantum scilicet omnia referuntur ad humanam indigentiam; non enim appretiantur res secundum dignitatem naturae ipsorum; alioquin unus mus, quod est animal sensibile, maioris pretii esset quam una margarita, quae est res inanimata; sed rebus pretia imponuntur secundum quod homines indigent eis ad suum usum."

primary binding force in human society. In fact they did not. Neither found it necessary, within the form of the commentary, to suggest that such a picture of society was inadequate. They both agreed that all humans have needs and that these needs must be continually satisfied through a system of reciprocal exchanges taking place according to proportional *contrapassum*. They both accepted Aristotle's statement that it is through this proportional exchange that citizens are united in the *civitas*.[65] As if this were not strong enough, Albert added the verb *salvatur* to the verb *commanet* found in the Grosseteste translation of this statement, with the result that in his commentary the very survival of the *civitas* is grounded in just and proportional exchange.[66] At one point Albert chooses the traditional metaphor of the human body to describe the organization of society around the central fact of exchange.[67] But he refers to this metaphor only once, perhaps because he recognized its inability to capture the dynamism and relativity of Aristotelian *contrapassum*. There is no head and no foot in Aristotle's model of exchange.

If there is a "heart" to Aristotle's model, it is located in the willingness of free men to return to one another the equivalent of what they have received. Men, he said, consider this return a kind of sacred duty.[68] There is an opening here for a discussion of the religious bonds of social communion, or the religious requirements for just economic requital, but it is utterly neglected by the commentators. Instead, using Aristotle's

---

[65] Aquinas (1969), 291a: "probat . . . quod justum commutativum contineat contrapassum secundum proportionalitatem, quia per hoc commanent cives sibi invicem in civitate quod sibi invicem proportionaliter contrafaciunt, prout scilicet si unus pro alio facit aliquid, alius studet proportionaliter facere pro eodem. Et manifestum est quod omnes cives hoc quaerunt ut eis proportionaliter contrafiat."
   Albert (Borgnet edn.), 355b: "In contrafacere enim proportionale, sicut dictum est, civitas commanere potest: civitas enim non commanet nisi indigentiae civium suppleantur." See also Kübel 342b.

[66] Arist. Lat. 26 (1972), 236: "In contrafacere enim proporcionale commanet civitas . . . Retribucione autem commanent." Arist. Lat. 26 (1973) 462: "per contrafacere enim proporcionale commanet civitas."
   Albert (Borgnet edn.), 355b: "in tali commutatione salvatur civitas."
   Borgnet 356a: "civitas in una libertate communicationis salvatur et commanet: et hoc debet intendere politicus."
   For the great social importance that Albert and Aquinas allow to trade, see Baldwin, *Just Price*, 63.

[67] Albert (Borgnet edn.), 357a: "Indigentia autem humana multa requirit et valde diversa, quae ab uno perfici non possunt: sed oportet quod a multis multa fiant. Propter hoc dicit Tullius, quod quemadmodum corporis humani aedificatur civitas. In corpore enim humano quamvis unum sit regens, scilicet cor, tamen multa membra officialia sunt in diversa officia distributa, quae ab uno perfici non possunt, ut mutuis obsequiis se invicem coadjuvent membra, et sic corpus in consistentia remaneat."

[68] [1133a1–5]. Arist. Lat. 26 (1973), 462: "Propter quod et graciarum sacrum prompte faciunt, ut retribucio sit; hoc enim proprium gracie. Refamulari enim oportet ei qui graciam fecit et rursus incipere ipsum graciam facientem."

figura, Albert accentuated the dynamism and mathematics of the exchange process:

Such a flux and reflux of grateful services [*fluxus et refluxus gratiarum*] holds the city together. Retribution is made proportionally, the returning services intersecting as diagonals intersect. As diagonals intersect, the service [*gratia*] of one is given to the other, and the service of the other is returned.[69]

Clearly the metaphor of the human body does not work here. We have instead the replacement of older social models with an incipient geometry of social relations. Furthermore, it was understood by both Aquinas and Albert in their role as commentators that at the center of this social geometry lay the measuring and commensurating continuum of money. It was money as *medium* that measured and so facilitated the proportional relation of all goods and services, money that served as an instrument of need equalization, money that linked diverse exchangers with diverse skills, money that was found at the locus of social interaction, and therefore money that actively bound together the community.[70]

COMMON VALUATION IN EXCHANGE

Although Aristotle maintained that money was instituted by law and by the community, he did not consider the role the community played in determining economic value. His limited model of direct exchange between two producers prohibited him from doing so.[71] It was within medieval economics that this critical line of thought in the development of price theory began to be pursued.[72]

Analyzing the relationship of the community to the individual was a widespread medieval intellectual concern. The elucidation and accept-

---

[69] Albert (Borgnet edn.), 356a: "Talis enim fluxus et refluxus gratiarum commanere facit civitatem: facit enim retributionem secundum proportionalitatem supradictam conjugatio gratiae secundum diametrum. Secundum enim diametrum gratia unius transit in alium, et gratia alterius retransit in istum."

[70] Jeremy Catto notes that both Albert and Aquinas, using the text of Book VIII of the *Ethics*, fastened on friendship as the binder of community and the basis of social *communicatio*. He neglects, however, both the discussions of Aristotle in which economic exchange is found at the center of human communication, and the enthusiastic acceptance, even embellishment of this idea by the early commentators. See Jeremy Catto, "Ideas and Experience in the Political Thought of Aquinas," *Past and Present* 71 (1976), 3–21. Further examples of the emergence of an exchange-centered social model in these commentaries can be found in Albert (Borgnet edn.), 360a: "Jam enim diximus, quod commutatione operum non existente, communicatio civium non erit." See also Aquinas (1969), 296a–b: "oportet omnia appretiari denariis, per hunc enim modum poterit esse commutatio rerum et per consequens communicatio inter homines; nummisma quidem adaequat res commutabiles sicut quaedam mensura faciens res commensuratas."

[71] For this reason this heading does not appear in chapter 2.

[72] See chapter 4 for a discussion of the influence of Roman and canon law on this development. See also Schumpeter, *Economic Analysis*, 60; Baldwin, *Just Price*, 20, 80; Ibanès, *Doctrine*, 41–43.

ance of communal rights and powers, and the balancing of these powers against those of the individual, were central to the political thought of Aristotle as well as to Roman law, canon law, and Christian theology.[73] The idea that a community could act as a whole was embodied in the important medieval legal construct of the corporation – the institutional cornerstone of the medieval university.[74] In the economic writings of Albert and Aquinas, however, the community as the prime determiner of market value was not considered as a formal "person," but rather as an *aggregate*, as the *mathematical product* of the multiplication of particular individuals and particular acts.[75]

The process of privileging the common over the personal in the determination of economic value was well underway in medieval legal thought before it made its appearance in these early commentaries.[76] From the early thirteenth century, the notion of common valuation played an increasingly important role in economic thought, and by the fourteenth century it began to appear as a component of proto-scientific thought as well.[77] It is important, therefore, to trace its philosophical beginnings in these early works, and to isolate this development by allowing it a separate category heading, one not found in the previous chapter on Aristotle.

A critical step in the elevation of the common was the depersonalization of the exchange process within scholastic thought. It derived, in part, from Aristotle's statement that in the equalization of profit and loss in economic transactions, no account was to be taken of the personality or the social condition of the participants.[78] Such a statement was consistent with the position current in both Roman and canon law. Albert made this point as strongly as it can be made. In the computation of equality in economic exchange, he wrote, whether the exchanger be an emperor, farmer, or priest, the differences in the dignity and status of buyer and

---

[73] Thomas Eschmann, "Studies on the Notion of Society in St. Thomas Aquinas," pt. I, *Mediaeval Studies* 8 (1946), 1–42; pt. II, *Mediaeval Studies* 9 (1947), 19–55. For the influence of Roman and canon law on the specifically economic construct of common valuation, see Baldwin, *Just Price*, 21, 28, 53; Baldwin, *Masters, Princes, and Merchants: The Social Views of Peter the Chanter and His Circle*, 2 vols. (Princeton, 1970), vol. I, 269; Noonan, *Usury*, 81–85.

[74] Eschmann, "Society," pt. I, 8–9.

[75] This is consistent with Aquinas' general conception of *communitas*. See Eschmann, "Society," pt. I, 28, 41.

[76] The history of the recognition of common valuation in exchange within legal thought is discussed in chapter 4.

[77] Langholm, "Scholastic Economics," 125; Ibanès, *Doctrine*, 41–42. For the reflection of this economic insight in scientific speculation, see chapter 7, 229–31.

[78] [1132a1–7]. Albert (Kübel edn.), 339b: "non respicitur ad conditionem personae, sed tantum ad quantitatem damni vel lucri, quod est in rebus ipsis, intenditur aequalitas rerum secundum quantitatem; sed hoc fit in commutationibus."

seller are to make absolutely no difference.[79] For Aquinas the idea of impersonality was no less important and was central to his definition of commutative justice.[80] The depersonalization of exchange went hand in hand with the acceptance of Aristotle's starkly mathematical conceptualization of the process.[81]

Aristotle had maintained that the varying need or demand for goods determined the price of goods in exchange. In keeping with his limited model of direct exchange between two producers, he personalized need as something that each exchanger brought to the exchange in various quantities. Scholastic thinkers emphasized even more than Aristotle that the need for any given object was determined by its utility – its usefulness to the exchanger. For example, where the Aristotelian text read "as many shoes must be exchanged for a house in proportion as the builder contributes more than the shoemaker,"[82] Albert commented, "as the builder exceeds the shoemaker in labor and expenses and in the utility of his work, so many more shoes are required to equal the house in value."[83] Although adding important detail, this comment still remained within the outlines of the Aristotelian text. But, as early as his first commentary, Albert began moving beyond Aristotle by expanding the concept of utility.[84] Where in Aristotle's scheme it is the usefulness to the individual exchanger that determines an artifact's value, in Albert's thought, it is the artifact's *common utility* that determines its value. Where Aristotle considered value in the context of individual exchange, Albert considered its formation in the context of a supra-personal marketplace.

Opus, the word Albert inherited to denote the common quality that was measured by money, became in his commentary at times *opus civilis*, or *opus in communitatis usum*.[85] Money moved from measuring personal use value to measuring the common quality that goods shared of being useful to the community.[86] The fullest statement of Albert's privileging of the community and the common in economic affairs resulted from a

[79] Albert (Borgnet edn.), 349b: "In talibus enim computationibus nullam differentiam facit personae vendentis vel ementis dignitas." See also Baldwin, *Just Price*, 62.
[80] Aquinas (1969), 284a: "in commutativa justitia attendatur aequale secundum arismeticam proportionem, manifestat per hoc, quod non consideratur ibi diversa proportio personarum."
[81] Lester Little (*Religious Poverty*, 33–34) stresses the Weberian formulation of the impersonality of money as an instrument, and the depersonalization of the exchange process that results from the introduction of money.
[82] Litzinger trans., vol. I, 423. Grosseteste, Arist. Lat. 26 (1973), 463: "Oportet igitur, quod edificator ad coriarium, tanta calceamenta ad domum vel cibum."
[83] Albert (Kübel edn.), 345a: "quod quantum aedificator excedit coriarium in expensis et labore et utilitate operis, tanta calceamenta aequantur domui."   [84] Ibanès, *Doctrine*, 38.
[85] Albert (Kübel edn.), 345a–b: "Et sic opus, quod omnia continet . . . est opus civilis."
[86] Albert (Kübel edn.), 344b: "conveniunt tamen in forma artis cuiusdam universalis quae disponit de omnibus politice, scilicet secundum quod serviunt communitati, et sic sunt appretiabilia."

linking of three separate arguments, all deriving from implications within the Aristotelian text. He interpreted Aristotle's statement that money was an artificial measure instituted by law to mean that money was specifically instituted by the community. He then linked the communal origin of money to the proper use of money as the common measure of all goods. Both common origin and common measure were then used to support his answer to the most difficult question: what, exactly, does money measure? He had already decided that money cannot measure things in their nature, only some accidental quality attached to all goods in exchange. Now he concluded that the quality money measured was not only common to all goods (*ad opus*) but was also defined commonly, i.e., in relation to the community (*ad operositatem communitatis*).[87]

Aristotle clearly understood that the demand for different goods and the usefulness of different products varied continually according to who needed what and when. He understood too that money as a continuous *medium* was the one instrument capable of measuring this ever-varying demand and use value. But medieval thinkers who took the idea of money as measure seriously began to question what, in reality, money price as measurement represented. They questioned as well the mechanism of price determination in the marketplace. Through their own experience they came to realize that prices did not vary as widely as the variation in personal need would indicate. The price of a product might change from day to day, but most often it remained constant to all buyers at any given time, regardless of their varying needs. Therefore, experience showed that Aristotle's equation of value to individual need did not work. Here the commentators did not hesitate to correct and add to the Aristotelian text. Their solution was that economic value, as measured by market price, is determined in relation to the community rather than to the individual.[88]

Although the commentators never actually replaced or explicitly criticized the Aristotelian model of direct exchange between two producers, in practice they continued to graft more complex, realistic, and accurate economic models onto the *figura* bequeathed to them. The measurement that price represented gained a certain "objectivity" in view of its being determined *secundum forum commune* rather than accord-

---

[87] Albert (Kübel edn.), 346a: "Et dicit, quod cum opera artium sint tam diversa, secundum veritatem suae speciei non possunt uno mensurari, sed secundum quod sunt *ad opus*, idest ad operositatem communitatis, sufficienter mensurantur per unum, et unum hoc est mensurans per suppositionem, idest per institutionem a civilitate, quae instituit numismata."

[88] For the Roman legal tradition that encouraged this observation, see especially *Digest*, 9.2.33; 35.2.63; Baldwin, *Just Price*, 20–29; Barry Gordon, *Economic Analysis Before Adam Smith: Hesiod to Lessius* (New York, 1975), 128–32. This question is considered in detail in chapter 4.

ing to an individual subject or situation.[89] Market price came to be seen as an aggregate product – the concrete (if ever-changing) numerical representation of a complex, supra-personal system. As the geometric *figura* of exchange was de-subjectified, it came to represent the marketplace as a kind of mechanism of equalization, in which the cross-conjunction of common estimation and common need "automatically" determined market prices.

A PROBLEM INTRODUCED BY THE EARLY COMMENTARIES

As scrupulous as Aquinas was in his treatment of the Aristotelian text, and as much as his commentary furthered its understanding, he was also unwittingly responsible for a critical misunderstanding on the part of certain of his readers. The misunderstanding was introduced during Aquinas' discussion of Aristotle's second form of particular justice, "corrective" or "rectificatory" justice (*iustitia directiva*). We have seen that Albert often referred to this form as *iustitia directiva in commutationibus et contractibus* because Aristotle applied it to the judge's *post facto* restoration of equality to inequitable economic transactions. Both Albert and Aquinas recognized that economic exchange took place between unequal exchangers, goods, and services. The only way for equality to be established between unequals was through the geometric process of *contrapassum*. But in the eyes of the judge called on to determine the justice of the result, all parties are equal. Since the participants are equal before the law, the judge reestablishes equality between them through an arithmetical rather than a geometrical process. He bisects the line of *lucrum/damnum* through addition and subtraction.

At times, Aquinas refers to the judge's reestablishment of justice as *iustitia directiva in commutationibus*. But here he unknowingly opened the door to a serious misinterpretation of Aristotle when he at other times referred to this form simply as *iustitia commutativa*, which can be taken to mean "commercial justice" or "justice in economic exchange."[90] Cer-

---

[89] Albertus Magnus, *Libri quarti sententiarum Petri Lombardi* (cited as IV Sent.), A. Borgnet (ed.), vol. XXIX of *Opera omnia* (Paris, 1894), d. XVI, art. 46, 638a: "Justum autem pretium est, quod secundum aestimationem fori illius temporis potest valere res vendita." For an insightful discussion of the terms "objective" and "subjective" as they apply to components of economic value, and particularly the transformation of the "subjective" element of human need into an "objective" element by the introduction of the concept of *common* need, see Baldwin, *Just Price*, 75. See also Noonan, *Usury*, 87.

[90] Aquinas begins the chapter on this second form of justice, (1969), 283a: "Postquam Philosophus ostendit quomodo accipiatur medium in justitia distributiva, hic ostendit quomodo accipiatur medium in justitia commutativa." For an analysis of Aquinas' creation of the term *iustitia commutativa*, see Gauthier and Jolif, *L'Ethique*, vol. I, 370–71. There is no doubt Aquinas understood that the arithmetical proportion governing *iustitia commutativa* did not cover the

tain future readers of Aquinas' commentary, fastening on his phrase *iustitia commutativa* and reading that this form was governed by arithmetical proportion, came to the erroneous conclusion that exchange transactions themselves were governed by arithmetical proportion.[91]

No misconception could be more damaging to Aristotle's economic argument or to an accurate perception of market exchange. It worked against the dynamic and relational model so central to *Ethics* v. It negated the self-regulating capacity of the marketplace to find its own values according to common need and common estimation (*secundum aestimationem fori*).[92] It did so because *iustitia directiva* (now often simply renamed *iustitia commutativa*) assumed the intercession of a judge consciously creating order by making decisions on the proper values of things. The mistaken conception that *iustitia directiva* (or the renamed *iustitia commutativa*) governed the exchange process worked into older, idealized conceptions of an economy subjectively ordered and subjectively controlled according to just values and ideal determinations. Again it must be stressed that Aquinas himself did not make this mistake: he merely introduced it by labeling *iustitia directiva* at times *iustitia commutativa*. His influence was so great, however, that from the time of his commentary this became the most common way to refer to this second particular form of justice.[93] The widespread currency of this phrase, and of the error attached to it, indicates that Aristotle's self-ordering geometry of exchange was not only foreign to certain thinkers, it was threatening to them – in the same way as the growing power of the marketplace to determine values was threatening.

We can marvel at the insight and understanding Albert and Aquinas brought to their commentaries, but we should not take either their economic thinking or their attitude toward the Aristotelian text as purely representative of their time. There were many who would not have been able to do justice to the *Ethics*, indeed, who would not have known what to make of it. The system it introduced was not only novel, it also brought a number of elements into the understanding of economic exchange that many perceived to be threatening to the social and religious order – perhaps with good reason. The intellectual model derived from the *Ethics* established the marketplace as a world governed by its

[91] mathematical form of the economic transactions themselves, but rather the *post facto* correction of those transactions by a third party or judge. Aquinas (1969), 290b: "in commutativa justitia lex attenditur solum ad differentiam nocumenti." See Baldwin, *Just Price*, 63.
[91] See chapter 4 for the conflict between geometric and arithmetic models of equalization in economic exchange.   [92] Albert, *IV Sent.*, d. XVI, 638a.
[93] An indication of the success of this phrase can be found in the Borgnet edition of Albert's *Ethics* commentary (348b), where the chapter covering *iustitia directiva* is introduced by the editors as "De justitia commutativa," even though Albert himself never used this term.

own laws and mathematics, forcefully creating its own order and equality, functioning independently of subjective or normative judgment, yet located at the connecting center of the *civitas*. The intellectual attempt to integrate this model of economic order with traditional models of natural order had profound repercussions within scholastic thought in the century following these first *Ethics* commentaries.

Chapter 4

MODELS OF ECONOMIC EQUALITY AND EQUALIZATION IN THE THIRTEENTH CENTURY

As important as Aristotle's economic ideas were to the formulations of Albert, Aquinas, and succeeding masters at Paris and Oxford, they constituted only a part of the textual inheritance on economic questions. In Aquinas' most thorough discussion of usury and just price in the *Summa theologiae*, Aristotelian insights from the *Ethics* and the *Politics* shared place with citations from the Bible, patristic authorities, post-patristic theological writings, Roman authors, and, importantly, Roman law and canon law. In varying degrees scholastic perceptions of money and market exchange were influenced by all these textual traditions.

In addition, scholastic economic thought was strongly influenced by its Christian setting – what its continuators took as their responsibility for the care and protection of souls. Economic determinations were tied not only to legal and ethical considerations but to the individual and his salvation.[1] The primary question scholastic thinkers asked concerning economic activity was not "how does it work?," but "what is permitted and what is not? what is sinful and what is not?"[2] Economic positions were often justified on moral grounds and framed in terms of the moral duty to protect the weak and to enforce economic justice.

Religious, legal, and ethical principles defining economic liceity were in turn tied to conceptions of the "natural order" of things and to Nature itself. Scholastic authors consistently defined usury not only as unethical but as "unnatural."[3] The *De peccato usure* of Remigius of Florence, written at the height of Florentine commercial expansion in the early

---

[1] Gabriel Le Bras, "Usure," in *Dictionnaire de théologie catholique*, vol. xv (Paris, 1950), cols. 2316–90, 2349: "La condamnation de l'usure s'appuie sur l'autorité sacré, la morale naturelle, l'ordre social, les droits positifs." See also Noonan, *Usury*, 2–3; Langholm, *Usury*, 13; Langholm, *Economics in the Schools*, 24.   [2] Le Bras, "Usure," col. 2351.

[3] Langholm (*Economics in the Schools*, 70) credits the theologian William of Auxerre (fl. 1230) with the grounding of usury theory in natural law. See Noonan (*Usury*, 47) for a discussion of St. Bonaventure's representative characterization of usury as "perversion of order." Aquinas' similar stance is discussed on 86–87.

fourteenth century, provides a clear example of the elaborate arguments used to demonstrate that usury and usurers violated not only the civil order but the very structure of God's universe.[4] In making this claim, scholastic thinkers could draw on Aristotle's definition of usury in the *Politics* as *maxime praeter naturam* in addition to a more popular literature of sermons and exempla in which usury was presented as a sin of cosmic proportions against the natural order.[5]

The habit of defining economic order in terms of natural order encouraged the linking of economic thought and experience to speculation about the structure of the natural world. This linkage was considerably strengthened by the fact that both scholastic economic thought and scholastic natural philosophy were grounded in conceptions of equality and equilibrium. Shared conceptions of equality, so rich in mathematical possibilities, thus provide a powerful lens through which to investigate the interface between economic and scientific thought. The evolving definition of equality (*aequitas, equalitas, aequatio, adaequatio*) within scholastic discussions of usury and just price over the course of the thirteenth century is the subject of this chapter.

While the ideal of equality as the proper end of all exchange remained constant over this period, the idea of what actually constituted equality changed as did the models used to describe the process of equalization. The direction of change was from a static, arithmetical model based on knowable values and knowable points of equality to a geometric model of equalization based on approximate values, proportional requital, and variable line-ranges of adequation; from an equality rooted in individual judgment to an equality viewed as the product of a supra-personal system. In chapters 6 and 7 I discuss how the emerging geometric model of equality in exchange influenced the formation of proto-scientific innovations within fourteenth-century mathematics and natural philosophy.

EQUALITY IN USURY THEORY

The simplest definition of a usurious loan is one in which the lender requires the borrower to repay more than the sum lent. Before the time of Charlemagne, Church law had forbidden only the clergy to take

---

[4] O. Capitani, "Il 'De peccato usure' di Remigio de' Girolami," *Studi medievali*, ser. 3, 6, Fasc. II (1965), 537–662, esp. 568, 638. A sample of Remigio's reasoning: usury is *contra regulam naturae* and, in particular, against the nature of the Heavens, because, while stars look small but are in reality large, the usurer (because of his wealth and his friendship with kings and high prelates) looks large but is in reality very small (*immo nichil*).

[5] For Aristotle's position, see *Politics* [1258b1–b8] and 86 in this chapter. For the treatment of usury as a sin against nature in sermons and exempla, see Jacques Le Goff, *Your Money or Your Life: Economy and Religion in the Middle Ages* (New York, 1990), esp. 49–54.

usury. In the ninth century the prohibition was extended to the general population. In Charlemagne's capitularies usury was defined simply as "where more is asked than is given."[6] The requirement for a perfect arithmetical equality in the loan contract was clear and without exception. Equity and justice demanded that one who lent five pounds could receive only five in return.[7] Such an arithmetical equality, so satisfying in its simplicity and balance, remained an ideal to which later writers on usury often returned.

With the great commercial expansion of the eleventh and twelfth centuries, questions concerning definitions of usury were raised with increasing frequency. For the first time in a collection of canon law, Gratian allowed the question its own heading in the *Decretum*.[8] St. Augustine, the first authority cited, defined the usurer simply as one who in a loan contract (*mutuum*) expected to receive back more than the amount lent. With this definition usury was extended to all manner of goods: wheat, wine, and oil in addition to money.[9] The second authority Gratian cited was St. Jerome, who used the word *superabundantia* to express the idea of an excess beyond the arithmetically equal in the loan contract.[10] All such *superabundantia* in temporalities were condemned as usurious.[11] Over the succeeding centuries this word appeared continually in discussions of usury to express the *unnatural* production of an excess where none was warranted, thus violating the requirement of natural equality in exchange.

In the third opinion cited, St. Ambrose supplied the simple rule: "usury is whatever exceeds the sum lent (or the principal) [*quodcumque sorti accidit*]."[12] Henceforth, this rule also had great currency. There was no discordance among the patristic canons cited by Gratian. the equality required in the loan contract was simple and *arithmetical*. There was no

---

[6] Noonan, *Usury*, 15.

[7] *Ibid.*; Le Bras, "Usure," col. 2333; Benjamin Nelson, *The Idea of Usury: From Tribal Brotherhood to Universal Otherhood*, 2nd edn. (Chicago, 1969), 4–5.

[8] The question had been considered by canonists before Gratian, particularly Ivo of Chartres. See Noonan, *Usury*, 17–18; Le Bras, "Usure," cols. 2334–37.

[9] St. Augustine cited in Gratian, *Decretum*, II.14.3.1, A. Friedberg (ed.), vol. I in *Corpus iuris canonici*, 2 vols. (Graz, 1959), col. 735: "id est si tu mutuum dederis pecuniam tuam, a quo plus quam dedisti expectes, non pecuniam solam, sed aliquid plus quam dedisti, sive illud triticum sit, sive vinum, sive oleum, sive quodlibet aliud, si plus quam dedisti expectes accipere, fenerator es."

[10] St. Jerome cited in *Decretum* II.14.3.2. See T. P. McLaughlin, "The Teachings of the Canonists on Usury," *Mediaeval Studies* I (1939), 81–147, esp. 96; Le Bras, "Usure," cols. 2337–38.

[11] The distinction continued to be made between multiplication and the production of superabundance in spiritual goods, which was promised by Christ, and the illicit creation of a superabundance in the loan contract. The role of this distinction in the dynamic of usury prohibition merits further investigation.

[12] St. Ambrose cited in *Decretum* II.14.3.3: "et quodcumque sorti accidit usura est; et quodcumque velis ei nomen imponas, usura est."

consideration of the circumstance or status of the participants which might justify the lender demanding one penny more in return than the sum lent. To put this Christian discussion in Aristotelian terms (as it was in the thirteenth century), the *mutuum* was governed by corrective or commutative justice (*iustitia directiva*), which required the restoration of a perfect numerical equality determined by the arithmetical process of addition or subtraction around a knowable point.[13] The concept of distributive justice with its more complex recognition of pre-existent inequalities, and the consequent necessity of geometrical or proportional equalization, was absent both in these patristic statements on usury and in the whole of Gratian's *Questio*.

Important additions to the definition of usury were made in the late twelfth century by the legally trained Popes Alexander III (1159–81) and Urban III (1185–87), after a century of rapid increase in the frequency and complexity of credit transactions. The very breadth of St. Ambrose's definition (*quodcumque sorti accidit*) invited the extension of usury theory to other commercial contracts involving credit. Urban III's judgment, *Consuluit*, exemplifies this extension and is noteworthy for a second reason.[14] Previously the most important scriptural texts used in condemnation of usury had come from the Old Testament.[15] Urban added Christ's command from Luke 6:35, "lend freely, hoping for nothing therefrom."[16] The use of this text, which soon became the most frequently cited anti-usury injunction, introduced the critical test of *intention* into the definition of usurious contracts.[17] By the thirteenth century, intention had become the purest test of usury; the question to be answered *in foro interno*, "did I intend inequality in the transaction?," a question which as conceived could be answered with a simple yes or no. In the century before the introduction of the test of intention, merchants had developed a variety of accounting techniques (using, for example, variable exchange rates for different coins in different markets at different times) to hide usury – a fact well known to the legists of the

---

[13] See chapters 2 and 3 for the mathematics of Aristotle's forms of justice.

[14] *Consuluit*, *Decretales Gregorii IX*, v.19.10, Friedberg (ed.), vol. II in *Corpus iuris canonici*, col. 814.

[15] The most important Old Testament texts are: Deut. 15:7–10; Exod. 22:25; Ezek. 18:5–9; Pss. 5:5, 14:5. See Le Bras, "Usure," col. 2337. For the difficulties Christian thinkers had in adopting the position found in Deut. 23:19–20, that it was permissible to exact usury from a stranger but not from a "brother," see Nelson, *Idea*, esp. 3–18.

[16] From *Consuluit*, in *Decretales Gregorii IX*, col. 814: "Verum quia, quid in his casibus tenendum sit, ex evangelio Lucae manifeste cognoscitur, in quo dicitur: 'Date mutuum, nihil inde sperantes': huiusmodi homines pro intentione lucri, quam habent, quum omnis usura et superabundantia prohibeatur in lege, iudicandi sunt male agere, et ad ea, quae taliter sunt accepta, restituenda, in animarum iudicio efficaciter inducendi."

[17] On the critical importance of intention in medieval usury theory from this time forward, see Jacques Le Goff, "The Usurer and Purgatory," in *The Dawn of Modern Banking* (New Haven, 1979), 25–52, esp. 44–51; Noonan, *Usury*, 20, 91–95.

time.[18] The new emphasis on *intentio*, where guilt was assessed in the internal forum of conscience, allowed clerical theorists to continue to assert an absolute ideal of arithmetic equality, even as the complexity of commercial transactions made such an equality increasingly difficult either to define or to enforce.[19]

While the theological position remained that a loan should be made out of charity and without hope of profit, realization grew within the more practical discipline of canon law that if the lender was to be denied a guaranteed reward he must also be protected from damages associated with the making of a loan. To ensure this protection, canon lawyers began to apply the Roman law concept of *interesse* to the loan contract.[20] In Roman law, *interesse* was a term recognizing damages occasioned by the breaking of a contract. Canon lawyers extended it to the loan contract to indemnify the lender for the failure of the borrower to meet his contractual obligations.[21] In the most commonly cited case, the lender was permitted to charge a penalty if the borrower failed to repay his loan within the agreed time.[22] Since the justification of *interesse* (or excess beyond the sum lent) from the beginning of a loan would mean the virtual dismantling of usury theory, medieval legists who acknowledged legitimate interest on a loan did so not as a profit but as a payment for damages (*non lucrum sed vitatio damni*), and not as a part of the loan contract itself but as an external factor, resulting (in the earliest cases) from a contractual failure on the part of the borrower.[23]

Whether intrinsic or extrinsic, the recognition of *interesse* meant that the violation of a strict arithmetical equality between sum lent and sum returned no longer infallibly constituted usury. In the Roman law of Justinian's time, where moderate usury was accepted and not condemned, the difference in the position of lender and borrower was allowed to express itself as a proportional difference between the amount lent and the amount to be received in return. Medieval theorists could never accept this kind of proportionality. There was more at stake here than the overturning of a purely economic position. Overturning usury theory

---

[18] See for example the complex discussion of a range of such contracts *in fraudem usurarum* by the near contemporary of Urban III, Bernard of Pavia, discussed in McLaughlin, "Canonists," 97, 112–25.

[19] For the relationship of the new emphasis on *intentio* to the actual complexity of loan contracts, see McLaughlin, "Canonists," 106–07. In the opinion of Le Goff ("Purgatory," 50–51), the contemporaneous development of the idea of Purgatory performed the similar function of permitting the maintenance of an idealized ethical economic stance in an increasingly complex economic world.

[20] McLaughlin, "Canonists," 140–41; Langholm, *Usury*, 50.

[21] Baldwin, *Masters*, vol. I, 282; McLaughlin, "Canonists," 140.

[22] McLaughlin, "Canonists," 140–43; Noonan, *Usury*, 109–12.

[23] Noonan, *Usury*, 106: "Interest is never thought of as payment on a loan; it is the 'difference' to be made up to a party injured by the failure of another to execute his obligations." For the development of this theory in canon law, see McLaughlin, "Canonists," 125–47, esp. 141–42.

would entail overthrowing the theological edifice of sin, restitution, and penance that had been built upon it, as well as the ideal of "natural" equality and justice at its core. However, as the reality of the monetized economy became more insistent and pervasive, usury theorists came to recognize that lenders often suffered economic damages that, in justice, required compensation. Canon lawyers gradually devised a solution to this problem: they continued to insist that a loan should be made without the hope of receiving back more than the sum lent, while recognizing specific cases in which the lender could licitly demand indemnification for loss.

One of the most liberal interpretations of this position was held by the influential canon lawyer Hostiensis (d. 1270).[24] At the beginning of the question on usury in his *Summa aurea*, he listed (in verse form!) twelve legitimate exceptions to the rule "nothing may be received beyond the sum lent," all the while insisting that the rule be strictly maintained in principle.[25] Hostiensis was on well-trod legal ground in accepting an increase beyond the arithmetically equal in compensation for actual damages suffered by the lender. Such legitimate compensation was known in canon law as *damnum emergens*.[26] He was in rarer company in his acceptance of an increase *ultra sortem* to indemnify a lender for the gain he *might* have made with his money had he not lent it. He wrote:

if some merchant, who is accustomed to pursue trade and the commerce of the fairs and there profit much, has, out of charity to me, who needs it badly, lent money with which he would have done business, I remain obliged from this to his *interesse*, provided that nothing is done in fraud of usury . . . and provided that said merchant will not have been accustomed to give his money in such a way to usury.[27]

Such indemnification for lost profit was known under the title *lucrum cessans*.[28] Hostiensis insisted on treating *lucrum cessans* as an external title and not as a case of legitimate usury, even though in such cases *interesse*

[24] Henry of Segusia or Hostiensis was a doctor of Roman law at Bologna before studying canon law at the University of Paris. The influence of Roman law is evident in his eventual position on legitimate titles to *interesse* in the loan contract. On Hostiensis' legal career, see Charles Lefebvre, "Hostiensis," in *Dictionnaire de droit canonique*, 7 vols. (1935–65), vol. v (1953), cols. 1211–27.

[25] The verse form had practical applications as a mnemonic device, indicating the closeness of Hostiensis' legal theory to current practice. Hostiensis, *Summa aurea* (Venice, 1574; reprint, Turin, 1963), Book v, "De usuris," col. 1623. For a discussion of Hostiensis' twelve categories, see McLaughlin, "Canonists," 125–44. For the general relation between canon law and civil law on the question of usury, see McLaughlin, "Canonists," 84–87.

[26] McLaughlin, "Canonists," 145.

[27] Hostiensis, *Commentaria*, v.16, cited and translated by Noonan, *Usury*, 118.

[28] McLaughlin, "Canonists," 145. Noonan (*Usury*, 118) claims that Hostiensis was "the first author to give unmistakable and full approval to a case of *lucrum cessans*."

could conceivably be calculated from the beginning of the loan. And despite his acceptance of *lucrum cessans* in principle, he continued to maintain that, since usury was in itself evil and unnatural, the idea of legitimate or moderate usury was a contradiction in terms.[29]

In Hostiensis' thought we see the tension that arose from the desire (or pressure) to maintain old ideals in a new and charged economic environment. This tension was reflected in conflicting models of equalization. The ancient ideal of a simple arithmetic equality remained in force, even while severely pressured by the logic of Roman law *interesse*, by the proportional implications of *lucrum cessans*, and, as Hostiensis clearly showed, by a new understanding of the productive combination of money and business acumen in the commerce and industry of the thirteenth century. The shift from an external to an internal test of usury through the focus on *intentio* allowed medieval lawyers and moralists to excoriate usury, and to insist on older rigorist ideals of strict arithmetical equality (*quodcumque sorti accidit*), while at the same time admitting "external titles" to increase (*damnum emergens, lucrum cessans*) and thus moving in practice toward a modified conception of proportional equalization.[30]

In a number of respects, the positions held by medieval theologians on usury differed from those of the lawyers. Although often informed and influenced by canon law on the subject, theologians were generally more conservative in respect to the justification of *interesse*.[31] The opinions of Thomas Aquinas, Hostiensis' contemporary, illustrate this distinction. Aquinas accepted the principle of *damnum emergens*, the lender's right to compensation for actual damages caused by the contractual failure of the borrower.[32] But he was considerably more sensitive than was Hostiensis to the philosophical implications of *lucrum cessans*.[33] He denied the lender's right to demand compensation for *possible* lost profit, since doing so involved selling what had only *probable*, rather than real existence. He wrote: "one should not sell something which one has not yet got and

---

[29] Hostiensis, *Summa aurea* v, col. 1630: "Quia omnis usura immoderata est, et ideo hodie omnis usurarius infamis est." Noonan (*Usury*, 50–51) speculates on the possibility that Hostiensis considered modifying the blanket condemnation of usury in his final writings on the subject.

[30] McLaughlin, "Canonists," 125 n. 357. For a discussion of the rich practical developments in usury theory taking place beneath the continued restatement of traditional and often abstract ideals, see Ovidio Capitani, "Sulla questione dell'usura nel medio evo," in Capitani (ed.), *L'etica economica medievale: testi à cura di Ovidio Capitani* (Bologna, 1974), 23–46.   [31] Ibanès, *Doctrine*, 41–42.

[32] Aquinas, *Summa theologiae* (*ST*), 61 vols. (New York, 1964–81), 2, 2, 78, 2, ad 1: "ille qui mutuum dat potest absque peccato in pactum deducere cum eo qui mutuum accipit recompensationem damni, per quod subtrahitur sibi aliquid quod debet habere; hoc enim non est vendere usum pecuniae, sed damnum vitare."   [33] Ibanès, *Doctrine*, 26–27; Noonan, *Usury*, 117.

which one may be prevented in many ways from getting."[34] Aquinas' distinction between what is "merely" probable and what is real was crucial to his position on equality in the loan contract. The denial of this distinction, the allowance of some degree of reality to the probable in economic life (and, later, to the probable in natural processes), would, in the generation after Aquinas, constitute one of the great changes in the scholastic conception of equality.

Aquinas' position on usury derived partly from Aristotle's strict definition of money as a measuring *medium*.[35] As a *medium*, money can serve only as a mid-term in exchange and never as an end-term, rented or sold in itself.[36] For Aristotle and for Aquinas, money has no natural value apart from its use to facilitate exchange.[37] The usurer who charges extra for the *use* of money is thus selling something that has no real separable existence, and, moreover, something that is not his to sell. In doing so he violates not only scriptural and legal injunction, but the equilibrium at the heart of nature's order.[38] Aquinas reveals the central place equilibrium holds in his philosophical position on usury in the opening words on this question in the *Summa theologiae*: "To accept usury from money lent is unjust in itself, for one party sells the other what does not exist, and this manifestly constitutes inequality which is contrary to justice."[39] At the center of the scholastic philosopher's opposition to usury was the belief that, in the words of Gabriel Le Bras, "En toute usure, il y a *inadaequatio*."[40]

From the middle of the thirteenth century, Aristotelian positions found an important place in theological arguments against usury.[41] The discussion of money in the *Politics* reenforced the traditional theological distrust and condemnation of *superabundantia*. Using the myth of Midas, Aristotle insisted that money in itself satisfied no natural human need.

---

[34] Aquinas, *ST*, 2, 2, 78, 2, ad 1: "Recompensationem vero damni quod consideratur in hoc quod de pecunia non lucratur non potest in pactum deducere, quia non debet vendere id quod nondum habet, et potest impediri multipliciter ab habendo." Note the distinction Aquinas makes between the probable and the real. The relationship between changing definitions of probability in economic thought and their reflection in medieval logic and science is a subject I return to in chapters 5 and 7.   [35] See chapter 2, 46–47.

[36] See Aquinas, *ST*, 2, 2, 78, 2, ad 5: "ille qui mutuat pecuniam, transfert dominium pecuniae in eum cui mutuat; unde ille cui pecunia mutuatur sub suo periculo tenet eam et tenetur integre restituere; unde non debet amplius exigere ille qui mutuavit."

[37] See Aquinas, *ST*, 2, 2, 78, 1, for Aquinas' extremely influential distinction between money and all fungible goods in this regard. See also Ibanès, *Doctrine*, 19–20; Langholm, *Economics in the Schools*, 241–44.

[38] This is St. Bonaventure's opinion as well and becomes characteristic of the theological tradition in usury theory from the thirteenth century. See Noonan, *Usury*, 47–48.

[39] Aquinas, *ST*, 2, 2, 78, responsio: "Dicendum quod accipere usuram pro pecunia mutuata est secundum se injustum, quia venditur id quod non est; per quod manifeste inaequalitas constituitur, quae justitiae contrariatur."   [40] Le Bras, "Usure," col. 2349.

[41] For a discussion of the place of Aristotelian insights in medieval usury theory, see Langholm, *Usury*, 54–69.

Since money satisfies no natural need, the desire for it is infinite, having no natural balance or bound.[42] Of all activity directed primarily to the acquisition of money, Aristotle believed usury was the most despicable and unnatural (*maxime praeter naturam*), because in the usurious loan, money, which was invented solely as an instrument of exchange, is made to generate itself, to give unnatural birth to itself.[43] Aristotle's distinction between the natural order and the acquisitive economic order fit well with the Christian condemnation of usury as a mortal sin, based on the theological definition of sin as perversion of order. Scholastics discomforted by the destabilizing dynamic of the economic order found they could use Aristotelian authority to deny, by definition, a multiplying power to money.

If, however, Aristotle's conservative position on money in the *Politics* served usury theorists well, his advanced discussion of money as an instrument of geometric proportionality in the *Ethics* created problems for them. Beginning in the twelfth century, the introduction of doubt, probability, and variable benefit into discussions of the loan contract threatened to bring the loan into the orbit of generalized economic exchange where value was determined relative to time, place, and person. Once in that orbit, of course, according to Aristotle's analysis in the *Ethics*, the geometric form of proportional requital would hold, rendering the doctrinal ideal of arithmetical equality not only untenable but unjust. Despite growing practical and intellectual pressures, neither canon lawyers nor theologians relinquished the traditional test of *quodcumque sorti accidit* in theory, even as in practice they came to recognize specific cases and "external titles" in which the traditional ideal of arithmetical requital could not be maintained.

EQUALITY IN JUST PRICE THEORY AND MARKET EXCHANGE

While there are similarities in the histories of scholastic just price and usury theory, particularly their grounding in ideas of equivalence, there are also striking differences. Over the course of two centuries the bounds

---

[42] *Politics* [1257b24–30]. Arist. Lat. 29, 1 (1961), 17: "Et infinite utique iste divitie que ab hac crimatistica . . . sic et huius crimatistice non est finis terminus, finis autem tales divitie et pecuniarum possessio."

[43] *Politics* [1258b1–b8]. Arist. Lat. 29, 1 (1961), 18: "autem vituperata iuste (non enim secundum naturam sed ab invicem est), rationabilissime odio habetur obolostatica, eo quod ab ipsa numismatis sit possessio, et non ad quod quidem acquisitum est. Translationis enim gratia factum est, tokos autem seipsum facit amplius, unde et nomen istud accepit. Similia enim parta generantibus ipsa sunt, tokos autem fit numisma numismatis; itaque et maxime preter naturam ista pecuniarum acquisitio est." For Aquinas' recognition of Aristotle's position, see *ST*, 2, 2, 78, 1, ad 3: "Et Philosophus naturali ratione ductus dicit in 1 *Polit.*, quod usuraria acquisitio pecuniarum est maxime praeter naturam."

of liceity in usury theory were expanded by the pressure of economic practice and by ever more acute observations of economic life. Despite the expansion in particular cases, however, in principle both lawyers and theologians maintained the traditional requirement for arithmetical equality in the loan contract. In regard to usury theory, the Church continued to impose the "what should be" of law, authority, and conscience on the "what is" of economic activity.

Until the middle of this century it was generally assumed that all aspects of medieval economic thought were similarly characterized by the determination to maintain ideal definitions, whether in opposition to, or in ignorance of, actual economic practice. Mixing preconceived ideas of pre-capitalist economic order with an exaggerated sense of the static and abstract nature of usury theory, historians imagined a medieval world innocent of economic calculation, where the actual workings of the market were only dimly perceived and where the rough give-and-take of exchange was denied validity in the name of corporative and religious ideals.[44] Working from such a picture, historians imagined that the just price (*iustum pretium*), so often mentioned by economic writers, was conceived of as an ideal price – a more or less objective measure of true value – determined primarily by the labor and expenses involved in a good's production and by its usefulness to the community. Such a price would allow laborers a modest surplus befitting their social position and permitting them to maintain that position. The just price thus conceived would necessarily be more or less fixed, independent of the fluctuations and vagaries of the marketplace. Seen in this way, just price theory was a sister to usury theory, another manifestation of the medieval desire to control economic activity in the name of religious ideals and social equilibrium.[45]

The problem with this neat vision of the just price is that there is ample textual support for a formulation of price and just price that opposes this tidy picture in almost every particular.[46] Raymond De Roover was one of the first (1958) to directly challenge the old view. He concluded that, in the majority of texts on this question from the thirteenth and fourteenth centuries, the just price "was simply the current market price."[47] In the decades since, historians have continued to question whether the just price ever existed, in fact or in theory, as a

---

[44] Raymond De Roover, "The Concept of the Just Price: Theory and Practice," *Journal of Economic History* 18 (1958), 418–34, esp. 419, cites numerous examples of this older interpretation and strongly criticizes it.

[45] See De Roover, "Concept," for a fuller characterization and critique of this school of thought.

[46] De Roover, "Concept," 421–26.

[47] De Roover, "Concept," 420. The positions of a number of other historians in substantial agreement with De Roover are provided below.

normative determination detached from market pressures. There remain, however, a number who question whether the "simple" identification of just price and market price fully conveys the meaning attached to this concept in medieval thought.[48] The debate over just price hinges on larger questions. To what degree were those who used the term "just price" conscious of the marketplace as a supra-personal system within which prices were determined? How well did they understand the workings of this marketplace and the impersonal factors determining prices? How could they attach the exalted concept of justice to the product of an economic process that functioned in some degree independently of personal will and judgment?

The medieval market never attained the heights of reification and idealization that it has in classical economic thought. Minute regulations regarding the size, weight, and quality of goods for sale, particularly manufactured goods, were issued for every market. Through the fourteenth century and beyond, princes and city councils maintained the right to regulate market prices in the name of the common good.[49] Maximum price edicts remained a traditional governmental response to economic crises.[50] As a rule, however, such interference was never more than temporary.[51] In the all-important area of grain, wine, and other agricultural products, prices were expected to fluctuate and were allowed to do so. Only when authorities recognized that the normal factors influencing market price were being unfairly manipulated by hoarders or profiteering resellers did they intervene.[52] The authorities were not alone in recognizing how markets worked in regulating prices. In every market the prices of agricultural products fluctuated seasonally in relation to their

---

[48] For a moderately stated position, see Samuel Hollander, "On the Interpretation of the Just Price," *Kyklos* 18 (1965), 615–34. Hollander objects to the idea that Aquinas held one consistent position on the determinants of just price. Ibanès (*Doctrine*, 36) holds that the Romanists supported a simple market price while canonists and theologians continued to deny the idea of a complete correspondence between price and value. De Roover rejected this distinction and failed to see any evidence of the existence of an "objective" just price reflecting medieval distrust of truly competitive prices. See Raymond De Roover, *La pensée économique des scolastiques: doctrines et méthodes* (Montreal, 1971), 24. Both Ibanès and Hollander cite Aquinas' statement in the *ST* (2, 2, 77, 1) – "si vel pretium excedat quantitatem valoris rei, vel e converso res excedat pretium, tolletur justitiae aequalitas" – to show that he conceived of situations where market price did not reflect just value. For a further discussion of this particular passage, see 96–99. See chapters 2 and 3 for more on the controversy concerning the place of *labor et expensae* in the economic thought of Thomas Aquinas and Albertus Magnus.

[49] See chapter 1 for the application of this right in the fourteenth century. See also *Ordonnances*, vol. I, 424–26 (1304–05); Miskimin, *Money and Prices*, 22; Britnell, *Commercialisation*, 2, 94.

[50] Attempts to control market prices in goods and wages came into play with renewed force after the market dislocations resulting from the Black Death. For France, see, for example, *Ordonnances*, vol. II, 366–78 (August, February, 1350); for England, see the series of *Statutes of Labourers* beginning in 1349. Royal attempts to control the market are discussed in chapter 1.

[51] Britnell, *Commercialisation*, 94. [52] Britnell, *Commercialisation*, 93–94.

bounty or scarcity, for reasons that were obvious to all.[53] The fact that market prices were established by market conditions in some measure independent of the needs of individual buyers and sellers was clear to everyone connected to the monetized marketplace, whether as producer or consumer.[54] From the twelfth century, medieval jurists consistently ruled that the market price of a good could be taken as its "just" price. They did so, even though they recognized that market prices were subject to wide variations over time and place and were determined in some degree independently of conscious human direction.

The roots of medieval legal thinking on market price and just price are found in Roman law.[55] In the *Digest* and *Codex* of Justinian's *Corpus iuris civilis*, the establishment of an equality between buyer and seller was recognized, without exception, as a dynamic process — a product of the conflicting desires of the buyer to pay as little as possible and the seller to charge as much as possible.[56] Not only was this conflict permitted by law; it was defined within the law as the *natural* state of affairs.[57] With the exception of outright fraud (e.g., selling at a greatly elevated price to someone incapable of good judgment), buyer and seller were permitted to outwit each other in order to obtain a price advantage.[58] As the *Digest* has it: "In sales and purchases it is naturally allowed [*naturaliter concessum est*] to buy a thing of greater value for a smaller price; and to sell a thing of lesser value for a greater price."[59] Despite the deception permitted in Roman law, there is no discussion there of a distinction between selling price and just value. A legally sufficient equality between cost and value was assumed to exist in all sale agreements entered into freely. This assumption was expressed in the often-repeated principle "a thing is worth what it can be sold for."[60]

Roman law contained one small exception to this rule. It appeared in

---

[53] De Roover, "Concept," 421; Kenneth S. Cahn, "The Roman and Frankish Roots of the Just Price of Medieval Canon Law," *Studies in Medieval and Renaissance History* 6 (1969), 3–52, 36–37; Baldwin, *Just Price*, 32–34; Postan, *Medieval Economy*, 21; Miskimin, *Money and Prices*, 21.

[54] Evidence for this statement is provided in chapter 1.

[55] See Cahn, "Roots," 43–51, for a discussion of Roman law influences on medieval just price theory.

[56] Baldwin, *Just Price*, 17: "The fundamental Roman law principle of sale and price was that of freedom of bargaining."

[57] *Digest*, 4.4.16.4: "in pretio emptionis et venditionis naturaliter licere contrahentibus se circumvenire." See also *Codex*, 4.44.10; 4.44.5; Cahn, "Roots," 12.   [58] Baldwin, *Just Price*, 17.

[59] *Digest*, 19.2.22.3: "Quemadmodum in emendo et vendendo naturaliter concessum est quod pluris sit minoris emere, quod minoris sit pluris vendere et ita invicem se circumscribere, ita in locationibus quoque et conductionibus iuris est." See Cahn, "Roots," 12.

[60] *Digest*, 36.1.16; 35.2.63; 9.2.33. In medieval jurisprudence, this principle appears repeatedly in glosses on the *Digest* and *Codex* as "Res tantum valet quantum vendi potest." See Langholm, *Economics in the Schools*, 93. Adding weight to this rule was the more general principle, central to Roman law, that a willing party is not injured. See *Digest*, 39.3.9.1; Langholm, *Usury*, 38.

two judgments, both limited to a sale of real estate by a minor at less than half the amount of the estate's "just price" calculated at the time of the sale.[61] In such a sale the minor seller was seen to have suffered excessive damage, and the buyer was left the choice of either furnishing the difference between the selling price and the "just price" or canceling the sale. Even within the judgments proposing this exception there is a spirited defense of the fundamental principle of free bargaining.[62] There is also, however, the contradictory implication that there exists a knowable "true price" for the land apart from its sale price, and that agreed prices are, in certain situations, not just.

Medieval lawyers greatly expanded this limited exception to the rule of free bargaining in Roman law. They did so under the heading *laesio enormis* (extreme damage).[63] In Roman law the rule protected only the underage seller of a piece of property.[64] In twelfth-century legal commentaries, the principle was gradually extended from real estate to all goods; then from sale by a minor to all sales in which less than one-half of the "true" value of the good had been received; and finally to include deceived buyers as well as sellers.[65] The new, expanded principle allowed that any payment exceeding one-half above or one-half below the "just price" provided either buyer or seller grounds to rescind the sale.

The question still remained how to determine a "just" price. Roman law at times recognized the necessity for a third party or judge to estimate the true value of goods, especially within the domain of public economic regulation in the late empire.[66] What criteria did judges use for these estimates of just price? The answer is not entirely certain, but the prevalent historical view is that the just price of Roman law was essentially the common or market price; in John Baldwin's words, "a normal and customary price . . . determined in commerce of free exchange which is regular and orderly."[67] The identification of just price with market price is conveyed in two nearly identical texts from the *Digest* that state: "The prices of things are determined not by their value and utility to individ-

---

[61] *Codex*, 4.44.8: "nisi minus dimidia iusti pretii, quod fuerat tempore venditionis." See also *Codex*, 4.44.2. See Baldwin, *Just Price*, 18–19, 22; Cahn, "Roots," 12–17.   [62] *Codex*, 4.44.8.

[63] The actual term *laesio enormis* did not get attached to the rule of one-half above or below the just price until the fourteenth century: Baldwin, *Just Price*, 18 n. 68.

[64] The lawyers of Justinian's time presumably believed that it was easier to determine a just price (i.e., objective price) for real estate than for other goods, based on the average yearly income of the property taken over a number of years. For the medieval continuation of this idea, see Baldwin, *Just Price*, 29, 53.

[65] Baldwin, *Just Price*, 22; Cahn, "Roots," 18–22. The canonists followed the lead of the twelfth-century Romanists in the treatment of this principle. *Laesio enormis* was first expressed in *Quum dilecti* of Alexander III in relation to a sale of real estate: *Decretales*, x.3.17.3, vol. II, cols. 518–19. It was again discussed in *Quum causa* of Innocent III, *Decretales*, x.3.17.6, vol. II, col. 520. See Baldwin, *Just Price*, 43–44.   [66] Baldwin, *Just Price*, 20; Cahn, "Roots," 32–33.

[67] Baldwin, *Just Price*, 20.

uals, but by their value determined commonly."[68] Medieval Romanists and canonists regularly cited and agreed with this opinion in their writings on just price. They recognized that price freely and commonly arrived at in the marketplace was the best guide to the determination of value. In this view, market or common value was seen as a *corrective* for the wide variations in individual needs and judgments. Similarly, in the determination of an unjust sale (*laesio enormis*), market price served as the corrective guide. The important place that findings of *laesio enormis* held in medieval legal thought reinforced the legal equation of common estimation, market price, and just price.

The rule of *laesio enormis* rests on the assumption that price determination is in some sense a mathematical problem, and that economic value can be represented as a numbered continuum, divisible as a line is divisible. The great thirteenth-century Romanists Azo, Accursius, and Odofredus, in seeking to make this rule practically applicable, added mathematical rigor to the understanding of price as a numbered continuum. With the extension of *laesio enormis* to cover all sales, a general rule had to be devised to cover the legitimate ranges for all prices. In order to define these limits precisely they applied medieval ratio theory.[69] Azo gave the rule: if the just price is ten and the lower legal limit of sale price is five, the upper limit should be fifteen.[70] This position was approved by Accursius, and from the authority of his gloss it became the standard application of the rule in both civil and canon law.[71] What is clear is that, by the middle of the thirteenth century, legists were routinely considering the just price not as a point but as a range in a continuum of value framed by an upper and lower limit.

To arrive at his position on just price, Accursius (d. 1263) synthesized positions from the *Code* and the *Digest*. He supported the general rule that, in a sale between individuals, price was an agreement arrived at through the process of free bargaining: *res tantum valet quantum vendi potest*.[72] But he gave greater weight to the opinion in the *Digest* that "The prices of things are determined not by their value and utility to individuals, but by their value determined commonly."[73] His resulting synthetic

---

[68] *Digest*, 35.2.63: "pretia rerum non ex affectu nec utilitate singulorum sed communiter funguntur"; *Digest*, 9.2.33: "pretia rerum non ex affectione nec utilitate singulorum, sed communiter funguntur." See Noonan, *Usury*, 81–99; Cahn, "Roots," 30–32.   [69] Baldwin, *Just Price*, 23.

[70] See Baldwin, *Just Price*, 23 n. 23, for Odofredus' mathematical determination from his commentary to *Codex*, 4.44.2.

[71] The canon lawyer Bernard Botone clearly reflects the mathematical interpretation of the Roman lawyers in his gloss on *Decretales*, x.3.17.6. On this, see Cahn, "Roots," 29 n. 55. See also Baldwin, *Just Price*, 45. Hostiensis also agreed with Azo's mathematical solution: Hostiensis, *Summa aurea*, Bk. III, cols. 943–44.   [72] Accursius, Gloss on *Codex*, 4.44.8; Baldwin, *Just Price*, 21.

[73] Gloss on *Digest*, 35.2.63. See Baldwin, *Just Price*, 28; Cahn, "Roots," 30; Langholm, *Economics in the Schools*, 260ff.

position, often repeated by future legists, was that the value of a commodity is determined by the price for which it can be *commonly* sold: *res tantum valet quantum vendi potest, scilicet communiter.*[74] In other words, for Accursius and the civil lawyers who followed him, the common estimation of value in the marketplace, i.e., market price, was the best guide to just price.[75]

Thirteenth-century canon lawyers arrived at a similar conclusion, emphasizing the element of common estimation (*communis aestimatio*) in the determination of just price and identifying this estimation with the current market price (*communiter venditur in foro*).[76] Out of this consensus grew the legal conception that economic equality (between the needs and interests of buyer and seller and between value and price) was the *product* of a supra-personal process of common estimation that occurred (under most conditions) naturally and without external intervention in the marketplace. In the words of the economic historian Odd Langholm, *communis aestimatio* "came to mean specifically the kind of estimate which the total community makes unintentionally through the impersonal working of the economy."[77] Not only was the just price achieved through a supra-personal process, as the product of a common estimation that shifted according to common need and scarcity, but the just price itself was understood within both canon and civil law as a shifting, relativistic determination.[78]

What did the theologians, concerned primarily with the religious ethics of exchange, make of this model derived by lawyers? There were many theological problems associated with a system in which equality and justice resulted from an impersonal and common process rather than from a conscious ethical decision on the part of the individual.[79] In the religious and philosophical universe of the thirteenth century, where the existence of order necessarily implied an active and intelligent orderer, where even the movement of the celestial spheres was seen to require the intervention of active intelligences, no other system conformed to the self-ordering, self-equalizing, relativistic model of the marketplace. How

---

[74] Accursius, Gloss on *Digest*, 35.2.63: "communi pretio aestimantur res. quod ergo dicitur res tantum valet quantum vendi potest, scilicet communiter." See Baldwin, *Just Price*, 28–29 nn. 75 and 76; Ibanès, *Doctrine*, 35–41; Langholm, *Economics in the Schools*, 93.

[75] See Baldwin, *Just Price*, 29, for a discussion of the meaning of current price in a medieval context.

[76] Baldwin, *Just Price*, 54; De Roover, *Pensée*, 55–56.

[77] Langholm, "Scholastic Economics," 125.

[78] Spicciani (*La mercatura*, 233) underlines that scholastics came to view economic value not in an absolute sense as an intrinsic and "quasi-metaphysical" property of the thing itself but as resulting from the human criterion of use value. See also Noonan's point (*Usury*, 85) that, with such a conception, just price is necessarily variable rather than fixed.

[79] Spicciani (*La mercatura*, 234) is one of the few writers who clearly recognizes the metaphysical problems that such a relativistic-mechanistic model presented to medieval thinkers.

could theologians who recognized the "unnaturalness" of an impersonal economic order within the larger *ratio divina* accept the legal equation of just price and market price?

The answer here is more complicated than it was in the case of the legists. From the earliest theological discussions of just price at the end of the twelfth century, a clear distinction was maintained between a price equality (just price) legally sufficient within the wide bounds of *laesio enormis*, and a price equality that conformed to the stricter moral requirements of divine law. These two requirements reflected and were reflected in two opposing models of equalization. The first, the older, which was at the heart of traditional usury theory, I call the *arithmetical model*. Although it developed long before the rediscovery of the *Ethics*, it fit well with Aristotle's arithmetically defined *iustitia directiva* found there.[80] It grew out of a sense of natural balance and divine order based on knowable points and perfections, and it was understood to be regulated by a *conscious* process of addition and subtraction toward the end of finding a knowable point of "natural" equality. It was represented by the ancient model of adding and subtracting weights to a scale in order to reach the perfect balancing *point* between gain and loss. Aristotle gave this model a mathematical dimension in *Ethics* v in his description of the judge who, after the fact of an unequal exchange, rectified inequality by bisecting the line of gain and loss to find the perfect mid-point.

The second form of economic equalization I designate the *geometrical model*. It derived from legal tradition and from practical observations of shifting prices in the marketplace rather than from congruence to theological conceptions of divine or natural order. It was a dynamic and self-regulating model in which equality was seen as the *product* of the cross-conjunction of opposing needs and benefits. It gained philosophical authority from the middle of the thirteenth century through the weight of Aristotle's concept of *contrapassum* – the recognition that geometric proportionality governed exchange.[81] Its geometrical form was reified through Aristotle's often-copied rectangular *figura proportionalitatis*, whose crossed diagonals represented the process of exchange equalization. Its elements were bound together by money as a connecting and measuring *medium*. Its dynamic and mechanical qualities were underlined by Roman law traditions that accepted free bargaining and recognized value as an aggregate product of common estimation in the marketplace. Within this geometric model, price lost its association with an ideal fixed point and came to be visualized as a range along a continuum of value.

---

[80] See chapter 2, 41–42.   [81] See chapter 2, 43–44, and chapter 3, 60–64.

Equality in just price theory and market exchange

The doctrine of *laesio enormis* provided yet another geometrical context for the concept of just price (and exchange equalization) in its positing of a mathematically defined range within which prices could fluctuate yet still remain "just." By the middle of the thirteenth century, the geometrical model of equalization had supplanted the arithmetical model in legal theory. But theologians and philosophers who were required to integrate their vision of economic order into their vision of divine and natural order had to be considerably more cautious.

The first theologians to discuss the contract of sale in the early thirteenth century were well aware that civil and canon law allowed deception within limits. They knew that, legally, all sales in which the price fell within the broad bounds of *laesio enormis* were considered final and just. They insisted, however, that the *lex divina* made no such allowances.[82] They therefore thought it quite possible for someone to knowingly buy for less than the just price or to sell for more. Such sales violated arithmetical equality and should thus, in their opinion, be subject to the same penalties as usurious loans.[83] According to their notion of equality, it was the responsibility of both buyer and seller to aim for and to achieve a knowable just price. Even though theologians of the first half of the thirteenth century never determined or even discussed how this just price was to be determined, they insisted that the smallest deviation from the point of just equality demanded restitution.[84]

The theological focus on equality continued unabated in the economic writings of Albertus Magnus and Thomas Aquinas. This is clearly seen in the opening words of Aquinas' response on just exchange in the *Summa theologiae*:

Considered in itself, the transaction of buying and selling is seen to have been introduced for the common utility of both buyer and seller . . . and therefore the contract should be instituted according to an equality of things between them. The value [*quantitas*] of things which come into human use is measured by their given price, and for this reason money was invented as [Aristotle] said. And therefore if the price exceeds the quantity of the value of a thing or, conversely, if the value of the thing exceeds the price, the equality that justice demands is

---

82 See Baldwin, *Just Price*, 68–71, for an overview of the theological position on price and the market before Aquinas. Langholm (*Economics in the Schools*) considers this question in relation to individual theologians at a number of points in his first five chapters. See also Baldwin, *Masters*, vol. I, 261–75.

83 Baldwin, *Just Price*, 69; Baldwin, *Masters*, vol. I, 265.

84 Baldwin, *Just Price*, 69 n. 127, quoting from the *Poenitentiale* of Thomas of Chobham: "In venditione autem secundum humanas si aliquis deceperit aliquem ultra medietatem iusti precii tenetur restituere illud quod ultra medietatem recipit. sed si minori quantitate decipit emptorem non tenetur restiturere. sed secundum legem dei si decipit emptorem in uno denario ultra iustum precium tenetur restituere."

destroyed. And therefore, to buy a thing for less or to sell a thing for more than it is worth is in itself unjust and illicit.[85]

For Aquinas, equality is at the center of economic exchange in three ways: (1) there is an assumed equality of need between buyer and seller before the sale (the buyer needs the good, the seller has decided he needs a particular sum of money to buy other goods – money serving as an instrument of equalization); (2) there is an assumed equality of benefit between buyer and seller as the result of the sale; (3) to satisfy justice there must therefore be an equality between the price paid and the value of the thing bought.[86] As he writes: "the paramount consideration in commutative justice is the equality of things."[87] But while his emphasis on equality in exchange is clear, his model of equalization is not. He never makes explicit in any of his writings how either the "equality of things" or the just price of a commodity in exchange is actually determined.[88] He is uncharacteristically vague on this point, especially in contrast to his teacher Albertus Magnus, who openly acknowledges the superiority of common estimation in the determination of value and specifically identifies the market price (*secundum aestimationem fori*) with the just price.[89]

It is difficult to believe that Aquinas' lack of clarity on this point resulted either from a lack of interest in or a lack of contact with the economic life of his time. In his discussion of the question of just price in the *Summa* he often demonstrates that he is a keen observer of current economic practices and is well aware of the mechanism of price determi-

---

[85] Aquinas, *ST*, 2, 2, 77, 1, responsio: "Uno modo secundum se, et secundum hoc emptio et venditio videtur esse introducta pro communi utilitate utriusque, dum scilicet unus indiget re alterius et e converso, sicut patet per Philosophum. Quod autem pro communi utilitate est inductum, non debet esse magis in gravamen unius quam alterius; et ideo debet secundum aequalitatem rei inter eos contractus institui. Quantitas autem rerum, quae in usum hominis veniunt, mensuratur secundum pretium datum; ad quod est inventum numisma, ut dicitur. Et ideo si vel pretium excedat quantitatem valoris rei vel e converso res excedat pretium, tolletur justitiae aequalitas. Et ideo carius vendere aut vilius emere rem quam valeat est secundum se injustum et illicitum."

[86] Note that this equality is not assumed, i.e., value is not perfectly identified with sale price. Note also that Aquinas is not talking here of a market price but only of an agreement between two individuals. Clearly in such personal agreements inequalities could result between price paid and value.

[87] Aquinas, *ST*, 2, 2, 77, 1, ad 3: "in justitia commutativa consideratur principaliter aequalitas rei." For Aquinas' use of the term *iustitia commutativa*, see chapter 3, 76–78.

[88] He came closest in *ST*, 2, 2, 77, 3, ad 4, when he deemed it just (if not perfectly virtuous) for a merchant to charge the high going price for corn in a time of scarcity, even knowing that the price would soon fall. De Roover (*Pensée*, 58) places particular emphasis on this statement in his argument that Aquinas equated the just price with the current market price. It is, I think, clear evidence that Aquinas understood that economic value is determined relative to scarcity and varies with time and place. The question remains why Aquinas failed to make this equation explicit.

[89] See chapter 3. Albertus Magnus, *iv Sent.*, d. XVI, art. 46, in (Borgnet edn.), 638: "Justum autem pretium est, quod secundum aestimationem fori illius temporis potest valere res vendita"; Baldwin, *Just Price*, 71.

nation in the marketplace. At one point he considers the case of a merchant who charges the high going price (*secundum pretium quod invenit*) for corn in a time of scarcity even when he knows that new supplies are coming that will soon cause the price to fall.[90] The case and the observation of commercial calculation upon which it is based demonstrate Aquinas' awareness that market value is determined relative to scarcity and need and that it varies with time and place.

Aquinas' commentary on Book v of Aristotle's *Ethics* demonstrates how well he comprehended Aristotle's analysis of exchange as a process of geometrical equalization.[91] It was in this commentary that Aquinas reintroduced the Augustinian example of the mouse costing less than the pearl in order to underscore the relativity of market price and the separation of economic value from natural value (*gradum naturae*) in the marketplace.[92] Aquinas thought this insight sufficiently important to repeat it in his economic discussion in the *Summa*.[93] Similarly, throughout the questions in the *Summa* Aquinas demonstrates his knowledge of decisions from both canon and civil law that legitimate the practice of free bargaining, recognize the commanding role of *aestimatio communis* in the establishment of market price, and equate just price with market price.[94] Why then did he refrain from identifying just price with current market price?

Recent historical opinion for the most part holds that Aquinas believed in the identity of just price and market price even though he failed to say so.[95] Another opinion denies this, asserting that Aquinas' concept of just price was determined by the "objective" factors of the labor and expenses involved in the production of a good rather than by the relativistic market

---

[90] Aquinas, *ST*, 2, 2, 77, 3, ad 4: "unde venditor qui vendit rem secundum pretium quod invenit non videtur contra justitiam facere, si quod futurum est non exponat. Si tamen exponeret vel de pretio subtraheret, abundatioris esset virtutis; quamvis ad hoc non videatur teneri ex justitiae debito." De Roover (*Pensée*, 58) places particular emphasis on this statement in his argument that Aquinas equated the just price with the current market price.   [91] See chapter 3, 62–63.

[92] See chapter 3, 70. See also Langholm, *Economics in the Schools*, 230.

[93] Aquinas, *ST*, 2, 2, 77, 3, ad 3: "sicut Augustinus dicit, pretium rerum venalium non consideratur secundum gradum naturae, cum quandoque pluris vendatur unus equus quam unus servus; sed consideratur secundum quod res in usum hominis veniunt." Baldwin (*Just Price*, 77) dates the Aquinas commentary on the *Ethics* to 1266, the questions on economics in the *Summa* to 1271–72.

[94] Aquinas, *ST*, 2, 2, 77, 1, ad 1: "Sic igitur habet quasi licitum, poenam non inducens, si absque fraude venditor rem suam supervendat aut emptor vilius emat; nisi sit nimius excesus, quia tunc etiam lex humana cogit ad restituendum, puta si aliquis sit deceptus ultra dimidiam justii pretii quantitatem."

[95] Despite Aquinas' failure to make the link explicit, Baldwin's general conclusion (*Just Price*, 76–77) is that, for Aquinas, the just price was "clearly equated" with the current price. So too, De Roover, *Pensée*, 56–58: "Thomas considerait comme juste le prix courant du marché." This was the general conclusion also of Noonan, *Usury*, 85, and Schumpeter, *Economic Analysis*, 93.

factors of scarcity, need, and common estimation.[96] Still others maintain that he had no unified position on just price but held conflicting models of value determination depending on whether he was considering sale between producers or sale by merchants.[97]

I find none of these positions satisfactory. I believe that Aquinas' failure to link just price with market price resulted not from a conflict of economic models but from his recognition that such a linkage brought with it serious theological, ethical, and metaphysical difficulties – difficulties imbedded in the geometric model of market equalization. One of the most important metaphysical principles for Aquinas is that order always implies the existence of an active, intelligent orderer. The equation of order with intelligence is so central to Aquinas' thought that it provides the basis for the proofs of God's existence with which the *Summa* begins. This principle is negated with the acceptance of an economic system in which equality is created as an accidental product of competing desires within an impersonal process, and where value (price) is detached from individual judgment. Similarly great difficulties accompany the acceptance of a shifting, relational estimation as a "just" solution to the problem of economic value, within a philosophical and theological value system centered on hierarchy and permanence. The legists did not focus on the theological implications of the geometrical model of equalization, but Aquinas was required to.

The ethical implications were equally disturbing. If just price equals common or market price and is divorced from individual judgment and direction, the individual's responsibility in economic activity is effectively eliminated. Furthermore, the recognition that ideal ends (equality and justice) could follow from the base motive of deception and the base desire to buy cheaply and sell dear severs the link between ethical order and natural order, a link Aquinas sought to strengthen at every point. The economic position Aquinas adopted in the *Summa* was crafted to take these serious philosophical difficulties into account.

Aquinas answered the Roman law recognition that deception in exchange was "natural" by insisting on the distinction between human and divine law in the definition of licit economic behavior. Human law, he wrote, is not concerned with virtue but with the maintenance of social

[96] Baldwin (*Just Price*, 77–79) considers the basis of the historical argument for objective determinants and rejects it. So too does Armando Sapori, "Il giusto prezzo nella dottrina di San Tommaso e nella pratica del suo tempo," *Archivo storico italiano*, ser. 7, 18 (1932), 3–56, esp. 24–30. I follow Baldwin, *Just Price*, 75, in placing the word "objective" in quotation marks, since I agree that need as the basis of market price was conceived as "common need" and as such was as much an objective consideration as the cost factors of labor and expense.

[97] For a nuanced statement of this position, see Hollander, "Just Price."

intercourse.[98] The civil law permits mutual deception within the broad range of *laesio enormis*. Divine law does not.[99] The legally permissible is not always the just. In parallel fashion, Aquinas does not deny that current market price provides a legally sufficient guide to value (he clearly recognized that it did), nor that in certain cases current price might conform to the highest standards of justice and virtue, nor that, even if too lax, common estimation might at all times serve as a useful indicator or bench mark by which to determine just price. But market price based on *communis aestimatio* cannot in itself guarantee an equality sufficient to merit the word "just."[100]

For Aquinas, a truly just equality between buyer and seller (which is to say a truly just price) requires individuals to consciously order each transaction to the end of *aequalitas iustitiae*.[101] He still, however, had not defined what this *aequalitas iustitiae* was in practical terms. It occurred, he said, when there was an equivalence between price and value. But how was this equivalence determined, and with how much precision could it be determined? To answer this Aquinas introduced into his theological and philosophical discussion a refinement of the legal rule of *laesio enormis*:

> The just price of things sometimes is not precisely determined [*quandoque non est punctualiter determinatum*], but rather consists in a certain estimate [*quadam aestimatione consistit*]. Therefore a small addition or subtraction does not seem to destroy the equality of justice.[102]

The modest phrasing of this conclusion disguises its great import. Here, even within the constraints of reconciling economic observations with theological requirements, Aquinas visualizes a true and just equality,

---

98 Aquinas, *ST*, 2, 2, 77, 1, ad 1: "Et ideo lex humana non potuit prohibere quidquid est contra virtutem; sed ei sufficit ut prohibeat ea quae destruunt hominum convictum, alia vero habeat quasi licita, non quia ea approbet, sed quia ea non punit."

99 *Ibid.*: "Sic igitur habet quasi licitum, poenam non inducens, si absque fraude venditor rem suam supervendat aut emptor vilius emat; nisi sit nimius excessus, quia tunc etiam lex humana cogit ad restituendum, puta si aliquis sit deceptus ultra dimidiam justii pretii quantitatem. Sed lex divina nihil impunitum relinquit quod sit virtuti contrarium."

100 This conclusion is supported by the argument that follows in the *Summa*, 2, 2, 77, 1, ad 2, where Aquinas concludes that the common is not identical with the "natural" (as it was taken to be in Roman law) but that the common way often leads down the (unnatural) road of sin. "Unde patet quod illud commune desiderium non est naturae sed vitii, et ideo commune est multis, qui per latam viam vitiorum incedunt." He specifically mentions here the common desire to buy cheap and sell dear. He concludes that individuals must work to overcome such common desires.

101 Aquinas, *ST*, 2, 2, 77, 1, ad 2: "Unde secundum divinam legem illicitum reputatur, si in emptione et venditione non sit aequalitas justitiae observata. Et tenetur ille qui plus habet recompensare ei qui damnificatus est, si sit notabile damnum."

102 *Ibid.*: "Quod ideo dico, quia *justum pretium rerum quandoque non est punctualiter determinatum, sed majus in quadam aestimatione consistit*; ita quod modica additio vel minutio non videtur tollere aequalitatem justitiae" (emphasis added).

one pleasing to God and consistent with the requirements of justice and virtue, not as a precise arithmetical *point* but as a *range* along a continuum of value. Moreover, estimation replaces knowing as the intellectual process through which this approximate range is recognized.

Though phrased in modest terms, the conception of the equalizing range had profound philosophical repercussions. In an intellectual tradition preoccupied with the concept of equality, changes in the definition of what constitutes equality, or in the understanding of how equality is achieved, would have multiple and expanding consequences. It was not only in the area of economic life that Aquinas pioneered the philosophical recognition of estimation and approximation. He introduced the concept of the legitimate range of variation (or "latitude" as he sometimes called it) into both his ethical thought and his natural philosophy as well.[103] With his acceptance of the notion of an equalizing range and his projection of this range onto the plan of nature, Aquinas began stepping away from an earlier philosophical model of natural activity fixed on knowable points of perfection, toward a more dynamic and fluid model accepting of approximation and estimation.

Once again the question is raised: if a new insight is found simultaneously in a number of spheres within one thinker's philosophical system, or within scholastic thought itself, how can the direction from which it came be determined? In Aquinas' case, the insight of the equalizing range could have come from many sources: observation of the physical world, authoritative texts, contemporary philosophical debate, social and economic experience, personal habits of perception and synthesis. Or, it could have come from all these nodes of influence (and more), taking shape as it was reflected between them. But while recognizing that the process is multi-directional, the evidence indicates that a critical influence in this case came from the direction of economic thought and experience. The recognition of an equalizing range based on estimation and approximation and the acceptance of that range as defining a "just" equality, appeared in medieval legal writings on price and value long before they appeared in Aquinas' economic thought or his natural philosophy.

The acceptance of estimation, approximation, and the equalizing range, seen here in its beginnings, had a remarkable future. In the century following Aquinas, these insights, along with a cluster of others that accompanied the geometric model of exchange, came to occupy a central place not only in economic speculation, but in proto-scientific speculation of the most forward-looking natural philosophers. Applied in the

---

[103] Aquinas' innovative use of the "latitude" in his speculation on ethics and natural philosophy is discussed in chapter 6.

context of natural philosophy, they became the instruments through which the image of nature was transformed.[104]

TWO MODELS OF ECONOMIC EQUALIZATION: HENRY OF GHENT AND GODFREY OF FONTAINES

The intellectual tension resulting from the attempt to reconcile increasingly sharp observations of the economic order with older conceptions of the natural order can be observed in conflicting definitions of economic liceity and equality within the writings of Henry of Ghent and Godfrey of Fontaines. Their opinions on economic questions are found mixed in among their quodlibetal questions written over the last quarter of the thirteenth century.[105] Both thinkers spent their early life in commercially active Northern cities (Ghent, Tournai, and Liège) and their later life primarily in Paris as teaching masters in theology. Both were independent and eclectic thinkers, although Henry drew more from an older, self-consciously conservative, Augustinian tradition, while Godfrey identified with the philosophical innovations of his contemporaries.[106] Both were strongly influenced by Aristotle's economic thought and the early commentaries on Aristotle by Albert and Aquinas, although, once again, Henry tended to follow the more conservative position on money and exchange expressed in the *Politics*, while Godfrey was fascinated by the positive and dynamic treatment of exchange in the *Ethics*. Finally, in the area of natural philosophy, both men made early and important contributions to the debate on the quantification of qualities, a debate which

---

[104] For the place of estimation and the equalizing latitude in fourteenth-century natural philosophy, see chapters 6 and 7.

[105] Henry of Ghent's first quodlibet is dated 1276. Henry was master of philosophy at Paris from 1270, master in theology from 1275, and wrote his quodlibetal questions over the period 1276–92. For a discussion of Henry's philosophical positions, see Stephen P. Marrone, *Truth and Scientific Knowledge in the Thought of Henry of Ghent* (Cambridge, Mass., 1985). Godfrey's quodlibets were written over the period 1285–96/7. For a dating of Godfrey's questions, see John F. Wippel, *The Metaphysical Thought of Godfrey of Fontaines: A Study in Late Thirteenth-Century Philosophy* (Washington, D. C., 1981), esp. xxi–xxxiii. For an explanation of the position of quodlibetal questions within scholastic debate, see Palemon Glorieux, *La littérature quodlibétique*, 2 vols. (Paris, 1935), vol. II, 9–50.

[106] Henry and Godfrey disagreed on a range of questions and subjects from theology to metaphysics to natural philosophy. In addition they were on opposite sides of the controversy surrounding the Condemnations of 1277 at the University of Paris. Henry was a force behind their promulgation; Godfrey was opposed. See Georges de Lagarde, "La philosophie sociale d'Henri de Gand et de Godefroid de Fontaines," in *L'organisation corporative du moyen âge à la fin de l'ancien régime* (Louvain, 1943), 57–134, esp. 57–60; Maurice De Wulf, *Un théologien-philosophe du XIIIe siècle: étude sur la vie, les oeuvres, et l'influence de Godefroid de Fontaines* (Brussels, 1904), 72, 81; Wippel, *Godfrey*, xix.

occupied a central place in the proto-scientific speculation of the four-teenth century.[107]

Strict arithmetical equality was the controlling principle behind Henry of Ghent's economic thought.[108] He championed an older, natural law definition of economic equity (*naturalis iuris aequitatem*) in opposition to contemporary legal opinion favoring the range of *laesio enormis* as a sufficient approximation of equality for the just price.[109] He insisted that the true basis of civil and canon law determinations on usury and just price should be the *natural law*, and that the form of just equality was rooted in nature (*aequitas naturalis*).[110] Therefore, he believed that theologians and philosophers were in a better position to discuss and define the legitimacy of exchange contracts than were lawyers.[111]

Since equality was a requirement of natural law, he believed that it must be enforced in contracts of buying and selling just as it was in contracts of loan.[112] From his analysis of previous authorities from Aristotle through Aquinas, he concluded that all unequal economic transactions violated natural law and were therefore in some sense usurious.[113] In order to enforce his arithmetical ideal of a knowable point of equality, he sought to extend the century-old test of usurious *intentio* from its recognized place in the *mutuum* to contracts of *emptio* (buying and selling), including, specifically, transactions undertaken by professional merchants.[114] He maintained the position, dating back to Augustine, that the

---

[107] For the place of Henry and Godfrey in this debate, see Maier, "Das Problem der intensiven Grösse," 30–43. For Godfrey's positions on the question of quantitative and qualitative change, see Sylla, "Godfrey," esp. 123–25.

[108] *Henrici de Gandavo: Quodlibet I* (*HG I*), Raymond Macken (ed.), vol. v in *Opera omnia* (Leuven, 1979), quest. 40, 221: "In isto ergo contractu emptionis et venditionis sic debet servari aequalitas inter mutuo dantem et recipientem, quod neuter plus recipiat quam det."

[109] *Henrici de Gandavo: Quodlibet II* (*HG II*), R. Wielockx (ed.), vol. vi in *Opera omnia* (Leuven, 1983), quest. 15, 97–100. Henry supports the existence of a single, ideal point of balance. For Henry's maintenance of this conservative economic position in the face of contemporary clerical opinion and practice, see his opinion on *redditus ad vitam* contracts, 109.

[110] *HG* 1.40, 219–20: "Dicendum ad hoc quod iustitia contractuum pertinet ad ius naturae, quod idem debet esse apud omnes gentes. Debet ergo in contractibus emptionis et venditionis aequitas naturalis observari, et penes illam determinari debet aequitas contractuum in iure positivo." Notice he is saying that positive law ought to insist on a true equality in contracts of buying and selling, not that it always does.

[111] *HG* 1.39, 217. After declaring usurious the contract of *redditus ad vitam* which the legists found acceptable, he wrote: "Nec debet forma huius contractus, in genere cuius contractus iniqui sit, peti a iuristis, sed magis a theologis et philosophis, cum sit iniquitas contra legem iuris divini et naturae, ut dictum est."

[112] *HG* 1.40, 221: "In isto ergo contractu emptionis et venditionis sic debet servari aequalitas inter mutuo dantem et recipientem, quod neuter plus recipiat quam det."

[113] *HG* 1.39, 210: "Hoc enim solum facit usuram, quod ultra datum aliquid speratur."

[114] On the extension of *intentio* to the merchant's activity, see Spicciani, *La mercatura*, 173–74.

propensity of merchants to seek personal advantage rather than equality exposed their occupation to the continual danger of sin.[115]

Henry's insistence on maintaining ideal standards in personal and commercial exchange might be taken as an indication that he was ignorant of the economic realities of his day. He was, in fact, an acute observer of economic matters, particularly sensitive to the important role of the merchant in his society.[116] He was one of the earliest thinkers to deny the ancient distinction between the honest labor of the farmer and artisan and the dishonest labor of the merchant.[117] While insisting that merchants maintain a strict equality between the value of their goods and the price they sell them for, he considered the labor involved in the transportation of goods as an improvement made to them, one that added real value and that therefore could be legitimately recompensed. He also recognized that the personal and professional qualities of the merchant added value to the goods he sold.[118] The merchant's expertise in knowing when and where goods were in short supply, his care in transacting his affairs, and even his professional reputation were all qualities that he could licitly translate into increased prices.[119]

A seeming contradiction between definitional rigorism and practical allowance also appears in Henry's attitude toward money and moneychangers. He insisted, citing Aristotle, that money's use is defined by its nature, and that money's proper nature, in turn, is defined by the reasons for its invention in the distant past.[120] According to Henry (again

---

[115] *HG* 1.40, 223: "Unde, cum pauci sint mercatores qui cum tanto studio servandae aequitatis vendant et emant, summe periculosa est emptionis et venditionis negotiatio."

[116] Henry disagreed with positions condemning all merchants (1.40, 224–25). Those who buy and sell for profit only and seek only to multiply their wealth (*ratione cupiditatis nimiae multiplicandi pecunias*) are condemned. Those who help bring the necessities of life to the community are praised.

[117] Spicciani, *La mercatura*, 179; *HG* 1.40, 227–28: "sicut faber potest carius vendere ferramentum quam emptum fuerit ferrum . . . sic mercator potest carius vendere merces translatas quam merces ipsae nudae emptae fuerunt."

[118] *HG* 1.40, 228: "Similiter in mutatione rei ratione temporis sicut in mutatione ratione loci . . . ita utile est ad acquirendum pecuniam esse expertum, qualia in quibus temporibus sunt rara et pretiosa; alia enim in aliis temporibus abundant. Et ideo empta vilius in tempore quo abundant, quando sunt vilia, custodita in aliud tempus quando sunt rara, iam per custodiam facta sunt pretiosa et vendi possunt carius, sicut dictum est de translatis per loca." See also Langholm, *Economics in the Schools*, 258.

[119] For the increase he considered permissible due to the merchant's reputation (1.40, 229): "alii cognoscentes eius industriam circa equos statim certificati per eum et eius industriam sunt quod equi illi valent pretium suum . . . et sic propter talem mutationem aestimationis circa valorem equorum bene potest iste statim equos suos aliquantulum carius vendere supra primum pretium equorum, vendendo opus industriae suae."

[120] See *HG* 1.39, 210, for the link between the nature of money and the reasons for its invention. His opinion here is identical to Aquinas'.

following Aristotle in the *Politics*), money was invented to facilitate exchange. Its only proper use is as a measure, a *medium*, a middle-term, a means of establishing equality and commensurability between things (*nummus sit medium sive regula sive pretium sive mensura in commutationibus*).[121] Money must always remain the mid-term in a sale and never the end-term. Consequently it must never be bought and sold as if it had value in itself. To violate this rule is to commit usury.[122] And yet despite Henry's traditional insistence on the ideal definition of money's proper use, which seems to ignore the actual uses of money in his own society, we find in his writings a precocious if limited defense of the profits of the normally despised moneychangers. They too, he writes, can charge for the labor they expend in computing and managing their complicated and necessary affairs.[123] Starting from the position of a pristine, definitional ideal, Henry created the most realistic space for the activities of merchant and moneychanger yet seen in scholastic theological speculation.[124] He accommodated new economic realities within a strict arithmetical definition just as earlier usury theorists found it possible to introduce proportionality by legitimizing *interesse* (*damnum emergens* and *lucrum cessans*) as external titles, while insisting on a usury prohibition arithmetically defined and strictly enforced. Despite the exceptions he allowed, Henry put himself forward as a spokesman for what he considered to be the traditional position on economic questions. It was in all cases illicit to resell something for more than the price originally paid, unless it had been changed or improved in some way.[125] Citing Augustine as had Aquinas in the *Summa theologiae*, Henry held that the intention of getting more for less was a sin in itself, even if that sin was shared by many or by all, and even if that sin was considered "natural" and so unpunishable in

---

[121] *Henrici de Gandavo: Quodlibet* VI (*HG* VI), G. A. Wilson (ed.), vol. X in *Opera omnia* (Leuven, 1987), quest. 22, 208. Also, I.39, 213: "nummus inventus est ad aequandam commutationes et in pretium commutati." Again, VI.22, 206: "quia res non frequenter ex se veniunt ad aequalitatem in commutatione, quia plus valet dolium vini quam raseria bladi, et non debet fieri, ut dictum est, commutatio nisi per aequalitatem, ideo ex necessitate numismatis inventus est usus." Henry's debt to Aristotle is clear here.

[122] *HG* VI.22, 203: "Pecunia medium est in commutationibus, et non terminus. Facere autem medium commutationum esse terminum ad maius lucrum, est usura." Henry calls the buying and selling of money against its proper nature *campsoria*, borrowing from William of Moerbeke's translation of the *Politics*. See *HG* I.39, 210.

[123] *HG* VI.22, 209: "sic nullo modo licet accipere amplius, nisi quatenus sua interest pro labore computandi et custodiendi numismata et cetera quae pertinent ad curam eius, dum tamen hoc faciat bona fide, alias autem nullo modo licet, sed est species usurae."

[124] Spicciani (*La mercatura*, 172) credits Henry with one of the earliest recognitions of "l'assurdita di una mercatura senza guadagno."

[125] *HG* I.40, 222: "Si vero vendat eam talem, non immutatam, sive statim sive postmodum carius, iniustus est et inaequalis propter factum inaequalitatis et iniustitiae, accipiendo plus ab emptore quam tradat eidem."

law.[126] If we examine Henry's argument in support of this position, we find that while ostensibly relying heavily on Aristotle's *Ethics* for its definitions, it in fact asserts a pre-Aristotelian ideal that harks back to early usury theory. It does so by simply eliminating Aristotle's fundamental concept of geometric proportionality in exchange.[127]

In his discussion of just exchange, Henry follows Aristotle in stating that there are two species of justice, *distributiva* and *directiva*. The first, *distributiva*, is mentioned once and never brought up again. He virtually dismisses its role in economic exchange.[128] By doing so, he avoids the highly problematic aspect of *geometric* proportionality so central to Aristotle's discussion of distributive justice, yet so difficult to reconcile with the arithmetical equality required in traditional usury theory.

After dismissing *iustitia distributiva* (by ignoring it), Henry considered equality and equalization in contracts of buying and selling solely in terms of Aristotle's second species of justice, *iustitia directiva*, governed by *arithmetical* proportion – subtracting from illicit gain or adding to loss to arrive at an ideal point of equality.[129] He recognized that *iustitia directiva* in Aristotle's (and Aquinas') discussion was limited to cases requiring remedy and equalization by an intervening third party or judge.[130] The only way he could maintain that *iustitia directiva* governed all economic exchange and still stay within Aristotle's definition was to extend the limited case of an intervening judge to cover all contracts of buying and selling.[131] How was this possible?

---

[126] *HG* I.40, 224: "Idcirco dicendum absolute quod talis peccat, et si omnes ementes et vendentes hoc intendunt, proculdubio omnes iniqui sunt, nec multitudo peccantium excusat." See Aquinas' similar position, 99, n. 100.

[127] The following argument closely follows Henry's presentation in question 40 of the first quodlibet, 219ff.: "Utrum emere vilius et in continenti vendere carius sit peccatum."

[128] *HG* I.40, 220: "Huius autem iustitiae, ut dicit PHILOSOPHUS in eodem, duplex est species. Una, quae est mercedum distributiva. Altera, quae est communicationum directiva, quae iniustitias in commutationibus factas et inaequalitates tentat aequare et in illis aequalitatem servare."

[129] *Ibid.*: "Et ad istam speciem [commutative justice] pertinet omnis commutatio emptionis et venditionis, in quibus inaequale lucri et damni reducitur ad medium." In the previous question (1.39, 212) he had established that commutative justice established equality according to arithmetical proportion.

[130] *HG* I.40, 221: "Unde, cum duo communicationem habentes inaequales facti sunt, ut uni videatur quod damnum passus est et alter lucrum indebitum a se abstulit, et super hoc dubitant, concurrunt ad iudicem, qui est quaedam animata iustitia et libra in pondere iustitiae ambos commensurans, apponens uni et subtrahens alteri, quousque sint aequales." This decision to limit the discussion of economic exchange to commutative justice is, as we have seen, unwarranted by the Aristotelian text but consistent with a particular reading of Aquinas' commentary. See the conclusion to chapter 3, 76–78.

[131] *Ibid.* Referring to the above exchange where a judge is required to restore equality he writes, "In isto ergo contractu emptionis et venditionis sic debet servari aequalitas inter mutuo dantem et recipientem, quod neuter plus recipiat quam det. Iste enim contractus propter necessariam commutationem rerum naturalium inventus est."

In a normal contact of sale there is no intervention by a judge – or is there? Henry's solution, more overstated but similar to the solution I have suggested for Aquinas, was that in every contract of sale, the participants themselves must serve as their own judges, *consciously* seeking to find the delicate balance of equality:

> And if equality is to be saved, all things given and accepted must be equal in value. And to do this, both buyer and seller ought to be judges, acting just as two arms of a scale and of animate justice, so that he who senses he has received the heavier weight of price might give some of it back to the other, so that equality is established, and so that they stand in relation to each other like arms of a scale, equally elevated and lowered.[132]

Henry's use of a mechanical scale to represent balance in buying and selling indicates the degree to which he recognized economic exchange as a process of equalization.[133] Note, however, that, at the same time as he uses a mechanical metaphor for exchange, he insists on the place of subjective judgment and ordering (*iustitia animata*) in establishing a true equality (just price). Henry's position itself represented a delicate (and perhaps unstable) balance. It closely resembles Aquinas' position in the *Summa theologiae* considered above. Both thinkers were conscious of a mechanism for price determination in the marketplace, and both sought to reconcile this mechanism with the ethical requirement of individual responsibility and the metaphysical requirement that order everywhere implied and required an intelligent orderer.

What is absent from Henry's picture is the dynamic geometry and crossed diagonals of Aristotle, or the legists' idea of price as a product of the crossed purposes of buyer and seller. They have been replaced by a conscious and judicious arithmetical process of addition and subtraction toward the end of equality. Henry disagreed with the rule from Roman law that a thing is always worth the price it can be sold for (*res tantum semper valeat pro tempore et loco, quantum vendi potest*). His reason was that, out of ignorance of the current price, or because of extreme need of the buyer, a price could be accepted by buyer and seller that was nonetheless

---

[132] *Ibid.*: "Et si debet servari aequalitas, aequale debet esse omnino in valore datum et receptum hinc et inde. Et in hoc *ambo debent esse iudices tamquam duo brachia librae et animatae iustitiae*, ut qui in pondere pretii sentiat se plus recepisse de eo quod est alterius, rescindat et reddat ei di suo, quousque fiat aequale, et sic stent quasi brachiis librae elevatis et depressis aequaliter" (emphasis added).

[133] Langholm (*Economics in the Schools*, 256) calls this "most likely the first clear statement of the principle of equilibrium in the history of economics." I cannot agree, given Aquinas' opening statement in his question on just price in the *Summa theologiae* (cited 95) and the many other statements on the subject of equality from lawyers and theologians writing before Henry. Moreover, I would argue that Henry's statement here is more retrograde than forward-looking.

unequal and unjust.[134] Henry sought to reformulate the rule of sale commonly put forward in the medieval glosses on Roman law. He stated that a thing is worth not what it *can* be sold for (*quantum vendi potest*) but rather what it *ought* to be sold for (*sed quantum vendi debet*).

At a number of points in his writings on economic questions, Henry recognizes the existence of markets in which the prices of goods are determined by an impersonal process relative to time, place, and other circumstances (*ratione loci et temporis et aliarum circumstantiarum*).[135] He also recognizes that the market value of a commodity is determined by its scarcity and based on its usefulness to humans rather than by its ontological or "natural" value.[136] At the same time, the example he uses, that of a dying horse, which, though less valuable than a healthy ass in monetary terms is nevertheless more valuable in its "nature and essence," reveals the strength of his sense of ontological or "natural order" and the difficulties involved in reconciling that sense with economic definitions of value.[137] In the last confused section of his discussion on price, Henry nevertheless tries to find an acceptable synthesis between the economic and natural order and between the conflicting notions of just price, value, and exchange equality proper to each of them:

In accordance with the equity of natural justice, a thing ought to be sold and bought for as much as it is worth and if someone wittingly sells it for more than it is worth at the time and place, or buys it for less, this is unequal, and it is unjust even if he is permitted to do so . . . For I do not say that a thing is always worth as much as it can be sold for at a given time and place, either because of ignorance or because of the buyer's necessity, but for as much as it ought to be sold for, so that the common saying: "A thing is worth as much as it can be sold for" is to be understood not in terms of what is possible in fact but in terms of what is possible in law, that is, for as much as it ought to be sold for at the place and time according to law.[138]

---

[134] In canon law these conditions provided exceptions to the principle that a price arrived at through free bargaining was just. As such they also provided grounds for a case of fraud in sale. Yet Henry used these exceptional cases to dispute the accepted legal definitions of just equality in economic exchange.

[135] *HG* VI.22, 207, where the price of eggs is to be determined "ratione loci et temporis et aliarum circumstantiarum."

[136] See for example *ibid.*: "scilicet rei substantiam, et eius valorem non in gradu et natura rei, sed ad usum hominum."

[137] *Ibid.*: "Quemadmodum licet in natura et essentia rei plus valet morbidus equus quam robustus asinus, ad usum tamen plus valet robustus asinus quam morbidus equus."

[138] *HG* I.40, 222: "Unde iuxta aequitatem iuris naturalis tantumdem res vendi debet et emi quantum valet, et si quis plus vendat eam scienter quam valet pro tempore et loco, vel minus emat, licet sic facere permittatur, proximo suo cum quo communicat, non contradicente, vel quia ignorat vel quia necessitas ad minus iustum recipere se compellit, hoc inaequalis est et iniusti. Non enim dico quod res tantum semper valeat pro tempore et loco, quantum vendi potest, vel propter inscientiam vel propter necessitatem ementis, sed quantum vendi debet, ut illud generale dictum: 'Tantum res valet, quantum vendi potest,' intelligatur non de potentia facti sed de potentia iuris,

Henry wants it both ways – to accept the fact of an impersonal, relativist market price, and yet to hold economic exchange within the bounds of individual judgment, rational order, and arithmetic equalization. But his mixing of ought and is, personal and supra-personal, arithmetical and geometrical, results in ambiguity rather than synthesis. The complexity of Henry's arguments concerning economic balance and equality indicate how connected they are to doctrinal, ethical, and metaphysical considerations at every point. These connections are perhaps most clearly seen in Henry's controversial position on the question of contracts *redditus ad vitam*.[139] Godfrey of Fontaines, as we will see, also wrote on this subject. An analysis of their conflicting views on this question throws into relief their conflicting concepts of equality and "natural" order.

It was common practice for individuals to give a flat sum of money to a monastery or other religious institution in exchange for the yearly return of a smaller fixed sum over the term of the donor's lifetime. Though this contract guaranteeing a return for life (*redditus ad vitam*) was generally held to be licit by canon lawyers, in Henry's opinion it was inescapably usurious and thus illicit. Once again Henry began his discussion of this economic question with definitions of usury and the proper use of money taken from Aristotle's *Politics* and *Ethics*. And once again, he specifically limited the form of justice governing this exchange to *iustitia commutativa* or *directiva*, determined arithmetically (*iuxta aequitatem arithmeticae*).[140]

In a contract *redditus ad vitam* Henry believed that the donor was in effect renting his money as one would rent a house for an annual return. Aquinas had provided the standard philosophical argument against the rental of money.[141] Renting money, like renting a house, implies the separation of the use of a thing from the thing itself. But money, unlike a

id est, quantum de iure pro loco et tempore vendi debet" (translation from Langholm, *Economics in the Schools*, 259). Langholm considers Henry a spokesman for the equation of market price and just price, and he cites the above as evidence for this. It seems to me that Henry is trying for much more here than a simple equation of market price and just price.

139 Henry's first quodlibetal question devoted to this question (1.39) is entitled "Utrum liceat emere redditus ad vitam." In his second quodlibet (question 15) he continued his analysis of this question ("Utrum licitum sit vendere redditus ad vitam"), specifically answering the criticisms that his first answer had generated. Langholm discusses Henry's position on this contract, but not in the context of equality. On this, see Langholm, *Economics in the Schools*, 273–75.

140 *HG* 1.39, 211–12: "Quod ut investigare possimus, notandum quod cum secundum PHILOSOPHUM in v Ethicorum iustitiae quaedam est pars distributiva, quaedam commutativa, commutativa est illa quae pertinet ad nostrum propositum. Cuius natura est, ut determinat ibidem PHILOSOPHUS, quod fiat per eandem quantitatem et dignitatem pretii rei datae et receptae, ut qui plus dedit de labore in cultura vinae, aliis eisdem existentibus, plus accipiat a paterfamilias in mercede, et qui minus, minus, et hoc iuxta *aequitatem proportionis arithmeticae*" (emphasis added).

141 In formulating this objection, Aquinas relies on the definition of money in Aristotle's *Politics*. We have already seen the central role this definition played in Henry's economic thought.

house, is inseparable from its use as a *medium* in exchange. Therefore its use cannot be rented as a separable entity. To do so is to violate equality by charging for what does not, in reality, exist.[142] Henry, always sensitive to the mathematical implications of natural equality, emphasized that such a contract went "directly against the equality of natural commutative justice, in which a perfect equality must be maintained between the sum given and the sum received."[143]

If for example someone gives a monastery one hundred pounds and in return receives fifteen pounds a year, in seven years he will receive the one hundred pounds he gave plus five pounds extra. The five extra pounds are clearly, in Henry's opinion, *ultra sortem*. The excess destroys the arithmetical equality required by natural law and is thus usurious. Furthermore, whoever gives the initial sum in a contract *redditus ad vitam* gives it in the hope of living long enough to receive something beyond it. But the hope and expectation of receiving more than the sum given is in itself clearly usurious.[144] Thus, despite the fact that this contract was in common use among religious institutions in his day, Henry's condemnation in his first quodlibet (1276) was unremitting.

In his second quodlibet Henry returned to this question. He acknowledged at the outset that his previous position had caused much displeasure and muttering against him. He asserted his right to hold economic positions that went against the decisions of both civil and canon lawyers, since the equity at the basis of economic exchange was founded on *natural law* (*aequitas iuris naturalis*) and was more properly in the sphere of theological speculation than legal decision. He was nonetheless careful to cite in this second quodlibet the authority of the canon lawyer Goffridus de Trano as well as Scripture in support of his position. Furthermore, he wrote that he now felt even more secure in his decision since important persons (*magni viri*) had personally told him they agreed with him. While such personal details are more likely to be found in quodlibetal questions than in other forms of scholastic discourse, still they are unusual here for their fullness of detail, and they attest to the great interest attending economic questions within the refined arena of late thirteenth-century scholastic debate.

---

[142] Aquinas, *ST*, 2, 2, 78, 1. For Aquinas' fullest statement of this position in *De malo*, 13, 4, and a discussion of its place in medieval usury theory, see Langholm, *Economics in the Schools*, 241–44; Ibanès, *Doctrine*, 19–20; Noonan, *Usury*, 51–57.

[143] *HG* I.39, 214: "Quod si fiat, accipit aliquid pro quo nihil dedit, et ita cum suo alienum tollit, quod est directe contra aequitatem naturalis iustitiae commutativae, in qua debet esse aequalitas omnimoda in pretio dati et recepti, ut dictum est." See also *HG* I.39, 215.

[144] *HG* I.39, 216: "Multo fortius ergo et in isto contractu quo redditus emuntur ad vitam, exspectando aliquid recipere ultra sortem, si forte supervixerit receptionem sortis. Sola enim spe et exspectatione plus recipiendi, etsi numquam aliquid reciperetur de sorte, usurae iniquitas in cupida voluntate committitur."

In this second question on the subject, Henry reinforced and clarified his earlier position that usurious intent was built into the nature of a contract *redditus ad vitam*. Recognizing, however, that the concept of natural law equity was central to this question, and indeed to all questions of economic activity, he attempted to clarify his strict definition of it. In doing so he revealed with great clarity the complex interplay of religious, ethical, metaphysical, and physical considerations underlying his conception of equality.

According to natural law, he wrote, there is but one *indivisible point* of true equality between buyer and seller. He translated this point of equilibrium into physical terms through the metaphor of the mechanical balance. The true mid-point of exchange is represented by the tongue of the scale, standing perpendicularly between the balanced arms.[145] This mid-point, although required by the logic of natural law (and necessarily consistent with the physical world governed by this law), was, he recognized, rarely achieved in human exchange. Since human knowledge is limited, we can only attempt to estimate this true point of justice. It exists as an ideal to be consciously sought by buyer and seller in good faith. We must therefore be satisfied with a small, divisible range of just price rather than insisting on an indivisible point.[146] Henry understood this permissible range to be quite small, as did Aquinas, and indeed much smaller than the legal limit of price variation allowed under the civil law rule of *laesio enormis*.[147] Clearly, he concluded, there is neither a point nor an acceptable range of exchange equality to be found in a contract of *redditus ad vitam* due to the uncertainties of life expectancy. Such a contract, therefore, violates the natural law requirement of arithmetical equality and is inescapably usurious.[148]

Godfrey of Fontaines was one of a number of thinkers who took exception to Henry's argument. His position reveals a newer theological attitude toward economic analysis, and, of great importance, an expanding conception of equality and equilibrium underlying this analysis. That

---

[145] *HG* II.15, 99: "aequitatem [aequitas] iuris naturalis, quae stat in medio indivisibili secundum naturam inter emptum et venditum, sicut lingula librae stat perpendiculariter inter brachia librae aequaliter ponderantia."

[146] *HG* II.15, 99: "(licet ex parte nostra medium illud est divisibile, quia nescimus singula ad unguem aestimare, et ideo relinquitur conscientiae ementium et vendentium, ne quis plus vendat quod suum est, quam bona fide credat pro tempore et loco debere valere, neque similiter minus emat)." Notice that both the point and the range are understood to change according to changes in the place and time of the contract. See chapter 6, n. 91, for Henry's use of this concept in the context of natural philosophy to describe a legitimate range or "latitude" in qualitative variation.

[147] Henry criticizes the laxity of the legal range of *laesio* since it is predicated on the unacceptable assumption that participants in economic exchange seek to deceive one another as to the nature of true value: *HG* II.15, 100.   [148] *HG* II.15, 100.

is not to say that Godfrey thought himself outside tradition in his economic thinking. His position on usury in the loan contract remained, in essence, traditional. Any contract was usurious in which the lender hoped or intended to receive more than the sum lent.[149] Godfrey insisted, as did Henry, that equality was the proper end of all economic exchange – it was their conceptions of equality that differed.[150]

In part this difference can be explained by Godfrey's greater acceptance of the canon law tradition in economic questions.[151] This is best seen in the central place he allowed the canon law solution of *communis aestimatio* in the determination of just price. Godfrey stressed both components of this idea: (1) value was a product of *common* judgment expressed in a market price that naturally varied according to time and place; (2) value was determined through a process of *estimation*.[152] Where Henry had recognized the role of human estimation in attempting to discover the real, pre-existent point of value equality, Godfrey identified value and value equality with the actual process of *aestimatio*. For him, *communis aestimatio* did not merely approach a pre-existent, arithmetically determinable just price; it *was* the just price.

Godfrey's redefinition of natural equality underlies his position legitimizing contracts *redditus ad vitam*.[153] In scholastic fashion, Godfrey began his quodlibetal question on this topic by stating the argument opposed to his position. He phrased it in Henry of Ghent's terms: such a contract is not licit because in all contracts of sale there must be equality between buyer and seller. In this contract, however, both parties hope to receive more in return than what they originally gave.[154] In response to

---

[149] *Godefroid de Fontaines: Quodlibet* x (GF x), J. Hoffmans (ed.), vol. iv in *Les philosophes belges*, 15 vols. (Louvain, 1924), quest. 19, 398–401.

[150] *Godefroid de Fontaines: Quodlibet* iii (GF iii), M. de Wulf (ed.), vol. ii in *Les philosophes belges*, 15 vols. (Louvain, 1904), quest. 11 (longae), 219: "debeat aequalitas secundum iustitiam conservari."

[151] De Wulf, *Un théologian-philosophe*, 72: "Sans avoir le titre de 'maître en décret' Godefroid est un juriste de premier ordre."

[152] GF iii.11, 220: "non vendat carius quam communius valere consuevit tempore solutionis, scilicet tempore pro quo vendit . . . vel parum plus vel minus, ita tamen quod non recedat notabiliter a medio secundum communem consuetudinem aestimando." Godfrey specifically cites the canon law decision *In civitate* [*Decretales Gregory IX*, v.19.6] in support of this position. Notice that, while depending on the standard legal definition of value, Godfrey cannot bring himself to accept the broad range of licit price represented by the legal principle of *laesio enormis*.
"Licet enim unicuique rei suum pretium imponere prout vult, dum tamen non excedat *notabiliter* pretium quod solet communiter aestimatione vel consuetudine statui" (emphasis added).

[153] The following argument is taken from *Godefroid de Fontaines: Quodlibet* v (GF v), M. de Wulf and J. Hoffmans (eds.), vol. iii in *Les philosophes belges*, 15 vols. (Louvain, 1914), quest. 14, 63–69, "Utrum licitum sit emere redditus ad vitam et recipere de redditibus emptis ultra sortem." I have tried as far as possible to maintain the structure of Godfrey's original argument.

[154] GF v.14, 63: "non sit licitum emere redditus ad vitam, quia in contractu emptionis debet esse aequalitas inter ementem et vendentem, sive inter rem quam dat emens et rem quam accipit a vendente. Sed in isto contractu emens dat minus ut plus recipiat, quia sperat rem quam emit esse plus valituram quam rem quam ipse dat; ergo et cetera."

this argument, Godfrey proposed that one might define such a contract as licit precisely because of the equality it represented. But the equality here would not be recognizable to Henry of Ghent. Rather, the equality Godfrey has in mind is an *equality of doubt*. Both buyer and seller are in equal doubt as to who will eventually give more and receive less in the bargain.[155] The argument between Godfrey and Henry was not over whether equality was the proper end of exchange – that was a given – but over the definition of equality.

Godfrey then stated his position. If the contract of *redditus ad vitam* is properly considered a loan contract, then it is illicit, since according to established authority, a contract of *mutuum* demands a strict arithmetical equality between sum lent and sum returned. If, however, such a contract is considered a contract of sale, then the demand for equality is satisfied by a *rational adequation*, that is, a rational agreement between buyer and seller concerning the equality of the exchange, based on estimation.[156] The equal status of the competing estimations is ensured by the equality of doubt. By defining rational adequation as an agreement based upon rational estimations, Godfrey substituted a dynamic *process* of value adequation for Henry's *a priori*, indivisible point of equilibrium.

Godfrey admitted that it was difficult to determine whether a contract of *redditus ad vitam* was a loan or a contract of sale. But once again he underlined his opposition to Henry by saying that, although he had doubts, he did not dare to condemn a contract that had been traditionally seen as licit and was in common use.[157] Where Henry felt it his duty to attack a common practice in the name of natural law, Godfrey felt it his duty to defend common practice and to demonstrate its accord with natural law. The great weight he allows community judgment and practice here is in line with his equation of *aestimatio communis* with true value.[158]

In a contract of *redditus ad vitam* one is not buying money, since to do so would clearly be against its nature and illicit in itself. One is buying instead the incorporeal right of receiving the money yearly (*ius per-*

---

[155] *Ibid.*: "Contrarium arguitur per contrarium, quia ille contractus videtur licitus in quo constituitur aequalitas inter ementem et vendentem. Sed ita contingit in proposito: nam aequaliter est dubium ex parte vendentis et ementis de plus vel minus recipiendo; ergo et cetera."

[156] *GF* v.14, 64: "Si autem sit contractus emptionis et venditionis, et fiat rationalis adaequatio inter rem emptam et pretium datum, est licitus contractus." Note that whereas Henry sought to bring all economic exchange under the arithmetical requirement for equality of the *mutuum*, Godfrey is concerned with limiting this to the loan contract.

[157] *GF* v.14, 65: "ideo non audeo iudicare illicitos contractus praedictos. Et ideo ad ostendendum quomodo possit esse licitus dictus contractus, declarandum est quomodo est ibi vere contractus emptionis et quomodo potest ibi fieri vere conveniens adaequatio."

[158] Privileging the community's judgment and estimation holds a prominent place in Godfrey's thought. See Lagarde, "Philosophie sociale," 61–69.

cipiendi), just as one can buy for a certain sum another's hereditary right to receive the annual fruits of a certain piece of land.[159] Natural law is in accord with the civil and canon law in accepting that all are free to alienate or sell what is their own. Seen in this light, the contract of *redditus ad vitam* is a contract of sale.[160] But where is the equality between price and value in such a sale? The great probability was that he who gave a certain sum of money to a monastery in exchange for an annual return would either live to receive back more or die leaving the monastery with more. There was not only no guarantee, there was effectively little possibility of establishing arithmetical equality between what was offered and what was received in return.

In response to this problem Godfrey simply asserted that such an arithmetical equality was neither required nor in fact possible to achieve in contracts of sale. Instead, in all contracts of *emptio*, equality is satisfied by a fitting *aestimatio*, made not according to any fixed value inherent in the things themselves, but according to the relative *proportions* of the usefulness of the things to the exchangers.[161] Godfrey then brought this process of estimation into sharper focus. Equality is satisfied if a number of buyers and sellers can be found willing to buy or sell at the estimate arrived at between an individual buyer and seller.[162] Again, the *sole* test of a just price is whether buyers (plural) can be found to accept it (*secundum quem ementes inveniuntur*).[163] There was truth in numbers, truth in the common, and truth in the *aestimatio* — a pure statement of value relativity in the marketplace, without any qualms that such relativity violated natural order.

Immediately after making this statement, Godfrey showed it was consistent with positive law on the subject, citing verbatim Accursius' dictum, "pretia rerum non ex affectu et utilitate singulorum, sed com-

---

[159] *GF* v.14, 65: "Ergo videtur quod possit dici in proposito quod redditus ad vitam, sive in pecunia, sive in quacumque re alia, sit ius quoddam percipiendi pecuniam vel talem rem; et est quid incorporale aliud ab ipsa re corporali. Quamvis ergo pecunia secundum se non sit emibilis, tamen ius percipiendi pecuniam vendi potest; sic etiam ius agendi ad certam pecuniae quantitatem potest vendi." Also, v.14, 66: "Nec hoc est illicitum; quia, cum emitur ius percipiendi triticum vel cum emitur fundus terrae, cum ex his statim possit haberi pecunia."

[160] *GF* v.14, 66: "Ex praedictis ergo videtur quod talis contractus sit contractus emptionis."

[161] *GF* v.14, 67: "est dicendum quod videtur posse fieri conveniens aestimatio, non quidem omnimode aequalitatis rei ad rem secundum se, sed *sufficientis proportionis* secundum quod in usum hominum natae sunt venire" (emphasis added). Note the link between the concept of proportionality and Godfrey's dynamic concept of equalization. Henry of Ghent, as we have seen, made a concerted effort to eliminate the problematic concept of proportionality from his discussion of equality and economic exchange.

[162] *Ibid.*: "ideo sufficit quod sit talis adaequatio secundum quam possit inveniri ut in pluribus emens, cum invenitur vendere volens, et hoc sive haereditarie sive ad vitam."

[163] *Ibid.*: "Etiam ita in venditione reddituum ad vitam non debet fieri secundum proportionem ad vitam ementis, quia hoc esset fieri adaequationem rei ad rem ut secundum se ad invicem comparantur, sed secundum illum modum secundum quem ementes inveniuntur."

muniter funguntur; et tantum valet res communiter quantum vendi potest."[164] But it is not enough to say that Godfrey's position resulted simply from his acceptance of the Roman and canon law tradition on economic questions. Although very knowledgeable in canon law, he was not speaking as a canonist here but as a philosopher and theologian. He had accepted a philosophical debate on this question that had been forcefully framed by Henry of Ghent in terms of natural law. Godfrey was not merely opposing the canon law solution to an ideal natural law solution. This would have accomplished little. He was demonstrating their congruence. He accomplished this not by forcing economic realities to conform to ideal definitions of natural order, but by expanding the definitions of natural order and equality to comprehend the dynamic of market exchange.

Godfrey made the link between natural law requirements and price agreement in the marketplace by arguing that economic agreements are reached through a process of estimation that is essentially *rational*, even if based on approximation and probability. Rather than seeing *redditus ad vitam* agreements as the unjust product of mutual deception, he saw them as the just product of various individuals of varying conditions with varying needs reasoning from self-interest.[165] The agreed price is thus a rational price. In this way Godfrey brought the dynamic process of price agreement into alignment with Aquinas' definition of natural law as "the participation of the rational creature in the eternal law."[166]

Both Thomas Aquinas and Henry of Ghent had insisted on a distinction between the "true" or "just" equality derived from natural law, and the practical market approximations of equality recognized by positive law. Their judgments concerning the requirements of natural law were derived from the image they held of the structure and logic of nature. The order that dominated their image of nature was everywhere and always an order imposed by mind. Within this vision, economic equality could be attained only through the conscious effort of individual exchangers to seek it. Equality as the product of the give-and-take of exchange, or price as the product of the supra-personal process of *aestimatio communis*, could not meet the test of natural equality. Godfrey denied this distinction. He

---

[164] *Ibid.* It is interesting that Godfrey makes an exception to this legitimization of market price in certain cases regarding food; see v.19, 68: "Cum enim alimenta sint necessaria ad vitam et pauperibus relinquuntur, non debent vendi nisi pro pretio quo probabiliter vita poterit sustentari, ne fame pereant contra intentionem relinquentis alimenta."

[165] *GF* v.14, 67: "ideo sufficit quod sit talis adaequatio secundum quam possit inveniri ut in pluribus emens, cum invenitur vendere volens . . . Et debent aliae circumstantiae considerari, sive quantum ad iuvenes, sive quantum ad senes, secundum quas propter necessitatem vel commoditatem vendentis, possunt inveniri tam iuvenes quam senes qui tales redditus emere velint."

[166] Aquinas, *ST*, I, 2, 91, 2.

did so not because he was concerned with practical rather than with philosophical implications in the manner of the legists, nor because his philosophical conception of a proper equality was less developed than that of his fellow theologians. He did so because he had expanded his conception of natural equality to bring it into line with the realities of market exchange.

Thomas Aquinas and Henry of Ghent were far from alone in recognizing that many of the active principles governing the economic order were outside and even opposed to traditional conceptions of "natural" order and equilibrium.[167] The Church's continuing position toward usury was in part a conscious effort to control economic activity and to bring it into line with natural law definitions of justice and equality. However, as the place of the monetized marketplace grew within society, and the recognition of its importance to the life of the *civitas* grew as well, traditional theological distinctions between natural order and market order, between natural equality and market equality, created intensifying intellectual tensions. How could something so central to the good of the community be unnatural? Godfrey of Fontaines was one of a number of thinkers of his time who resolved this tension by recognizing market exchange as a system governed by its own proper logic and possessing its own proper equilibrium – but a logic and equilibrium that were in themselves rational and natural. In doing so they expanded and reshaped the image of natural order to comprehend the logic of the marketplace.

At the cusp of the fourteenth century, a conception of nature incorporating the geometric model of market equalization was taking shape. The new model was concerned more with process than result, more with relations than perfections. It operated within ranges and mathematical limits rather than according to a precise numbered plan. It was defined more by the expanding and contracting line than by the point, and was therefore better described by geometry than arithmetic. The result was a conception no less rational than the earlier one, even though more accepting of estimation, approximation, and probability. It was within this new conception of nature that scientific thought emerged.

[167] Langholm, *Price and Value*, 87.

Chapter 5

EVOLVING MODELS OF MONEY AND MARKET EXCHANGE IN THE LATE THIRTEENTH AND FOURTEENTH CENTURIES

The scholastic analysis of money, exchange, and market value, already well underway in the thirteenth century, continued to evolve over the fourteenth century, with the commentaries on Aristotle's *Ethics* and *Politics* providing the most important loci for this discussion.[1] In this chapter I discuss the advances made within each of the six categories of connection between scholastic economic thought and proto-scientific speculation. I remind the reader that these categories are heuristic constructs. My use of them to isolate the core insights underlying the scholastic model of economic exchange should not obscure the consistent, even necessary connections between these insights in the writings considered in this chapter. Dynamic equilibrium, relativity, the measuring continuum, and common valuation all work together within the new social geometry of the fourteenth century.

EQUALITY, THE MEAN, AND EQUALIZATION IN EXCHANGE

Through the fourteenth century, the controlling principle in economic thought remained equality. Without exception, thinkers defined economic exchange as a process of equalization. In the restricted case of the loan or *mutuum*, ideal equality continued to be determined in most cases

---

[1] Editions of fourteenth-century commentaries cited in this chapter are: Jean Buridan, *Quaestiones in decem libros ethicorum Aristotelis ad Nicomachum* (Oxford, 1637); Buridan, *Quaestiones super octo libros politicorum* (Oxford, 1640); Walter Burley, *Expositio in Aristotelis ethica Nichomachea* (Venice, 1500); *Geraldus Odonis: Expositio in Aristotelis ethicam* (Brescia, 1482; Venice, 1500); *Maistre Nicole Oresme: Le livre de politiques d'Aristote*, Albert Douglas Menut (ed.) (Philadelphia, 1970); Oresme, *Le livre de ethiques d'Aristote*. Manuscripts of cited commentaries: Aegidius Lessinus (?) *Quaestiones super libros ethicorum*, ms. Vat. lat. 2172 (1r–59r); Albert of Saxony, *Expositio super libri ethicorum*, ms. Bib. Maz. 3516 (84–156); Walter Burley, *Expositio super octo libros politicorum*, ms. Balliol 95 (161r–232r); Gilles d'Orleans (Aegidius Aurelius), *Super ethicam*, ms. BN lat. 16089 (195r–237v); Guido Terrenia, *Super ethicam*, ms. BN lat. 3228 (1r–55v); Henricus de Frimaria, *Sententia totius libri ethicorum*, ms. Vat. lat. 2169 (1–284v); Richard Kilvington, *Quaestiones morales super libros ethicorum*, ms. Vat. Ottob. lat. 179 (25r–59r).

arithmetically – the sum lent determined the sum that could be required in return, with no allowance for proportionality.[2] However, in the broader realm of economic exchange that included buying (*emptio*) and selling (*venditio*), traditional requirements for arithmetical equality came under increasing challenge. Thinkers began to question the knowability and even the existence of definable points of exchange equality. They began to formulate alternative models of economic equalization based no longer on knowable points but on lines or legitimate ranges, no longer on arithmetical addition and subtraction but on the acceptance of proportional and (in Aristotelian terms) geometrically determined requital in exchange.

The main sources of economic theory in this chapter are taken from Aristotelian commentaries, but there were a number of thinkers working outside the commentary tradition who made important early contributions to the emerging geometric model of equalization. The two most influential were Peter John Olivi (1248–98) and John Duns Scotus (c. 1266–1308). Of these two, Olivi's analysis was first, both in time and in the range and acuity of his economic observations. Olivi's contributions reveal with particular clarity the complex play of forces in medieval economic thought. Of all scholastic writers on economic questions, he was perhaps the most willing to confer rationality and liceity on current economic practices. At the same time, he was a leader of the rigorist party of Franciscans – sworn to poverty, opposed to Franciscan ownership of property in any form, convinced that the preoccupation with the temporal and the sensible was the path toward error.[3] He was a critic of Aristotle because he saw in Aristotelian empiricism an overvaluation of the things of this world, and yet his own observations of economic life were unsurpassed in their accuracy and attention to detail.[4]

Until recently, Olivi's economic thought was known primarily through its appearance, at times copied word for word, in the fifteenth-

---

[2] Alexander Lombard's early fourteenth-century treatise on usury provides a restatement of this position. See Lombard, *Un traité de morale économique au XIVe siècle: Le "Tractatus de usuris" de maître Alexandre d'Alexandrie*, A.-M. Hamelin (ed.) (Louvain, 1962). For the central place of arithmetical equality in usury theory in the fourteenth century, see Lombard, *Un traité*, 79.

[3] The standard biography of Olivi is F. Ehrle, "Petrus Johannis Olivi: sein Leben und seine Schriften," *Archiv für Litteratur und Kirchengeschichte des Mittelalters* 3 (1887), 409–552. For Olivi's place within the party struggles of the Franciscans, see David Burr, *Olivi and Franciscan Poverty: The Origins of Usus Pauper Controversy* (Philadelphia, 1989). Despite his rigorist position, Olivi recognized that poverty should be practiced relative to person and position. Rather than impose exterior restrictions on use, he sought a solution based in personal intention. On this, see Langholm, *Economics in the Schools*, 349.

[4] Olivi's attitude toward Aristotle and empiricism is discussed by David Burr, *The Persecution of Peter Olivi* (Philadelphia, 1976), esp. 12–16; O. Bettini, "Olivi da fronte ad Aristotile," *Studi francescani* 55 (1958), 176–97. For the contrast between Olivi's acceptance of current mercantile practices and his severity on the position of Franciscan poverty, see Spicciani, *La mercatura*, 184–86.

century sermons of St. Bernardino of Siena (1380–1444). Bernardino did not publicly acknowledge his debt to Olivi, and it is only modern scholarship that has uncovered it. Perhaps the best way to illustrate Olivi's precocity is to note that his economic writings have been regarded by twentieth-century economic historians as extraordinarily sophisticated and forward-looking, even when they were mistakenly thought to have been composed by St. Bernardino, a century and a half after their actual date of formulation.[5] While Olivi's writing is remarkable for its perception and understanding of actual economic practice, the more historians discover about the range and sophistication of economic thought in the thirteenth century (and particularly the influence of Roman and canon law on that thought), the less novel some of his insights appear.[6] As we saw in chapter 4, the primary elements of the new approach to market equalization (clearly visible in the work of Godfrey of Fontaines) were the recognition of the role of *aestimatio* and *probabilitas* in the determination of price, the acceptance of mutual agreement as the prime determinant of "just" equality in exchange, and the integration of these essentially indeterminate factors with the scholastic requirement of a governing *ratio*.[7] One can see echoes of this cluster of ideas in Olivi's precocious definition of *capitale* and in his striking, if conditional, acceptance of a multiplying power in money.[8]

---

[5] See for example Raymond De Roover's assessment in *San Bernardino of Siena and Sant' Antonino of Florence: The Two Great Economic Thinkers of the Middle Ages* (Boston, 1967). De Roover had considered San Bernardino "one of the greatest economists of all time," until he was made aware that Bernardino had copied his most important statements on value and price determination almost word for word from a manuscript he possessed of Olivi's *Tractatus de emptionibus et venditionibus, de usuris, de restitutionibus*. On this, see De Roover, *San Bernardino*, 19. Olivi's authorship of this manuscript is established in D. Pacetti, "Un trattato sulle usure e le restituzioni di Pietro di Giovanni Olivi falsamente attribuito a fr. Gerardo da Siena," *Archivum franciscanum historicum* 46 (1953), 448–57. For reasons why San Bernardino would be unable to credit Olivi publicly for his insights, see 124–25. For a bibliography of writings by and about Olivi, see Servus Gieben, "Bibliographia Oliviana," *Collectanea franciscana* 38 (1968), 167–95.

[6] For two differing assessments of Olivi's originality, see the very positive assessment of Giacomo Todeschini, "Oeconomica Franciscana II: Pietro di Giovanni Olivi come fonte per la storia dell'etica-economica medievale," *Rivista di storia e letteratura religiosa* 13 (1977), 461–94. For a position stressing Olivi's debt to thirteenth-century economic thought more than his originality, see Julius Kirshner and Kimberly Lo Prete, "Peter John Olivi's Treatises on Contracts of Sale, Usury, and Restitution: Minorite Economics or Minor Works?," *Quaderni fiorentini* 13 (1984), 233–86, esp. 259. Similarly, see Kirshner, "Storm Over the 'Monte Comune': Genesis of the Moral Controversy over the Public Debt of Florence," *Archivum fratrum praedicatorum* 53 (1983), 219–76, 236. Langholm (*Usury*, 43–44, 116) also leans more toward tradition than innovation in his assessment of Olivi's thought. Spicciani (*La mercatura*, 192–93) recognized considerable originality in Olivi's treatment of usury and market value, while at the same time recognizing his debt to the legal tradition.   [7] See chapter 4, 110–14.

[8] All citations from Olivi in this work are taken from the following two editions of his economic questions: (1) *Quodlibet* I, quests. 16 and 17, and *De contractibus usurariis*, in Amleto Spicciani (ed.), "Gli scritti sul capitale et sull'interesse di Fra Pietro di Giovanni Olivi," *Studi francescani* 73 (1976),

Equality, the mean, and equalization in exchange

Olivi's concept of *capitale* descended from a long-debated question within canon law: whether the lender who could have invested his money in commerce at a probable profit should be indemnified for the loss of that probable profit (*lucrum cessans*). As we have already seen, the influential legal scholar of the middle of the thirteenth century, Hostiensis, generally accepted an indemnification for *lucrum cessans*, while Aquinas denied it by making a strict distinction between what was merely probable (the profit expected from investment) and what was real (the actual sum lent).[9] Olivi asked the question: if someone intends to invest his money in trade for a profit, and instead, out of charity, lends the money to a friend in need, can he expect back from his friend not only the sum lent but in addition the profit he lost by not investing in trade? Olivi's answer to this question was an unqualified yes: the borrower was responsible for indemnifying the lender for his loss of "probable profit," and for restoring a "probable equivalence" (*probabiliter equivalens*) to the loan contract. The required indemnity *ultra sortem*, however, applied only when the money lent had been previously intended for profitable investment in trade. Only in this case could the probable profit the money would have earned be considered as existing, in some sense, as a superadded value (*valor superadiunctus*) within the loaned money itself.[10]

The question of greatest interest from the perspective of investigating connections between economic and scientific thought is how Olivi found it possible to bring his dynamic and seemingly open-ended concept of *capitale*, with its base in probability and its acceptance of superadded value, into a scholastic scheme still preoccupied with the ideal of

---

289–325; (2) *Tractatus de emptionibus et venditionibus, de usuris, de restitutionibus*, in Giacomo Todeschini (ed.), *Un trattato di economia politica francescana: il "De emptionibus et venditionibus, de usuris, de restitutionibus" di Pietro di Giovanni Olivi*, fasc. 125–26 (Rome, 1980). An earlier partial edition and Italian translation of Olivi's *Tractatus* can be found at the conclusion of Spicciani, *La mercatura*.

[9] See chapter 4, 84-86, for a discussion and bibliography on this question. Acceptance of *lucrum cessans* continued to be sporadic. Olivi's fellow Franciscan, Alexander Lombard, writing nearly a generation after him, rejected it as a legitimate title to *interesse*. So too did the important fourteenth-century canonist Johannes Andreae. For more on the history of this doctrine, see Kirshner and Lo Prete, "Olivi's Treatises," 270–73; Kirshner, "Storm," 253; Noonan, *Usury*, 120–26; Langholm, *Usury*, 98.

[10] Olivi, *Tractatus*, 85: "is cui prestatur tenetur sibi ad probabiliter equivalens, seu ad preservandum ipsum a damno probabilis lucri, tum quia illud quod in firmo proposito domini sui est ordinatum ad aliquod probabile lucrum non solum habet rationem simplicis pecunie seu rei, sed ultra hoc quamdam *seminalem rationem* lucrosi quam communiter capitale vocamus, et ideo non solum habet reddi simpliciter valor ipsius sed etiam *valor superadiunctus*" (emphasis added). Olivi (*Tractatus*, 84) recognized that *lucrum cessans* was the logical continuation of *damnum emergens*, which was generally accepted as a title to interest.

equality (not profit) as the governing principle of exchange.[11] Olivi recognized that there was a problem here. In the *De contractibus usurariis* he examined moral questions raised both by his conception of *capitale* and by the practice that was common in the commerce of his day (on which he based his conception) of charging for the use of productive capital. He posed this question: if I lend a merchant *capitale* at interest so that he can in turn invest it and profit from it, is it usurious if I receive interest on my loan? When I receive interest *ultra sortem*, my gain is guaranteed, while the merchant, because of the uncertainty of trade, could lose on the transaction. Can such an exchange satisfy the requirements of equality? Furthermore, by agreeing to pay me back more than the sum (*capitale*) I lent, the merchant has, in effect, bought the right to whatever profit he can make from my money. But since his profit is in the future he has no way of making a rational decision as to whether or how much he will benefit from the transaction, and both equality and rationality are essential to proper, non–usurious economic transactions. Olivi asked whether this transaction was clearly usurious, given these built-in uncertainties and inequalities.[12] In good scholastic fashion he first argued that it was. He began with Aquinas' objection to any equalization concerning the merely probable. The lender by selling probable future profit was selling what in fact did not exist – something clearly against natural law and the requirements of equality. Moreover, the lender was selling what was not his to sell, i.e., the industry of the borrower, which was the real productive cause of any profit from the loan, rather than the money itself.[13]

Olivi's response to these traditional objections involved both a critical reevaluation of the question of probability and a refinement of all the elements we have seen in Godfrey of Fontaines' response to Henry of

[11] For a fuller discussion of the history of the concept of *capitale* and the distinction between *capitale* and all other uses of money, particularly money as measure, see Kirshner and Lo Prete, "Olivi's Treatises," 267–72 and nn. 65–68. Kirshner and Lo Prete are careful to demonstrate the strict limits of this concept for Olivi. It is noteworthy that Todeschini attributed Olivi's position on *capitale*, as well as his acceptance of *lucrum cessans*, to his full comprehension of the implication of the idea of *equalitas proportionum* in exchange ("Oeconomica II," 490–94).

[12] Olivi, *De contractibus usurariis*, 321–22: "Si vero ponatur casus, qui in quibusdam terris inter mercatores saepe contingit, in quo quidem capitale currit in mercando ad periculum tradentis ita quod, quandocumque inde in mercando vel alio modo inculpabili perditur, ammittitur ei qui tradidit capitale; lucrum tamen est ei fixum et certum, quia mercator, cui traditum est illud capitale, emit totum futurum lucrum praedicti capitalis tanto pretio quantum probabilitas futuri lucri potest ante proventum lucri rationabiliter aestimari, videtur quibusdam hic esse usuram quadruplici ratione."

[13] Olivi, *De contractibus usurariis*, 322: "id quod non est in aliquo suo causaliter vendit ac si esset causaliter in illo seu proventurum ex illo, eandem inaequalitatem committit quae reperitur in contractu usurae . . . scilicet quod ibi venditur id quod non est ac si esset, aut quia quod non est suum venditur quasi suum: sed praefatum lucrum non provenit causaliter ex praedicto capitali, sed potius ex mercantis industria et actu." For the authority of this objection, see Aquinas, *ST*, 2, 2, 78, and chapter 4.

Ghent on the question of contracts *redditus ad vitam*.[14] Olivi declared, in opposition to Aquinas, that the *probability* of profit has a certain *real* existence and value in itself. His acceptance here of approximation and probability in an economic context and his integration of probability into the controlling scholastic construct of rational adequation mark a great shift in the philosophical conception of equality and equalization. Though worked out within economic thought and motivated by the desire to bring actual economic practice into alignment with philosophical definitions of rationality, this shift would have momentous implications for the redefinition of equality and equalization in other realms of scholastic thought, including natural philosophy. The integration of probability and estimated ranges into rational models of equilibrium is one of the great achievements of proto-scientific thought in the fourteenth century.[15]

Olivi recognized that the merchants of his day expected to make a profit when they borrowed a sum of money to invest in trade. They therefore expected to repay the lender more than the sum they borrowed (*ultra sortem*) in recognition of the profit they considered probable on the loan. Olivi, following the merchants of his day (and in contrast to Aquinas), allowed this probability of profit a real, if discounted, value. Rather than condemning the buying and selling of "merely" probable value as being contrary to nature, he reasoned that since merchants routinely estimated the value of probable profit and measured it in monetary terms (the amount *ultra sortem* they were willing to pay for the loan), then it was licit for them to buy and sell it for a monetary price.[16] In this view, probable profit has a real and measurable existence as a kind of fructifying power within the *capitale* itself.[17] Olivi further recognized that merchant borrowers paid less for the probable profit attached to the borrowed capital (i.e., the interest they agreed to pay on the loan) than they believed they would make in actual profit by investing it. In other words, he recognized that merchants had learned how to rationally discount the probable.[18] The borrower, being knowledgeable in the art of

---

[14] See the conclusion to chapter 4, 110–14.   [15] See chapter 7, 214–20.

[16] Olivi, *De contractibus usurariis*, 323: "appretiabilis valor probabilitatis seu probabilis spei lucri ex capitali illo per mercationes trahendi. Ex quo enim haec probabilitas habet aliquem valorem aliquo temporali pretio appretiabilem, potest licite illo pretio vendi."

[17] *Ibid.*: "Ergo praedictum interesse probabilis lucri quodammodo causaliter et quasi seminaliter continebatur in praedicto capitali in quantum capitali, alias enim non posset licite exigi." In contrast, money as the price and measure of goods was conceived by Olivi in the Aristotelian sense as a fixed measure set by law. See Kirshner and Lo Prete, "Olivi's Treatises," 267.

[18] Olivi, *De contractibus usurariis*, 323: "ex quo probabilitas illa minori pretio venditur quam lucrum ex mercationibus capitalis credatur suo tempore futurum et valiturum, constat quod in eius venditione semper creditur probabiliter quod emptor eius sit finaliter lucraturus seu plus quam in emendo dedit habiturus." On this, see Langholm, *Usury*, 135–36.

trade (so Olivi assumed), had learned to estimate the probable profit to be made from borrowed capital and, weighing the risks involved, could be expected to buy it for a rational price. In Olivi's vision, the guarantee of equality and rationality in this type of open-ended loan, in which greatly varying degrees of risk and uncertainty are involved, is provided solely by the voluntary mutual agreement to the terms of the loan.[19] Mutual agreement implies mutual recognition of gain, which, in itself, provides the basis of equality in exchange.

For Olivi, the multiplying or fructifying power in *capitale* is not present in money *per se*, but only in money intended for investment in commerce. In his view, this multiplying power can be conveyed to money (in itself definitionally sterile) through the energy, industry, and skill of the commercial borrower. His concept of *capitale* grew out of his recognition that bringing the world of commerce within the bounds of philosophical rationality required the intellectual imagination of a new form of equalization, one based on multiplication, estimation, and probability. Such a form existed in the marketplaces of the towns of southern France and Italy (where Olivi lived most of his life) *before* it found expression in scholastic economic thought. Olivi makes this clear when he notes that the concept of productive capital was in common use (*communiter capitale vocamus*) before his decision to provide it with a philosophical rationale.[20] With the great expansion of commerce in the thirteenth century, and the equally great successes of the calculating merchant, the rationality of an exchange system built around probability and superadded values impressed itself on the minds of observers – even observers as far removed from practical commercial activity as mendicant friars. The task Olivi and other thinkers accepted was to formulate new models of explanation, consistent with the scholastic requirements of *ratio* and *equitas*, capable of comprehending the new forms of rationality they perceived in economic activity.

The combining of old and new can be seen in Olivi's choice for a term to represent the multiplying power added to *capitale* through the industry and skill of the merchant. He chose *rationes seminales*, a philosophical term pertaining to potentiality in matter with a long pedigree in Christian theology and natural philosophy stretching back to St. Augustine.[21] By

---

[19] Olivi, *De contractibus usariis*, 324–25: "dicendum quod immo causaliter, seu aequivalenter aut praevalenter ex ipso educitur, pro quanto scilicet futurum lucrum suarum mercationum iam quasi esse in ipso praesupponitur et tamquam iam praesuppositum venditur et emitur; et certe ipse emptor cum sit in arte mercandi et lucrandi industrius et voluntarius non emeret illud lucrum nisi bene sciret illius emptionem probabiliter esse sibi lucrosam." [20] Olivi, *Tractatus*, 85.

[21] *Ibid.*: "sed ultra hoc quamdam *seminalem rationem* lucrosi quam communiter capitale vocamus, et ideo non solum habet reddi simpliciter valor ipsius sed etiam *valor superadiunctus*" (emphasis added). See Langholm, *Economics in the Schools*, 371–73.

using a traditional concept to describe a potentially destabilizing economic power, he brought the rationality of economic life into line with traditional philosophical definitions of rational order. In applying to economic activity a term traditionally used to make sense of nature and natural activity (from Plato onward), Olivi specifically linked the new dynamic of economic order to the order of nature.

In works spanning more than two decades, the intellectual historian Michael Wolff has investigated Olivi's transference of terms and concepts between the spheres of economic and physical thought.[22] He has focused on the concept of productive capital, particularly Olivi's explanation of how the productive labor of the merchant is transferred to money itself, turning sterile coin into fruitful *capitale*. Wolff believes that it was while searching to explain the capacity of money to absorb and hold the *virtus* of the commercial investor that Olivi came upon the physical analogy of objects set in motion by persons who strike or throw them.[23] In a series of questions exploring the transference of *virtus* in this physical context (again described through the term *ratio seminalis*) Olivi arrived at what is often considered the first formulation of the concept of *impetus* in the Latin West.[24] On the basis of verbal and formal similarities between the concepts of *capitale* and *impetus*, Wolff argues for a "functional tie" between Olivi's speculations in economic ethics and his forward-looking contributions in the realm of scholastic natural philosophy.[25] Behind both sets of concepts Wolff sees a new "paradigm of dynamic causality" at work – a paradigm that fits very closely the proto-scientific developments in natural philosophy considered in the concluding chapters of this work.[26]

In his *Tractatus de emptionibus et venditionibus*, Olivi presents his clearest statements on the relationship between reason, agreement, value, and adequation in exchange. Since he recognized that economic value and price are determined with respect to human utility and not with respect to absolute or essential value, he maintained that in every agreement on price there is, *ipso facto*, an equalization of utility between buyer and seller.[27] The equality of justice in exchange is satisfied when both parties believe they are acquiring the same quantity of utility. Implicit in Olivi's

---

[22] Michael Wolff, *Impetustheorie*; Michael Wolff, "Mehrwert."
[23] Michael Wolff, *Impetustheorie*, 178–88, 245.
[24] *Petrus Johannis Olivi: Quaestiones in secundum librum sententiarum*, Bernhard Jansen (ed.) (Quaracchi, 1922), quests. 23–31, esp. q. 31: "An omnia, quae educuntur de potentia materiae, sint ibi prius secundum suas essentias seu secundum rationes seminales"; Michael Wolff, *Impetustheorie*, 184–94.
[25] Michael Wolff, "Mehrwert," 419–23, esp. 419.   [26] Michael Wolff, "Mehrwert," 422.
[27] Olivi, *Tractatus*, 55: "valor et pretium rerum venalium potius pensatur in respectu ad nostrum usum et utilitatem quam secundum absolutum valorem suarum essentiarum. Item equitas iustitie commutative est ut quantum utilitatis mihi confert, tantum idem tibi conferam."

scheme is the conception of market exchange as a system in which human utility (*utilitas*) and need (*indigentia*) are continually measured and quantified by changing market price. In effect, utility and need function here as variable (intensible) qualities attached to goods in exchange, expanding and contracting according to differences in time, place, person, and condition. I use the word "intensible" here because this is the term that came to be commonly used in scholastic natural philosophy to denote a quality capable of expansion and contraction along a continuum of value. Both *utilitas* and *indigentia* in Olivi's economic thought fit this definition. I emphasize the recognition of the quantification of qualities by price in Olivi's economic thought because of the important role such quantification played in the proto-scientific speculation of the generations following Olivi.

Olivi fully realized that the equality reached in exchange was based only on estimates, albeit rational ones.[28] It is in the realm of the analysis of the limits of approximation and estimation that he made some of his most acute observations:

The judgment of the value of a thing in exchange seldom or never can be made except through conjecture or probable opinion, and so not precisely, or as if understood and measured by one indivisible point, but rather, as a *fitting latitude* [*sub aliqua latitudine competenti*] within which the diverse judgments of men will differ in estimation. And such a latitude will therefore contain various *degrees* [*gradus*] and little certainty, and much ambiguity attached to the estimates, with some greater and some less [emphasis added].[29]

Henry of Ghent's knowable, mathematical point of equality is completely abandoned here. The fluidity and dynamism in Olivi's equation of approximates is unmistakable, as is the language of latitudes, lines, and degrees used to describe it. With the "latitude" of just price, Olivi philosophically integrated the implicit legal range of *laesio enormis* into the scholastic requirement for *equalitas*.[30]

Although Olivi's thinking can be taken as indicative of the high level of market understanding existing in the late thirteenth century, it is

---

[28] Spicciani, *La mercatura*, 194.

[29] Olivi, *Tractatus*, 53: "Item tertio sciendum quod huiusmodi pensatio valoris rerum usualium vix aut numquam potest fieri a nobis nisi per coniecturalem seu probabilem opinionem, et hoc non punctualiter, seu sub ratione seu mensura indivisibili in plus et minus, *sed sub aliqua latitudine competenti*, circa quam etiam diversa capita hominum et iuditia differunt in extimando. Et ideo varios gradus et paucam certitudinem, multumque ambiguitatis iuxta modum opinabilium in se includit, quamvis quedam plus et quedam minus" (emphasis added).

[30] For the history of the use of the word *latitudo* in scholastic debate, particularly its use by Aquinas and St. Bonaventure in discussions of the mean of virtue, see chapter 6, 183ff., and O. Lottin, "La connexion des vertus chez Saint Thomas d'Aquin et ses prédécesseurs," in his *Psychologie et morale aux XIIe et XIIIe siècles*, 6 vols. (Gembloux, 1957), vol. III, pt. 2, 197–252, esp. 223–28.

extremely difficult to gauge his influence on economic thought before the popularization of his ideas in the sermons of San Bernardino more than a century later. (Even here their true source remained uncited.) Due to Olivi's identification as a leading theoretician of the spiritual Franciscans, a year after his death (1299) a number of his works were condemned and burnt for political and doctrinal reasons having nothing to do with his economic positions. His reputation for religious orthodoxy continued to suffer after his death, culminating in 1319 when his entire corpus was censured and declared forbidden.[31] No succeeding economic thinker, including San Bernardino, could admit to reading or being influenced by Olivi's writings.[32] Nevertheless, one can find evidence of Olivi's thought "peeping out" in the work of later theorists, particularly that of his fellow Franciscans.[33] Two of Olivi's most important insights concerning the process of market equalization – insights that were particularly applicable to the emergence of scientific thought – were essentially repeated by Duns Scotus, the most influential thinker of the generation following Olivi.

Duns Scotus, like Olivi and like Godfrey before him, legitimized the idea of equality in exchange *secundum rectam rationem* defined according to human use rather than the inherent value of the things exchanged.[34] Duns Scotus also recognized that equalization was a dynamic process determined through mutual agreement. Since exchange was rooted in necessity, both parties had to be willing at times to give more or take less than what they might perceive to be the just value of the things exchanged. Without such a willingness to give and take, agreement, though necessary to life, would be difficult to attain.[35] While eliminating the old ideal of

---

[31] Burr, *Olivi and Poverty*, 125–26; De Roover, *San Bernardino*, 19.

[32] For the tangled history of the transmission of Olivi's economic thought, see Spicciani, *La mercatura*, 181–89, and for a discussion of Olivi's place in economic history, 181–219; Langholm, *Usury*, 26.

[33] The word *spunti* is Spicciani's in regard to Olivian insights in the work of Duns Scotus, *La mercatura*, 135. Olivi's influence is more certain on the Franciscan general Geraldus Odonis. Both Langholm and Spicciani see Odonis' treatise on economic contracts (Siena, U.V.8) as leaning heavily on Olivi's work; Langholm, "Scholastic Economics," 124; Spicciani, *La mercatura*, 188.

[34] All citations from Duns Scotus unless otherwise noted are from John Duns Scotus, *Quaestiones in quattuor libros sententiarum* (*Opus Oxoniense*), L. Wadding (ed.), vol. XVIII of *Opera omnia* (Paris, 1894), d. xv, q. 2, 255–357. Duns Scotus, *Sent. iv*, 282b–83b: "Quod domini rerum juste eas permutant, si sine fraude servent aequalitem valoris in commutatis secundum rectam rationem . . . quia frequenter res, quae in se est nobilior in esse naturali, minus est utilis usui hominum, et per hoc minus pretiosa . . . Et propter hoc additur, *secundum rectam rationem*, attendentem scilicet naturam rei in comparatione ad usum humanum, propter quem fit commutatio ista."

[35] Duns Scotus, *Sent. iv*, 283b–84a: "Quandoque autem relinquitur ipsis commutantibus, ut pensata mutua necessitate reputant sibi mutuo dare aequivalens hinc inde, et accipere; durum est enim inter homines esse contractus, in quibus contrahentes non intendant aliquid de illa indivisibili justitia remittere sibi mutuo, ut pro tanto omnem contractum concomitetur aliqua donatio."

strict arithmetical equality, Duns Scotus added a religiously inspired guide to just exchange unmentioned by Olivi. He wrote that in all probability the requirements for just equality will be satisfied if both parties do for the other what they would like to have done for themselves.[36]

The most direct link between Duns Scotus and Olivi was their shared concept of a "latitude" (*latitudo*) of just price. Duns Scotus specifically criticized Henry of Ghent's position that there was one knowable indivisible *point* of equality in exchange.[37] Rather, he maintained, equality was determined within a great *latitude* (*magna latitudo*) of equivalence between the amounts of utility attached to things in exchange. Commutative justice was satisfied if price fell within the *degrees* (*in quocumque gradu*) of this latitude.[38] In considering the degree of the extension of this latitude, Duns Scotus related it either to the legal limit of *laesio enormis*, or to previously established community limits on the just range of price.[39] Clearly both Olivi and Duns Scotus conceived of value as an extendable continuum measured by money.[40]

Duns Scotus' conception of equalization included elements of intervening subjectivity, particularly his association of just agreement with the golden rule. Olivi hesitated less in seeing the market as a *mechanism of equalization* that ordered itself independently of the intention of its individual participants. The clearest illustration of this recognition is found in his response to the old question of whether it was licit to charge a very high price for grain in times of great scarcity. Olivi's answer was yes.[41] But rather than making a distinction between market forces and concern for the social good as Aquinas had, Olivi showed them to be mutually reinforcing. Rather than setting limits on market-determined behavior, or defining it as antisocial, Olivi deemed it rational and constructive. He made the striking point that greater damage to the common

---

[36] Duns Scotus, *Sent. iv*, 284a: "Et si iste est modus commutantium, quasi fundatus super illud legis naturae: *Hoc facias alii, quod tibi vis fieri*, satis probabile est, quod quando sunt mutuo contenti, mutuo volunt sibi remittere, si secundum aliquid deficiunt ab illa justitia requisita."

[37] Duns Scotus, *Sent. iv*, 283b: "Ista autem aequalitas secundum rectam rationem non consistit in indivisibili, sicut dicit quidam Doctor."

[38] *Ibid.*: "quod justitia commutativa respicit, est *magna latitudo*, et intra illam latitudinem non attingendo indivisibilem punctum aequivalentiae rei et rei, quia quoad hoc, quasi impossibile esset commutantem attingere, et in quocumque gradu circa extrema fiat, juste fit" (emphasis added).

[39] *Ibid.*: "Quae autem sit ista latitudo, et ad quantum se extendat, quandoque ex lege positiva, quandoque ex consuetudine innotescit." See Langholm, *Economics in the Schools*, 410.

[40] Both Olivi and Duns Scotus use the terms *latitudo* and *gradus* to describe the continuum of value measured by money. Both as well were early contributors to the proto-science of *calculationes* in which the *latitudo formarum* became the primary conceptual measurement of qualities. This connection is considered further in chapter 6.

[41] Olivi, *Tractatus*, 58, does make a strong distinction (also made by Aquinas) between a legitimate price rise due to an increase in common need (market price), and the illegitimate increase of price to take advantage of a particular individual's particular need.

good would be caused if prices did *not* rise during famine. It was precisely the rise in price, pegged to the rise in scarcity, that induced the possessors of the grain to sell to the community rather than hoard for their own use during times of famine.[42]

In Olivi's vision, the community was served rather than damaged by the natural workings of the market, and the common good was served rather than damaged by the desire for wealth on the part of individual producers. While such a view was unusually blunt in the context of scholastic economics, and quite out of touch with the Aristotelian distrust of the desire for riches, it was consonant with current understanding and practice. There is considerable evidence that the role of rising prices in the diminution of hoarding in times of famine was well understood outside the university – both by officials responsible for victualing the growing cities and by peasants and merchants supplying the market – before it was enunciated by Olivi.[43] It is no surprise that, where scholastic economic thought proves most penetrating, it is most closely tied to current economic practice and understanding. Only scholastic thought, however, was called on to synthesize practice and principle; it alone was capable of interpreting current practice in the light of ideal philosophical models of form and activity.

As we turn to consider fourteenth-century additions to the concept of market equalization found in the commentaries on Aristotle's *Ethics* and *Politics*, it is important to stress the essential conservatism of the commentary tradition and the continuing predominance of the early commentaries of Albert and Aquinas within that tradition. Olivi's economic writings were remarkable for their time in conveying a sense of the author's direct observation of current economic practice. This immediacy is seldom present in the Aristotelian commentaries. Again and again, possible new avenues are dropped and discussions cut short when they approach problems and questions not previously considered within the tradition. Often the same bare examples, the same analyses, indeed the same phrases we have already examined in the work of Aristotle, Albert,

---

[42] *Ibid.*: "quia si tunc pretium non habet augeri, hoc ipsum esset in preiuditium boni communis. Quia habentes non sic prompte communiter non habentibus et egentibus res huiusmodi vendere vellent; et ideo minus bene communi egestati provideretur." Langholm (*Economics in the Schools*, 361) calls this position "an early (and in scholastic economic sources perhaps the first) clear expression of an *instrumental* approach to the theory of price."

[43] Ibanès, *Doctrine*, 22–23. See *Ordonnances*, vol. I, 424–26, and chapter I for a discussion of laws concerning market prices and hoarding. The *Didascalicon* of the twelfth-century thinker Hugh of St. Victor contains a precocious recognition (II.23) that "the pursuit of commerce . . . commutes the private good of individuals into the common benefit of all." See Hugh of St. Victor, *The "Didascalicon" of Hugh of St. Victor: A Medieval Guide to the Arts*, Jerome Taylor (tr.) (New York, 1961), 77.

and Aquinas are found in the later works. As a general rule, novelty was not a priority, nor was it the purpose of the commentaries in the educational program of the medieval universities.

Jean Buridan (c. 1295–c. 1358) provides one of the great and notable exceptions to this rule. All of his surviving writings derive from lectures delivered during his tenure as teaching master in the Faculty of Arts at Paris where he spent his entire career. He was perhaps the most prolific and influential interpreter of Aristotle in the fourteenth century, the author of commentaries on the *Nicomachean Ethics* and the *Politics*, as well as almost all of Aristotle's many works on natural philosophy.[44] Buridan's insights and innovations as a natural philosopher are well recognized by historians of science and are discussed at many points in chapters 6 and 7 in this book.[45] His insights into the economic order and the workings of the marketplace are equal in acuity and innovation to his insights into the workings of nature. In five of the six categories of economic/scientific speculation considered in this study, either Buridan's analysis, or that of his student at Paris, Nicole Oresme, represent the culmination of four-teenth-century university thought on the subject. As natural philos-ophers, academic administrators, and economic observers without peer, Buridan and Oresme provide the clearest evidence of the connection between economic life and abstract speculation, between social experi-ence and the emergence of scientific thought.

Between Olivi and Duns Scotus at the cusp of the century and Buridan and Oresme at mid-century stand a number of commentators whose work, while often adhering to the Aristotelian text, nevertheless reveals a more sophisticated concept of market equalization than that found in the early commentaries of Albert and Aquinas, and considerably advanced over Aristotle's limited model of direct exchange between producers. Of these commentators, the closest heir to Olivi and Duns Scotus was their fellow Franciscan, Geraldus Odonis (1290–1349?).[46] Odonis' *Scriptum*

---

[44] A brief biographical sketch of Buridan is given in chapter 1, 31–32. For a list of Buridan's manuscripts, see Edmund Faral, "Jean Buridan: notes sur les manuscrits, les éditions, et le contenu de ses oeuvres," *Archives d'histoire doctrinale et littéraire du moyen âge* 15 (1946), 1–53. For a more extensive treatment, see Michael, *Buridan*, vol. II.

[45] For a concise statement of Buridan's contributions to scientific thought, see Ernest Moody, "Jean Buridan," in *Dictionary of Scientific Biography*, C. C. Gillespie (gen. ed.), 18 vols. (New York, 1970–90), vol. II, 603–08.

[46] In addition to his membership in the Franciscan Order, Odonis shared with Olivi a background as student and teacher in the economically active cities of southern France and in Paris. In contrast to Olivi, Odonis was an active administrator in the Franciscan Order, rising, in 1329, to the position of minister-general during a period of intense opposition between the Spirituals and the Conven-tuals. Again in contrast to Olivi, Odonis, on balance, opposed the Spirituals. Despite their differences, it is clear that Odonis was familiar with Olivi's *Tractatus* and used it, at times verbatim, in the writing of his own treatise on economic contracts (Siena U.V.8). On this subject, see Langholm, *Usury*, 46; Langholm, *Economics in the Schools*, 512–13; Spicciani, *La mercatura*, 188. For

super libros ethicorum was completed in the years immediately preceding his election as minister-general of the Franciscan Order in 1329.[47]

There are many places in the Aristotelian text where equality in exchange is specifically discussed. At these loci we find Odonis grafting the acceptance of estimation, approximation, and agreement (as found in the writings of Olivi, Duns Scotus, and Godfrey of Fontaines) onto the Aristotelian model. His largest correction of the Aristotelian text concerned the role of the judge in the equalization of gain (*lucrum*) and loss (*damnum*). In Aristotle, the judge as *iustitia animata* served as the embodiment and guarantor of justice in exchange. It was the judge's role to intervene in contracts where one of the parties had gained at the expense of the other and to restore equality by bisecting the line of *lucrum/damnum*. As we have seen, it was the role of the judge as intervening orderer in the misnamed *iustitia commutativa* that at times inhibited the conception of economic exchange as a self-regulating process.[48] Odonis, in his gloss on Aristotle, sought to show that the intervening judge was *not* a necessary component of equalization in exchange, and consequently, not necessary to the establishment of commutative justice.

If commutative equality is seen as a point midway between the line of gain and loss, then the judge is constantly needed because of the difficulty of attaining this point and the many circumstances that might influence its determination. But if commutative equality is seen as a line or latitude, as the *ipso facto* result of mutual agreement based on rational estimation, then, assuming the absence of fraud, the judge is rarely or never needed. The law, noted Odonis, allowed license in bargaining and trading.[49] For this reason, there is no injustice in an agreement made voluntarily and mutually, despite the possible eventuality of uneven gain and loss (*commutatio fuit legitima et utrobique voluntaria*).[50] Later commentators extended

---

Odonis' biography, see the article on Gurial Ot (Geraldus Odonis) by C. V. Langlois in *Histoire littéraire de la France*, vol. XXXVI (Paris, 1927), 203–25.

[47] The edition of Odonis' *Scriptum super libros ethicorum* I cite from (Brescia, 1482) is unpaginated. I have substituted my own page numbers (1–44) starting from the beginning of Book V. In addition, each citation is cross-referenced to the paginated edition, *Geraldus Odonis: Expositio in Aristotelis ethicam* (Venice, 1500). For the dating of Odonis' commentary, see James J. Walsh, "Some Relationships Between Gerald Odo's and John Buridan's Commentaries on Aristotle's 'Ethics,'" *Franciscan Studies* 35 (1975), 237–75.   [48] Discussed in chapter 2, 42–43.

[49] Odonis, *Ethics* V, lect. 7, 21a (1500, 103rb): "puta in vendere et emere in quantis vel in quot commutationibus aliis lex dedit licentiam commutandi."

[50] *Ibid.*: "in commutatione voluntaria nec lucrum nec damnum sunt iniusta postquam data est legis licentia. et postquam ab utraque parte commutatio est voluntaria. Puta si hodie duo commutarent vinum et bladum cras vilesceret vel caresceret vinum, unus haberet lucrum sine hoc quod iniustum faceret, et alter damnum sine hoc quod iniustum pateretur. quia commutatio fuit legitima et utrobique voluntaria."

this important departure from Aristotle in the direction of eliminating the subjective orderer.[51]

As if to demonstrate by philosophical argument that free agreement between exchanging parties was more to be trusted as a guide to equality than the decision of a judge, Odonis interjected a discussion into his commentary on Book v found neither in Aristotle nor in earlier commentaries at this point. He asked a question of particular concern to Franciscan writers: whether a well-regulated society rested more on the virtuous activity of men or on the strength and enforcement of positive law. As one might have guessed from Odonis' earlier valorization of free agreement in exchange, he concluded that the good man was more rational, just, divine, and to be prized than the good law.[52] In this as in so many of his economic decisions, Odonis was in substantial agreement with his fellow Franciscans Olivi and Duns Scotus. This unanimity raises two related questions: (1) what role did Franciscan theology play, particularly its emphasis on the freedom and primacy of the will, on the evolution of "economic" insights such as the validation of estimation and mutual agreement in the determination of just equality? (2) Was there a position on economic exchange, generally sympathetic to the give and take of the marketplace and so favorable to capitalist acquisition, that can be labeled specifically "Franciscan"? Clearly Franciscans are found among the most sensitive and understanding of economic observers, particularly in their willingness to recognize agreement as a dynamic process of equalization. However, the long existence of parallel arguments in Roman and canon law and their acceptance and development by non-Franciscan thinkers, including Dominicans and secular masters such as Godfrey of Fontaines, Jean Buridan, and Nicole Oresme, argue against the existence of a specifically "Franciscan economics."[53]

Certainly, in a tradition as intentionally synthetic as scholastic thought, Franciscan theological positions influenced Franciscan definitions of economic liceity. Scholastics sought to integrate economic insights with

---

[51] Buridan, *Ethics* v.10, 409: "in voluntariis commutationibus aut distributionibus nullum accidit lucrum aut damnum contra justitiam; hoc addito, quod neuter in pretium ponat illud cujus ipse non est dominus, quod dico ad excludendum usuras." For the influence of Odonis' commentary on Buridan, see Walsh, "Odo and Buridan."

[52] Odonis, *Ethics* v.11, 21b–24a (1500, 103rb–104ra). For a discussion of Franciscan voluntarism, see Bonnie Kent, *Virtues of the Will* (Washington, DC, 1995), 94–149, 236.

[53] Todeschini argues the case for specifically Franciscan theological influences on the economic thought of Olivi and later Franciscans in his "Oeconomica II," esp. 464–66. Kirshner does not believe in the existence of a specifically "Franciscan economics" or paradigm. In a series of articles on contemporary reactions to the common Italian institution of the *monte* or forced city loan in the fourteenth and fifteenth centuries, Kirshner continues to criticize any neat categorization of economic position in terms of religious affiliation. See Kirshner, "Storm," esp. 228–31; Kirshner, "The Moral Theology of Public Finance: A Study and Edition of Nicholas de Anglia's *Quaestio disputata* on the Public Debt of Venice," *Archivum fratrum praedicatorum* 40 (1970), 47–72.

theological requirements as intently as they integrated models of economic activity with models of nature. Thus the Franciscan emphasis on the primacy of the will would have facilitated the acceptance of free agreement as the sufficient guarantee of equality in exchange, and would have worked, in a general sense, toward the acceptance of the Roman law position allowing free bargaining in the determination of price and value. We can see this in the thought of Odonis, who often surpassed Duns Scotus in the acuity of his economic observations.[54] Where Duns Scotus had maintained that a general adherence by both parties to the golden rule served as the guarantor of equality in exchange, Odonis returned to the geometric language of Aristotle. He recognized that each party sought from the other a benefit proportional to the benefit offered.[55] This was not yet a clear acceptance of the Roman law recognition of deception in exchange as "natural," but it was a step beyond Duns Scotus, and a long way from Henry of Ghent's insistence that all parties seek the true point of equality in their exchanges.[56] Olivi had made the point a quarter-century before Odonis: all buyers wanted to pay less than the value of the thing bought, all sellers wanted to charge more.[57] Thus, in Olivi's view, the equality of mutual voluntary agreement superseded the desire for unequal advantage on the part of the exchangers – the desire for inequality was transformed into equality through the dynamic process of free and informed exchange. Ignorance, inexperience, even a compelling personal need could all undermine the equalizing process by undermining the necessary requirement of voluntary agreement. All were therefore grounds for invalidating a contract.[58] But Olivi maintained that, if the voluntary agreement to exchange was not taken as proof of equality, and if after the exchange complaining parties were free to go to a judge for equalization, then the entire exchange process, so necessary to society, would be damaged by overlegalization and complication.[59]

Other fourteenth-century commentators on the *Ethics* went further

---

[54] See Langholm, *Economics in the Schools*, 508–33, for a detailed study of Odonis that reveals his great powers of economic observation and the boldness of his economic thought.

[55] Odonis, *Ethics* v, lect. 9, 28a (1500, 105rb): "illi comuniter querunt per alios proportionale beneficium sibi contraferri. volentes sibi serviri sicut et ipsi aliis servierunt . . . quia istud est quasi naturale homini."

[56] *Ibid.*: "si isti beneficiantes non male querunt remunerari. et benefitiatus non contrafaciat aliquod benefitum. sequitur quod erit servitus inter cives. videbitur enim quod *servientes sint aliorum servi*." I emphasize the self-regulating aspect of this agreement, but it should be noted that Odonis insisted on the responsibility and good faith of the parties to an exchange. For the Roman law position, see *Digest*, 19.2.22.3, and see chapter 4, 90–92.

[57] Olivi, *Tractatus*, 61: "Nam ex utriusque partis libero et pleno consensu incohatur et rectificatur. Ita quod emptor vult sibi plus rem emptam quam pretium eius, et venditor econtrario."

[58] Langholm, *Economics in the Schools*, 358.

[59] Olivi, *Tractatus*, 60: "Sed communis consensus et consuetudo vult quod non omnis excessus in huiusmodi sit restituendus . . . quia alias infinite querimonia et litigia ex huic orirentur."

than Odonis toward bringing Aristotelian geometric *contrapassum* into line with the Roman law of sale. Jean Buridan, writing more than a decade after Odonis, followed him in limiting the judge's role in establishing equality, but surpassed him with the assertion that a "just" exchange has taken place even in cases where one party has gained an unequal advantage. In his commentary on Aristotle's *Ethics*, Buridan argued that profit and loss occur commonly in buying and selling without being *contra iustitiam*. The quest for personal advantage is the natural condition of just exchange.[60] Here, even in a scholastic commentary in which economic principles had to conform to philosophical principles, Buridan was altering the Aristotelian model in order to bring it into line with practical observations and considerations. Odonis and, even more, Buridan were beginning to formulate a conception of a market system so self-ordering that equality could result as a *product of willed inequalities*. None of this is found in Aristotle or the thirteenth-century commentaries.

If the role of the judge or subjective orderer was diminishing as a principle of market order and equalization, the role of money as an instrument of equalization in exchange was growing and taking its place. In the literature of the period, money was most often represented as a destroyer of justice, social order, and religious ideals.[61] The biographies of St. Francis graphically express his refusal to even touch money because of his fear of its corrupting influence. While this distrust remained pervasive in society, by the late thirteenth century, writers, including Franciscans, began to articulate a counterposition based on the growing intellectual appreciation of money as an instrument of order and equalization in exchange.

One can see this development in a group of *Ethics* commentaries dating from the late thirteenth and early fourteenth centuries.[62] Henry of Freimar, the author of the most thorough of these commentaries, noted that since

---

[60] Buridan, *Ethics* v.10, 409: "lucrum vel damnum si accidant, non sunt contra justitiam, immo sic in commutationibus est vera mercatura, secundum quam communiter utraque pars accipit quod magis est utile sibi, quod si non ita fiat, hoc accidit commutationi." Notice the value given to purely practical considerations here.

[61] For an excellent discussion of this common theme and constant subject of fierce satire, see Yunck, *Lady Meed*, esp. 63–69. See also chapter 1.

[62] For a general description of this group of commentaries, available at present only in manuscript, see Langholm, *Wealth and Money*, 28–30, 85–87. They were particularly insightful in their analysis of the properties of money and its equalizing role in exchange. For example, Vat. lat. 2172, 38va: "In commutatione debet servari equalitas . . . ad rem et non ad nummisma non potest semper equalitas servari . . . sed per nummisma salvatur ista equalitas." For discussions about the date, probable author, and possible "averroist" inclination of this manuscript, see Gauthier, "Trois commentaires 'averroistes.'" O. Lottin offers a conflicting interpretation in "A propos de la date de certains commentaires sur l'Ethique," *Recherches de théologie ancienne et médiévale* 17 (1950), 127–33.

money as a numbered *medium* allowed one to recognize immediately the relationships between the values of different goods, it became the means by which diverse things were "reduced to equality."[63] The heightened sensitivity of fourteenth-century theorists to the central role of money as equalizer, not only in exchange but also in the life of the community, can be clearly seen a decade later in Odonis' deductive proof of its necessity: without exchange there would be no intercourse between men; without the equalization of goods there is no exchange; without commensuration of value according to moneyprice, there is no equality.[64]

In Aristotle's *Ethics* it was the judge who was given the primary task of "reducing" things to equality. It is striking how similarly the role of the two – the judge as subjective equalizer, and money as objective instrument of equalization – were described in the fourteenth-century commentaries. Odonis talked of the judge as the *medius* and finder of the *medium*, who equated the unequal by *reducens ad equalitatem et ad medium*.[65] He used identical words to describe money's function as the measure of excess and defect, once again toward the end of *reducens ad equalitatem et ad medium*.[66] Although both had the same function, it is clear that the direction in fourteenth-century theory was toward the elimination of the judge in the normal course of exchange and his replacement by the mechanism of market agreement with money as the instrument of equalization. The model was shifting from the personal to the instrumental, from mind to mechanism as the basis for establishing order and equality.

MATHEMATICS AND THE GEOMETRY OF EXCHANGE

Throughout the fourteenth century, commentaries on Book v of Aristotle's *Ethics* involved mathematical discussions of exchange, particularly in explaining the distinction between arithmetic and geometric proportionality, continuous and non-continuous proportions, and the function

---

[63] Henry of Freimar (Henricus de Frimaria), *Ethics* v, 148ra: "Sed nummismata cognoscimus certam quantitatem equalem commutabilis. Eo quod per ipsam determinamus quantitatem precii et valoris rei commutabilis." *Ibid.*, 148va: "[Cognoscimus] inquantum una res superexcedat aliam in valore . . . ad hoc ut ad equalitatem reducantur." Langholm, in particular, has emphasized the importance of this commentary from the first decade of the fourteenth century and its influence on the later commentaries of Buridan and Oresme. See Langholm, *Wealth and Money*, 24–28, 85; Langholm, *Price and Value*, 59–60, 113–14; Langholm, *Economics in the Schools*, 536–49.

[64] Odonis, *Ethics* v, lect. 9, 31b (1500, 106ra): "hec mensura et pretium rerum nummisma sit necessarium probat tali deductione quia non existente commutatione rerum non erit communicatio hominum. Non existente autem equalitate rerum non existet commutatio. Non existente vero commensuratione secundum precium et nummisma non existet equalitas."

[65] Odonis, *Ethics* v, lect. 7, 19a (1500, 102vb): "iudex est medius et inventor medii. iudex enim adequat inequalia reducens ad equalitatem et ad medium."

[66] Odonis, *Ethics* v, lect. 9, 29a (1500, 105va): "mensurans per consequens superhabundantiam et defectum et aliqualiter reducens ad equalitatem et ad medium secundum quod sunt appreciata ad estimationem et comparationem nummismatis."

of Aristotle's geometric *figura proportionalitatis* in exchange. Where Aristotle and the early commentators had given the simplest mathematical examples of economic proportionality, a number of fourteenth-century commentators, Geraldus Odonis in particular, chose examples that did some justice to the complexity of economic calculation in everyday life.[67] In one case Odonis spoke of two merchants who had invested unequally in *unum capitale*, and he analyzed the proper mathematical way to split the profit of 300 pounds. In another, Odonis fit the calculation of spoils from a captured castle into a proportional equation. The count A had brought with him 100 men, the baron B had 50 men. The spoil was divided into two unequal parts. The count's part (G) was twice the other (D). Overall, Odonis' discussion of geometric proportionality in the context of just exchange occupies more than three pages of his commentary.[68]

Aristotle gave great weight to the role of arithmetic equalization in economic life. Indeed, he states at one point that *iustitia directiva* was applicable to all economic transactions between individuals assumed to be equals: buying and selling, loans, pledges, deposits, and rents.[69] While he undoubtedly meant that such justice was applicable only after the fact of an unjust exchange, in the confused text that has come down to us, this point is lost. Walter Burley (c. 1275–c. 1345), yet another mathematician/natural philosopher to comment on the *Ethics*, provided clarification here.[70] Although he generally followed the Aristotelian text very closely, in this case he pointed out its inconsistencies and asserted the necessity of recognizing the primary place of *contrapassum* and geometric equalization in just exchange.[71] Jean Buridan also went back to the Aristotelian text to explain the inconsistencies that led some to hold on to

---

[67] See Aquinas, *Ethics* (1969), v, 280b, and chapter 3, 59–60. Aquinas' simple example stated that Socrates who worked two days should get twice as much in wages as Plato who worked only one day.   [68] Odonis, *Ethics* v, lect. 6, 10a–14a (1500, 100rb–101rb).

[69] Aristotle, *Ethics* v.2 [1130b30–1131a4].

[70] For a biography of Burley that demonstrates once again the fourteenth-century scholar as an active participant in the social and political life of his time, see Conor Martin, "Walter Burley," in *Oxford Studies Presented to Daniel Callus* (Oxford, 1964), 194–230. It was while living in the household of the powerful ex-treasurer and chancellor of England Richard de Bury that Burley wrote his commentary on the *Ethics* (1338). See Courtenay, "London *Studia*," 136–37; Lowrie J. Daly, "The Conclusions of Walter Burley's Commentary on the *Politics*, Books I to IV," *Manuscripta* 12 (1968), 79–92. For the conservatism of Burley's approach as a commentator, see S. Harrison Thomson, "Walter Burley's Commentary on the *Politics* of Aristotle," in *Mélanges Auguste Pelzer* (Louvain, 1947), 557–78. A listing of Burley's *Ethics* manuscripts can be found in Langholm, *Price and Value*, 91.

[71] Burley, *Ethics* v, 83ra: "ostendit in quibus et qualiter est verum quod iustum commutativum et contrapassum sunt idem et intendit quod iustum commutativum continet in se contrapassum non secundum equalitatem sed secundum proportionem. Contra superius dictum est quod in iustitia commutativa medium accipitur non secundum proportionem geometricam que consistit in equalitate proportionis: sed secundum proportionalitatem arismetricam que consistit in equalitate quantitatis."

the mistaken idea that the process of exchange involved arithmetical equalization.[72] He stated flatly that it was geometric *contrapassum* involving proportionality that was the basis of justice in exchange.[73] Indeed, Aristotle's geometric *figura* of exchange continued to fascinate fourteenth-century commentators, and it regularly appeared as a drawn and labeled figure accompanying the text.[74]

In the commentaries of the fourteenth century, as the idea of a knowable point of just equality faded, and as the importance of the judge as an external, subjective equalizer in normal exchange also receded, the perceived economic role of an arithmetically achieved equality dwindled as well. As it did, the understanding grew that market exchange was governed primarily by geometric equalization.[75] In the first decade of the fourteenth century, the commentator Henry of Freimar argued that equality could be considered either absolutely (*de medio rei*) or relatively (*de medio quo ad nos*), and that, while arithmetical proportion pertained to the equalization of things considered absolutely in themselves, geometry governed the equalization of all things considered relatively and proportionally, i.e., all goods in exchange.[76] As soon as one considered the *process* of valuation and admitted the variable factor of *indigentia*, one entered the realm of geometric equalization.[77] The development of the concept of just equalization through mutual agreement and *aestimatio* from the late thirteenth century furthered the equation of economic value and relativity and thus of economic exchange and geometry.[78]

---

[72] Buridan, *Ethics* v.10, 408.

[73] Buridan, *Ethics* v.14, 425: "Sic igitur patet, quod contrapassum proportionabile est iustum in omni commutatione, et non contrapassum aequale quantitative."

[74] The commentary of Albert of Saxony (1316–90) focuses on Aristotle's *figura proportionalitatis* and provides labeled diagrams: *Expositio super libri ethicorum* v, 122r–123v. For the place of this work in the commentary tradition and its reliance on the earlier commentary by Walter Burley, see Georg Heidingsfelder, *Albert von Sachsen: sein Lebensgang und sein Kommentar zur nikomachischen Ethik des Aristoteles* (Münster, 1927), esp. 83–91. The mathematical implications of Aristotle's forms of equalization are discussed as well by the English mathematician and calculator Richard Kilvington in his *Quaestiones morales*, esp. 53rb–55vb.

[75] Though the existence of such a point was denied in regard to most economic transactions, the majority of theologian-theorists continued to insist on the reality of such a point in the loan or *mutuum*.

[76] Henry of Freimar, *Ethics* I, 57va: "quia arismetica quidem considerat medium secundum absolutam quantitatem rei, geometria autem solum secundum plenitudinem proportionis et non secundum quantitatem rei absolutam ut magis patebit in [Book] v de iustitia distributiva."

[77] Odonis, *Ethics* v.12, 27a (1500, 104vb): "Et cum hoc est hic proportionalitas quodammodo geometrica facta cumparatione ad indigentiam maiorem vel minorem ut videbitur supra textum."

[78] Henry of Freimar, *Ethics* v, 147va–147rb. Henry was one of the strongest spokesmen for the "objective" factors of labor and expenses as components of economic value, so that the producer could be assured of being adequately recompensed in exchange. This in no way lessened his understanding of the market variability of price according to scarcity and need, or his understanding of *contrapassum* as a geometric process. See Langholm, "Scholastic Economics," 123; Langholm, *Price and Value*, 107.

The change of emphasis from an economic system governed by arithmetic equalization to one governed by geometric equalization paralleled the developing conception of the marketplace as a self-ordering system – a mechanism of equalization at the heart of the *civitas*.[79] The subjective orderer was replaced by the conception of a dynamic process, where equality was the product of intersecting needs and purposes, represented by the crossed diagonals of Aristotle's geometric *figura proportionalitatis*. The conjunction point of equality was accomplished not through personal aim, or the judge's order, but through agreement and the impersonal instrument of money. The Franciscan general Odonis, who was also a noted mathematician and theoretician of the continuum,[80] made the interesting comment that the Aristotelian *figura* of proportionality governed exchange even though the majority of individuals involved knew nothing of diagonals and had no sense of themselves as being part of a geometric figure.[81] In Odonis' view, geometry and geometric form underlay market exchange on a level deeper than, and independent of, human knowledge and intent. In this sense it resembled the geometric form that natural philosophers like Nicole Oresme were beginning to discover underlying all relations in nature.[82]

The majority of the surviving commentaries on Aristotle's *Ethics* in the fourteenth century were written by scholars with strong interests in natural philosophy: Walter Burley, Gerald Odonis, Henry of Freimar, Richard Kilvington, Jean Buridan, Nicole Oresme, Albert of Saxony. All of these writers took the geometry of Aristotelian *contrapassum* seriously. Perhaps the strongest indication of this is the recognition by a number of authors that *contrapassum* within the geometric *figura* could provide proportional equality even between *numerically incommensurable* values. Nicole Oresme, Jean Buridan, and the English mathematician and Calculator Richard Kilvington all recorded in their commentaries to Book v that no proportion can be found between the *diameter* of the square and its sides considered arithmetically, but considered geometrically as a line, propor-

---

[79] Walter Burley was among those who connected geometric *contrapassum* with a dynamic social conception of the *civitas*. Burley, *Ethics* v, 83r: "iustum commutativum continet illud per quod cives commanent sibi invicem in civitate et per quod salvatur civitas: quia leges ordinantur ad hoc quod civitas salvetur: et quod cives ad invicem commaneant. Sed per hoc quod cives proportionaliter compatiuntur: commanent cives sibi invicem in civitate. ergo iustum commutativum continet contrapassum secundum proportionem."

[80] Odonis' independent position on the real existence of indivisibles in the continuum is discussed in John Murdoch and Edward Synan, "Two Questions on the Continuum: Walter Chatton (?), OFM, and Adam Wodeham, OFM," *Franciscan Studies* 26 (1966), 212–88.

[81] Odonis, *Ethics* v, lect. 9, 30b (1500, 105vb): "Exponit iterum modum contrapassionis. que omnino vult fieri secundum dyametrum et cum centum mille homines dyametrum nescientes sciunt valde bene modum commensurandi contrafaciendi et contrapatiendi."

[82] For developments in this geometrical understanding, see chapter 7, 202–07.

tion can always be found.[83] By considering the mathematical advantages of geometric commensuration within their discussion of money and exchange in *Ethics* v, the commentators indicate a connection between money's capacity to render commensurable the seemingly most diverse goods and services, and the geometric line's capacity to render commensurable even numerically incommensurable values.[84] As a geometric model replaced an arithmetical model in the comprehension of economic exchange so too did a geometric model supplant an arithmetical one in the comprehension of nature's form and activity within fourteenth-century natural philosophy. As it did so, the geometric line as an instrument of relation and commensuration assumed an ever-greater role in proto-scientific speculation.[85]

MONEY AS *MEDIUM* AND AS MEASURE

It was Aquinas who first attached the word *instrumentum* to money.[86] By the early fourteenth century this notion had found its way into literary culture. Brunetto Latini, the friend of Dante who translated many ideas derived from Aristotle's *Ethics* into popular literary form, wrote in his *Li livres dou tresor*:

Et deniers est ausi comme justice, sans ame, por ce k'il est un mi par quoi les choses desigaux tournent a egalité; et puet on baillier et prendre les grans choses et les perites par deniers. Et il est uns estrumens.[87]

Within university culture, the thinker who most fully realized the implications of money as impersonal instrument was Nicole Oresme (c.

---

[83] Oresme, *Ethiques* v, 285: "Et generalment proporcionalité est en toutes quantités, combien que aucunes ne puissent pas bien estre mesurees par nombres; si comme il appert en geometrie des proporcions qui sont entre choses incommensurables, et son appellées irracioneles ou sourdes." Oresme is here glossing Aristotle's statement that proportionality extended past the relationship of unit numbers to all things that could be numbered [1131a30–32]. On this subject, see also Kilvington, *Quaestiones morales*, 55vb.

[84] Buridan's comment comes in v.11 of the *Ethics*, which compares the finding of the *medium* in *iustitia distributiva* and *iustitia commutativa*, 414: "Dicitur [iustitia distributiva] autem medium secundum geometricam proportionem equalitas proportionis aliquorum duorum vel plurium; et dicitur *geometrica*, quia potest reperiri in continuis secundum nullum numerum commensurabilibus: verbi gratia, sumamus quadratum A, et quadratum B, dicemus autem quod proportiones diametrorum illorum quadratorum, quanticunque sint quadrati, ad latera eorundem sunt aequales, licet diameter sit incommensurabilis costae."

[85] Nicole Oresme in particular was fascinated by the question of mathematical incommensurability and with line as an instrument capable of relating incommensurables. He dedicated great effort and a number of his most important works in natural philosophy to this subject. The relationship between money as a measuring line and the geometric line as an instrument of relation is discussed in chapter 6, 184–94, and chapter 7, 205–11.

[86] Bridrey, *Théorie de la monnaie*, 110–11; Langholm, *Wealth and Money*, 67–68.

[87] Brunetto Latini, *Li livres dou tresor*, Francis J. Carmody (ed.) (Berkeley, 1948), vol. II, 29, 2, 199.

1320–82), the great fourteenth-century mathematician, geometer, and natural philosopher.[88] Along with his philosophical speculation on subjects that would now be termed mathematical and "scientific," Oresme wrote vernacular translations and commentaries on both the *Ethics* and the *Politics* of Aristotle at the request of King Charles V of France, and he authored a unique and widely read autonomous treatise on money, the *De moneta*, or *Tractatus de origine, natura jure, et mutacionibus monetarum*.[89] In this treatise Oresme writes that money is an instrument most subtly and rationally devised.[90] It is "an instrument ordered toward the end of living well,"[91] "an instrument for exchanging natural riches,"[92] a measuring tool that "permits all things to be measured together,"[93] and, finally, "an instrument of equalization" (*instrumentum equivalens*) or "balancing instrument," whose working created the basis for just exchange.[94]

Although Oresme's consciousness of money as an instrument was particularly acute, he generally remained within an older tradition of Aristotelian commentary in his recognition of its role in facilitating proportional equalization.[95] Walter Burley's commentary on Book v of the *Ethics* provides a representative example of fourteenth-century attitudes on the subject of money as measure.[96] If Burley's treatment of this subject was remarkable in any way, it was not for its novelty but for its

---

[88] A brief biography of Oresme focusing on his administrative experience within the bureaucracies of university, Church, and royal government is given in chapter 1, 29–31.

[89] Oresme, *De moneta*. Oresme soon translated his *De moneta* from Latin into French: *Traictie de la première invention des monnoies de Nicole Oresme*, L. Wolowski (ed.) (Paris, 1864). The *De moneta* is discussed in chapter 1.

[90] Oresme, *De moneta*, 4: "subtilitati sunt homines usum invenire monete, que esset instrumentum permutandi adinvicem naturales divicias." As authority for his appreciation of money as instrument, Oresme cited Cassiodorus' exclamation: "O inventa prudencium!" (chapter 11, p. 17). For Oresme's debt to previous commentators in the formation of his economic thought, see Langholm, *Wealth and Money*, esp. 18, 80ff.; Bridrey, *Théorie de la monnaie*, 411–30.

[91] Oresme, *Politiques* 1, 65: "un instrument ordoné a bien vivre."

[92] Oresme, *De moneta*, 5: "moneta est instrumentum permutandi divicias naturales."

[93] Oresme, *Ethiques* v, 297: "monnoie qui faites toutes choses estre mesurées ensemble."

[94] Oresme, *De moneta*, 10: "Moneta siquidem est *instrumentum equivalens* permutandi divicias naturales" (emphasis added). Johnson translates "instrumentum equivalens" as "balancing instrument." *Traictie*, 20, has: "monnoie est l'egal instrument à permuer les Richesses naturelles." The implications of the everyday use of this tool on developing ideas of measurement and relation in fourteenth-century natural philosophy are pursued in chapter 6.

[95] For two statements out of many that preceded Oresme on this subject, see Odonis, *Ethics* v, lect. 9, 29a (1500, 105va): "mensura omnium venalium. mensurans per consequens superhabundantiam et defectum et aliqualiter reducens ad equalitatem et ad medium secundum quod sunt appreciata ad estimationem et comparationem nummismatis"; Guido Terrenia, *Ethics*, 44vb: "diversorum non potest fieri adaequatio nisi eorum valor appreciari possit nisi enim possit estimari quam valet bladum plusquam calciamentum non possit estimari inter ea equalitas."

[96] For an analysis of Burley's place in the discussion of money as measure, see Langholm, *Price and Value*, 93. For a concise biography of Burley that concentrates on his scientific and mathematical achievements and his place in the science of *calculationes*, see the article by John Murdoch and Edith Sylla in *Dictionary of Scientific Biography*, vol. ii, 608–12.

repetition of basic tenets. Following Aristotle, Burley comments on the great diversity existing among goods in exchange. This diversity can be brought into relation and equation only if one common measure is established in comparison to which all things find their measurement. This common measure is money.[97] Money was invented so that it could be the *medium* and measure of all goods in exchange (*medium et mensura omnium commutabilium*).[98] Following Aquinas, Burley recognized that money is an artificial and relative measure, incapable of measuring things in themselves or according to their natures. The *natural* measure of things in exchange is human need (*indigentia humana*), which is common to all commodities. And it is this common quality of *indigentia* that money is capable of measuring.[99] Thus, in Burley's understanding, money measures the common quality of human need attached to all goods in exchange.[100] Burley's firm grasp of the Aristotelian conception of money as measure should be considered in light of his position as an influential figure among the early Merton Calculators, and as a natural philosopher whose work was characterized by a fascination with qualitative measurement.

The questions of what money measured and how money measured engaged scholastic thinkers. Although everyone recognized that money measures and quantifies *indigentia* (following Aristotle and the early commentators), they also came to recognize that price often measured *indigentia* in relation to yet other qualities inhering in commodities. One of the first economic questions to provoke this recognition concerned the just price and the permissibility of the merchant's profit. Was it licit to re-sell something for more than one had paid for it? If so, where did this *superabundantia* – this superadded value – come from? This question was critical to the maintenance of a conception of equality in exchange.[101] From the late thirteenth century, the answer generally given is two-tiered: to re-sell for more is to violate "natural equality" *unless* the re-sold commodity has been improved through labor. Although self-consciously traditional in economic matters, Henry of Ghent was one of a number of

[97] Burley, *Ethics* v, 83va: "ostendit quod omnia illa quorumpotest esse commutatio debent esse comparabilia: quia omnia illa que comparantur habent aliquod unum in quo vel secundum quod comparantur et narrat quod est illud in quo omnia comparabilia comparantur quod est mensura omnium commutabilium: quis est nummisma."

[98] Burley, *Ethics* v, 83vb: "ad hoc enim inventum est nummisma ut sit medium et mensura omnium commutabilium: ut per nummisma sciatur quod commensurabilium plus valeat et quod minus."

[99] Burley, *Ethics* v, 84ra: "omnia commutabilia commensurantur per indigentia naturaliter et per nummisma secundum conditionem hominum . . . secundum veritatem tamen et secundum proprietatem ipsarum rerum impossibile est res tam differentes commensurari."

[100] *Ibid.*: "sed solum per comparationem ad indigentiam hominum possunt contineri sub una communi mensura. et ideo nummisma non commensurat res commutabiles ex sui natura. sed per institutionem et hoc per comparationem ad indigentiam humanam."

[101] Ibanès, *Doctrine*, 35–36.

thinkers who permitted merchants to expand the arithmetical equality between buying price and re-selling price by charging for their labor of transportation, which he considered a legitimate improvement to the goods sold. He recognized, moreover, that adding to the value of this labor were the *mental qualities* that the successful merchant brought to his task.[102] By knowing where and when to buy and sell, the merchant's "expertise" and "care" had value in themselves – value which could be measured, translated into monetary terms, and legitimately added to the cost of the commodity.[103] Even something as ephemeral as the merchant's good reputation could be translated into monetary terms and licitly added to the resale price.[104]

Duns Scotus, who disagreed with certain of Henry's economic positions (notably Henry's belief in a just point of arithmetical equality in exchange), nevertheless agreed with him that the merchant's personal qualities could be valued in monetary terms and licitly added to the re-sale cost of his goods.[105] Both Duns Scotus and Henry expanded the notion of money as a measure of qualities, and justified the merchant's profit in this manner, because they had come to recognize the useful service that merchants perform for the community. Duns Scotus maintained that an addition to the re-sale price was licit under two conditions: (1) if the merchant was performing a useful service to the community and (2) if the added price corresponded to the merchant's "diligence, prudence, care, and risks."[106]

Thus, by the end of the thirteenth century, both Henry of Ghent and Duns Scotus shared in the understanding that the merchant's personal qualities could be measured by money and translated into monetary terms in the marketplace. Both men were also among the first to introduce questions concerning the measurement of qualities and qualitative change into their natural philosophy.[107]

---

[102] See Spicciani, *La mercatura*, 179.

[103] *HG* 1.40, 228: "ita utile est ad acquirendum pecuniam esse expertum, qualia in quibus temporibus sunt rara et pretiosa; alia enim in aliis temporibus abundant. Et ideo . . . per custodiam facta sunt pretiosa et vendi possunt carius, sicut dictum est de translatis per loca." For a fuller account, see chapter 4, 139–40.

[104] *HG* 1.40, 229: "alii cognoscentes eius industriam circa equos statim certificati per eum et eius industriam sunt quod equi illi valent pretium suum . . . et sic . . . bene potest iste statim equos suos aliquantulum carius vendere supra primum pretium equorum, vendendo opus industriae suae."

[105] Spicciani, *La mercatura*, 226.

[106] Duns Scotus, *Sent.* IV, 317a: "Primum est quod talis commutatio sit utilis Reipublicae. Secundum est, quod talis juxta diligentiam suam et prudentiam et sollicitudinem et pericula accipiat in commutatione pretium correspondens."

[107] For Duns Scotus' influence on the Oxford Calculators in the area of qualitative measurement, see Clagett, *Science of Mechanics*, 206; Steven J. Livesey, "Mathematics *iuxta communem modum loquendi*: Formation and Use of Definitions in Heytesbury's *De motu locali*," *Comitatus* 10 (1979), 9–20, esp. 10. In addition to Duns Scotus, Livesey also discusses Henry of Ghent, Godfrey of

Money as medium *and as measure*

Peter John Olivi was the economic thinker whose writing most closely reflected the thinking in the marketplace around him. Olivi, like Henry of Ghent, recognized that without a profit commensurate to his labor and experience, the merchant would not perform his service to the community.[108] However, in his detailed discussion of wage variations he went beyond any of his predecessors in the area of money's measurement of qualities. According to Olivi, there were a number of factors that led to wage variations between different occupations such as lawyer, doctor, soldier, architect, and laborer. The largest price differential is caused by relative scarcity: there are many more laborers than doctors; therefore their labor is less valued. Other factors include the risks and skill and even the responsibility involved in the performance of the task.[109] The architect earned more than the laborer even though the laborer might work harder, because the architect was being paid additionally both for his refined skill and his greater responsibility for the outcome of the project.[110] The duke or leader receives a higher stipend than his knight, who in turn receives more than the common foot soldier and so on, in part because the one holding the higher position is required to give greater skill (*peritia*), energy (*industria*), and mental attention (*sollicitudo mentalis*) to his task. Furthermore, Olivi recognized that the labor and long study required to develop certain skills added further to their rarity and thus increased their market value in relation to those requiring less preparation and study.[111] In Olivi's dynamic vision, personal qualities and skills were

Fontaines, and Walter Burley as early contributors to the scholastic debate on qualitative intension and remission. Duns Scotus actually gives a brief summary of opposing opinions in the debate on qualitative change in his commentary to Lombard's *Sentences* (*Opus Oxoniense*), bk. 1, d. XVII. Duns Scotus' understanding of the just price as an intensible range or "latitude" discussed in chapter 5, 126 (*Sent.* IV, 283b) provides another link between his economic perceptions and later philosophical and mathematical insights underlying the science of *calculationes*. For more on this subject, see the discussion of the history of the measuring "latitude" within natural philosophy in chapter 6, 192–94.

[108] Olivi, *Tractatus*, 63: "Item nisi essent honorabiles et fide digni, non eis a diversarum terrarum gentibus prout expedit huic offitio crederetur. Si etiam non essent pecuniosi non possent grandes et caras merces prout terris expedit providere. Ex his autem aperte concluditur quod lucrum predictis circumstantiis competens inde possint et debent reportare. Et quo ulterius sequitur quod *usque ad aliquam mensuram congruam possunt suarum mercium pretium augere*" (emphasis added).

[109] Olivi, *Tractatus*, 56: "Tertio observat laborem vel periculum et industriam adductionis rerum vel obsequiorum. Nam merces vel operosa obsequia, que cum maiori labore et periculo communiter adducuntur vel fiunt plus ceteris in pretio paribus ponderamus."

[110] Olivi, *Tractatus*, 57: "Et ideo fossori aut cesori lapidum quamquam plus corpore laboranti, non tantum pretium datur sicut architectori qui cum altiori peritia et industria fossori aut cesori precipit et demonstrat agenda." The idea that variations in skill of the producers were to be proportionally recompensed in exchange was found as well in Aristotle's *Ethics* [1133a12–14].

[111] Olivi, *Tractatus*, 57: "Quia si ad altiora offitia debite exigenda exigitur maior peritia et industria et amplior sollicitudo mentalis, et etiam multo et diuturno studio ac peritia et labore, multisque periculis et expensis communiter acquiritur peritia et industria talis; et etiam quia pauci sunt et rari sunt ad hoc idonei, et ideo maiori pretio reputantur."

continually being translated into changing monetary values as they were measured against shifting need and utility in the marketplace.

Olivi connected the measurement of mental and personal qualities by money directly to his key economic concept of the licit range of value equalization in exchange – what he called the "latitude" of just price (*latitudine iusti pretii*).[112] Thus in his economic thought Olivi combined the measurement of qualities with the concept of the *latitudo* or measuring range divided by degrees, a combination that was later found often in proto-scientific speculation, particularly in the measuring schemata of the Calculators. In all this, Olivi, while going into greater economic detail than his predecessors, was barely scratching the surface of the actual gradation of skills and personal qualities routinely measured by money in the marketplace of the late thirteenth century.

The Aristotelian text of *Ethics* v.5, was so insightful and mathematically sophisticated regarding the question of money as measure that most fourteenth-century commentators were, like Burley, content to explicate its subtleties, repeat the additions of Albert and Aquinas, and add their own minor refinements in language or presentation. The major exception to this cautious approach was provided by Jean Buridan.[113] Buridan was not directly associated with the intellectual movement of *calculationes*, but he made important contributions in a wide range of scientific fields and was, perhaps, the most influential figure in natural philosophy of his century. His analysis of money as a measure of qualities points up further connections between the scholastic understanding of the function of this instrument and systems of measurement developing within fourteenth-century natural philosophy.

Aristotle wrote in Book v that society was held together by the proportional requital of goods and services [1133a1–5].[114] Albertus Magnus had emphasized the geometry of this process by stating that services were exchanged according to the model of "intersecting diagonals."[115] Buridan delved deeper into the question: how could two different services be compared and commensurated, and, more puzzling, how could a monetary quantity (numbered price) be commensurable with an

---

[112] Olivi, *Tractatus*, 64: "Ex latitudine vero quia sicut ars et industria artificis sibi licite fit lucrosa, sic industria mercatoris in rerum valore et pretio prudentius examinando, et ad subtiliores minutias iustum pretium producendo, potest licite sibi valere ad lucrum et maxime cum in hoc, salva *latitudine iusti pretii*, aliis communiter prosit" (emphasis added). For the place of the measuring "latitude" in fourteenth-century proto-scientific thought, see chapter 6.

[113] For a biography of Buridan that indicates his considerable experience in the political, social, and economic life of his time, see Edmund Faral's article in the *Histoire littéraire de la France*. Similarly, see Michael, *Buridan*, vol. I, 160–237. For biographical information centering on Buridan's scientific interests, see Moody in *Dictionary of Scientific Biography*. [114] See chapter 2, 51–52.

[115] Albertus Magnus, *Ethics* (Borgnet edn.), 356a: "facit enim retributionem secundum proportionalitatem supradictam conjugatio gratiae secundum diametrum." See chapter 3.

incorporeal entity such as a service rendered?[116] He began his answer by noting that on first inspection the measurement of services by price appears impossible. Then, to illustrate that such measurements were both possible and common, he chose (what appears to us) an extreme example of its kind, but one that made sense in the context of his society.

Buridan presented his argument to an imagined doubter: suppose someone gave you ten pounds and you in return gave him the acknowledgment *grates Domine*. You will say, no doubt, that the ten pounds you received was worth more than the acknowledgment you gave, and that there was no equality in the transaction. But consider that all value in the marketplace is relative and is estimated according to human need. Imagine that the giver of the ten pounds is very rich and so has little or no need of money but great need of honor. Imagine also that the recipient of the money is a man of great honesty and goodness. In such a situation, said Buridan, the poor man's simple *grates Domine* provides a fit and proportional adequation for the rich man's ten pounds.[117] In this elegant example not only is a price found for a seemingly immeasurable service, but the price is "just," i.e., represents an equalization of value between buyer and seller. And once again, the attachment of a numbered price to personal qualities is involved: in this case, it is the poor man's qualities of *bonitas* and *honestas* that have been calculated and recompensed in the exchange. The equation is remarkably flexible. The relativistic estimation of equality by the buyer shifts naturally as he is richer or poorer, and as he is more or less in need of the service he is buying. Economic exchange is built on the shifting estimation, measurement, commensuration, and equalization of seemingly unrelatable and immeasurable entities – including subjective human qualities.

Buridan's treatment of Aristotle's arithmetically determined *iustitia directiva* shows similar ingenuity. In the Aristotelian system, in cases of unjust exchange of any sort, it was the judge's role to step in and equalize the gain and loss of the two parties. As we have seen, Aristotle conceived of this operation in terms of restoring equality to a single *line* of *lucrum/damnum* that had been wrongly divided into two unequal parts. He

---

[116] Buridan, *Ethics* v.14, 423: "non potest quantitatis aequalitas observari inter ea quae non sunt secundum quantitatem commensurabilia; sed iustum invenitur ubi commutabilia non sunt commensurabilia secundum quantitatem, ut sic gratiarum actiones reddantur pro data pecunia."

[117] Buridan, *Ethics* v.14, 425: "Et si dicas, quod hoc est impossibile, quoniam si aliquis dat tibi decem libras, et tunc dicas ei, *grates Domine*, tunc non valet tantum dictum tuum quantum decem librae. Ad hoc respondeo, quod valor rerum aestimatur secundum humanam indigentiam, possibile est igitur, quod dans decem libras est valde dives, ita quod nihil vel modicum indiget pecunia, sed magis indigeret honore, et forte recipiens est magnae bonitatis et honestatis, ita quod dans decem libras multum honoratur, si ille recipiens dicat, *grates Domine*: immo quandoque si solum dignetur recipere donum illud, propter quod sic optime fit adaequatio secundum proportionem ad mensuram valoris."

gave considerable emphasis and detail to this model of line equalization, and a drawn line often appeared in the medieval manuscripts of the *Ethics* at this point.[118]

I have already noted that the role of the judge in the normal process of exchange was reduced in the commentaries from the fourteenth century and often replaced by money as an instrument of equalization. Buridan, while limiting the role of the judge, at the same time carefully analyzed the model of measurement and equalization associated with him. He criticized the Aristotelian model of *lucrum/damnum* for its imprecision. Aristotle had said that all exchanges could be reduced, after the fact, to one line of gain and loss requiring bisection and equalization by the judge. Buridan countered that in every exchange there was not one, but a minimum of two measuring lines involved because each party carried his own line of *lucrum/damnum*. Gain and loss could not be established universally and absolutely, but had to be considered relatively and proportionally to the differing potential for each person to gain and lose from the transaction. Equality was to be found not through the bisection of a line, but through the proportional relation of two or more lines, each line proper to an individual in a given exchange.[119]

Buridan recognized that his position positing two lines of *lucrum/damnum* was open to the objection that there cannot be two points of just equality in one exchange. He replied that two different shades of darkness in two separate subjects could each be a *medium* between white and black. This was possible as long as whiteness and blackness were considered relative to the potential of each subject for these qualities, rather than absolutely: "the darkness in Socrates is the *medium* between white and black only in so far as Socrates can innately be white or black, and not in so far as Plato can be white or black."[120] In the same way, said

---

[118] Aristotle, *Ethics* [1132a23–b15], and see chapter 2. For fourteenth-century comments on the line of *lucrum/damnum*, see Burley, *Ethics*, 81vb: "ista nomina scilicet damnum et lucrum proprie dicuntur in rebus possessis: sed . . . illa nomina transferuntur ad injurias personales unde quando passio est mensurata secundum mensuram iustie tunc illud quod est plus vocatur lucrum: et illud quod est minus vocatur damnum." See also Odonis, *Ethics* v.10, 16b (1500, 102rb): "sciendum quod bonum et malum restringi possent ad lucrum et damnum."

[119] Buridan, *Ethics* v.10, 410: "quod plus debito et minus debito sunt duo iniusta, hoc quidem ex parte facientis, illud autem ex parte patientis. Et ita in operatione iusta sunt duo iusta secundum unum ex parte facientis et alterum ex parte patientis . . . propter quod oportet in operatione iusta duo esse iusta et duo esse iniusta modo predicto."

[120] *Ibid.*: "quod nullum est inconveniens duo fusca ejusdem gradus inveniri in duobus subiectis, et tamen utrumque esset medium albi et nigri. Et etiam forte dici posset, quia illa duo fusca non essent media albi et nigri secundum eandem particularem rationem albi et nigri, sed secundum hoc subiectum esset innatum esse album aut nigrum fuscum in eo esset medium albi et nigri . . . Fuscum igitur in Socrate non esset medium albi et nigri, nisi secundum quod Socrates est innatus esse albus et niger, et non secundum quod Plato potest esse albus aut niger." (He cites *Physics* v as his authority here.)

Buridan, Socrates has had a just transaction with Plato and vice versa if at the end both have a *medium* between plus and minus insofar as each could possibly have had plus or minus, and not, as Aristotle had maintained, insofar as one had plus and the other minus.[121]

Buridan's language here is very close to the language found in contemporary debates within natural philosophy concerning the quantification of qualities. His method of establishing equality through the proportional relation of two or more measuring lines represented much more closely than did Aristotle's scheme a form of equalization found in the writings of the Oxford Calculators. Similarly, his use of an expanding and contracting line to measure the values of gain and loss proper to each subject is similar to the use made by the Calculators of the "latitude of qualities" (*latitudo qualitatum*) – their prime conceptual instrument of measurement.[122] Furthermore, Buridan explained his economic model in language reminiscent of Calculator debates on the measurement of qualities, using their favorite example of varying degrees of the quality of whiteness to represent the varying degrees of *lucrum* and *damnum*.

Buridan was not the first to link money as a measure of qualities to the measurement of particular physical qualities in the manner of the Calculators. The first such statement I have found occurs in the *Ethics* commentary of Henry of Freimar written in the first decade of the fourteenth century. Money, said Henry, is the measure of all things according to their human usefulness in the same way as whiteness is the measure of all colors.[123] But having connected money as measure to qualitative measurement, Henry went no further with the analogy. Buridan actually built on the connection between money and whiteness to analyze in greater detail the problem of determining economic gain and loss. In doing so he reveals the levels of connection (and the flow of insights) between his observations of natural activity and his observations of economic life, both of which were framed by logical and philosophical concerns defined by his intellectual culture.

After correcting Aristotle's model of arithmetical equality, Buridan turned to the geometry of *contrapassum*, and particularly the role of *indigentia* as the natural measure of things in exchange. Rather than existing as something in itself attached directly to commodities, *indigentia*,

---

121 Buridan, *Ethics* v.10, 410: "si Socrates iuste commutat Platoni, ipse neque plus habens debito, neque minus, habet medium pluris et minoris, in quantum ipse posset habere plus, et posset habere minus: et similiter Plato medium habet, in quantum ipse posset habere plus, et posset habere minus, et non in quantum esset possibile quod unus eorum habeat plus, et alter minus."

122 See, for example, John Dumbleton's use of drawn, proportionally divided lines to measure and relate qualitative latitudes in chapter 6, 187–90.

123 Henry of Freimar, *Ethics* v, 148ra: "Ergo relinquitur quod nummisma sit mensura omnium rerum commutabilium per ut veniunt in usum humanum sic etiam album est mensura omni coloris."

he maintained, was a measure of the *relationship* between commodities and individuals. To illustrate this he noted that *indigentia* seemed to have an inverse relationship to the things it measured, for the more one had of anything, the less one needed it.[124] Therefore one cannot measure *indigentia* arithmetically as numbers are measured by the number one, or wine is measured by the quart (that is, according to the general rule for measurement given by Aristotle in *Metaphysics* x). Rather, as with all relationships, it is necessary to measure *indigentia* proportionally – geometrically. As *indigentia* is itself a proportional measure, so too is money. Whatever relation a given commodity has to human *indigentia*, it has the same relation to money as the numbered measure of *indigentia*.[125]

Buridan searched for an example to illustrate the proportional and relativist system of measurement attached to money in exchange. He found it in the world of nature – in a mathematical relationship at the cutting edge of scientific speculation in the fourteenth century, and in an area where Buridan made his most forward-looking contributions as a natural philosopher – the physics of motion. *Indigentia*, he wrote, measures value in the same way that we measure motion by time or space. Measurement in both cases is geometrical rather than arithmetical. In such cases, the measure does not have to be a unit of the measured. He then illustrated the proportional relationship between *indigentia* and value by comparing it to an equation between two motions (A and C) measured by their two times (B and D), so that $A/B = C/D$, and if B is twice D, A will be twice C.[126] With this surprising choice of an illustrative case, Buridan provides evidence of a two-way connection between the mathematical understanding of relative value measurement in the marketplace and geometrical conceptions of measurement in natural philosophy, particularly in the fertile scientific area of kinematics.

---

[124] Buridan, *Ethics* v.16, 430: "sed indigentia non patitur secum res voenales, quia cum habemus eas non indigemus eis." Buridan was concerned to refute objections to *indigentia* as measurement based on this inverse relation of the measure to the measured.

[125] Buridan, *Ethics* v.17, 433: "valor pecuniae mensuratus fuerit secundum proportionem ad humanam indigentiam, omnia commutabilia poterunt appreciari secundum proportionem ad pecuniam; qualem enim proportionem habebunt ad humanam indigentiam, talem proportionem habebunt ad pecuniam humanae indigentiae proportionatam."

[126] Buridan, *Ethics* v.16, 431: "alia [i.e., indigentia] est mensura secundum similitudinem proportionis, verbi gratia, sicut motum mensuramus spatio vel tempore, nam si motus A, sit in tempore B, et motus C in tempore D, arguimus sic: sicut A ad B, ita C ad D; igitur per locum A transmutata proportione, sicut A ad C, ita B ad D: sed tempus B est duplum tempori D, ergo motus A est duplex motui C; et ita etiam potest exemplificari si motus mensuretur spacio."
  "Dico igitur quod indigentia humana mensurat commutabilia secundum similitudinem proportionis, non secundum aequationem quantitatis."

146

RELATION AND THE RELATIVITY OF VALUE IN EXCHANGE

From the late thirteenth century, the language of the marketplace and particularly the language of relation and relativity expanded into all areas of life and thought.[127] Peter John Olivi showed both knowledge of and characteristic independence from Aristotelian doctrine in his analysis of relativity in exchange.[128] At the beginning of his *Tractatus de emptionibus et venditionibus*, he remarked (following St. Augustine) that value can be assessed in two ways: either according to an ontologically defined nature, or according to human use. According to natural order, a mouse is valued higher than bread since it is higher in the scale of perfection, while according to human use, bread is valued higher. Since buying and selling are "ordered" (*ordinantur*) to human use, the second is the mode proper to the marketplace.[129] Olivi then isolated three factors in market use value. The first, which San Bernardino later labeled *virtuositas*, added a partially non-relational dimension to value by stressing the actual qualities of the commodity that made it better or more useful for a given task.[130] Olivi's second and prime component of market value concentrated on the relationship between supply or *raritas* and human need (*indigentia*), recognizing that *indigentia* is a quality that varies in relation to the difficulty of obtaining certain goods at certain times.[131] Olivi said that it was this second factor of *raritas* that explained why things more needed by humans, like water or air, cost less than things less needed but rarer and

---

[127] In logic this expansion is witnessed by the acceptance of the concept of *relatio* or "being in relation" as a primary category of being. See Gordon Leff, *Medieval Thought: St. Augustine to Ockham* (Baltimore, 1962), 277: Mark Henninger, *Relations: Medieval Theories 1250–1325* (Oxford, 1989). No modern author I have read on this subject recognizes the possible influence of monetization (i.e., money as an instrument of relation, the marketplace as a self-ordering system of relation, the relativity of values in exchange) or other social and economic developments on the new ontological status given relation in the fourteenth century.

[128] For a general discussion of Olivi's use of Aristotle, see Bettini, "Olivi da fronte ad Aristotile"; Burr, *Persecution*, 28, 75. See also Kirshner and Lo Prete, "Olivi's Treatises," 247.

[129] Olivi, *Tractatus*, 52. Spicciani, *La mercatura*, 190–92, discusses this aspect of Olivi's economic thought.

[130] In regard to this component of value, the word *virtuositas* is found in St. Bernardino's hand in the margin of the manuscript of Olivi's *Tractatus*. Bernardino used this term when he wrote his own economic sermons, copied in large part from Olivi's manuscript. Once again note that these components of value copied by Bernardino were taken by economic historians to be remarkably perceptive for the middle of the fifteenth century – a century and a half after their actual formulation. On this, see Olivi, *Tractatus*, 11; De Roover, *San Bernardino*, 19–20.

[131] Olivi, *Tractatus*, 53: "Secundo modo, secundum quod res ex sue inventionis raritate et difficultate sunt nobis magis necessarie, pro quanto ex earum penuria maiorem ipsarum indigentiam et minorem facultatem habendi et utendi habemus." Bernardino called this second factor *raritas*. Although Aristotle had failed to connect *indigentia* to factors of supply, this connection was being made by fourteenth-century commentators, notably Henry of Freimar, probably independently of Olivi's work. See Langholm, "Scholastic Economics," 123–24.

harder to obtain, like gold or balsam. Olivi's addition of the concept of *raritas* to the factor of *indigentia* represented an advance over the Aristotelian analysis of value.[132]

Olivi's third factor addressed an element completely unconsidered by Aristotle: the relativity of personal preference – the varying attraction of different goods to different buyers.[133] Here the element of the subjective will so important to the Franciscans came into play. Olivi asserted that the greater or lesser subjective appeal (*beneplacitum*) of one thing over another was "not a small element of value."[134] The label San Bernardino attached to this third element, *complacabilitas*, expresses well Olivi's sense that the relation between commodity and subject was a factor in itself. All three factors, but especially this third, underline the importance of "fit" or "congruence" as a determinant of value.[135] Not only did Olivi recognize more fully than any thinker before him the importance of this factor in the marketplace, but he was one of the earliest thinkers to exhibit similar forms of relational valuation in both his physical and his religious thought.[136]

The definition and minute explication of the *ordo rerum* was a traditional preoccupation of Christian thought. In the marketplace of the thirteenth century, however, the *ordo rerum* was redefined – away from the longitude of perfection and toward the latitude of relation. From the late thirteenth century, one sees the continual repetition in many forms of St. Augustine's recognition that economic value was determined relatively *quoad nos* and not considered hierarchically according to an ontology of inner natures, or as Oresme put it, "a leur naturel valeur ou

---

[132] For the introduction of the factor of supply to Aristotelian *indigentia*, see Langholm, "Scholastic Economics," 123; Langholm, *Price and Value*, 113–16.

[133] Olivi, *Tractatus*, 53: "Tertio pensatur secundum magis et minus beneplacitum nostre voluntatis in huiusmodi rebus habendis."

[134] *Ibid.*: "Uti enim prout hic sumitur est rem in facultatem voluntatis assumere vel habere, et ideo non modica pars valoris rerum utibilium pensatur ex beneplacito voluntatis, sive plus vel minus complacentis in usus huius vel illius rei; et istam habendam ad nutum: iuxta quod unus equus est gratior uni et alius alteri, et unum ornamentum sive iocale est gratius uni quam alteri." The importance of subjective desire as a determinant of value is downplayed in modern economic theory. See De Roover, *Pensée*, 48–49; De Roover, *San Bernardino*, 18–20.

[135] The first suggests a fit between object and requirement; the second a fit between scarcity and need; the third a fit between the object and the subjective desire of possession.

[136] Giacomo Todeschini ("Oeconomica II," 491) makes an interesting observation regarding Olivi's use of the word *idoneitas* (fitness, competence) in a religious context: "La 'idoneitas' che altrove era misura di perfezione attiva o contemplativa dell'uomo evangelico, diviene ugualmente misura della capacità retribuibile dell'operatore economico." For an example of commensuration and gradation in Olivi's Franciscan rule, see Burr, *Persecution*, 16. For the reflection of Olivi's relativism in his precocious insights into the causes of projectile motion, see Clagett, *Science of Mechanics*, 517; Michael Wolff, "Mehrwert," 418–19. Relational thinking shared by fourteenth-century natural philosophers was at the heart of theoretical advances in the analysis of motion. See chapter 7, 240–45.

parfeccion."[137] The "just price" came to be understood by all economic writers as a relative price.[138]

Even thinkers who had succumbed to the habit of relational thinking recognized that relativistic market order was alien to a Christian *ordo* conceived hierarchically and concerned with eternal values. This is apparent in a series of sermons written by Oresme while he was serving as dean of Rouen Cathedral (1363–77) and Bishop of Liseux (1377–82) toward the end of his life. Beginning in the first sermon he warned against valuing wealth and social position above inner virtue and charity. He voiced the old theme that gold corrupted balance and justice.[139] He indicated his fear that the concern for gaining wealth had supplanted the concern for gaining salvation.[140] Money provided no satisfaction for the soul.[141] Merchants were condemned for their venality,[142] and poverty was valued over wealth.[143] Money, Oresme insisted, did not and could not measure the inner nature of things, though men often acted as if it did.[144] In short, Oresme emphatically contrasted the order of the spirit to the order of the market, spiritual wealth to its material counterpart.[145] All of these themes, of course, had long traditions behind them.

At the same time, there are many instances within these sermons in which the religious sphere is invaded by relativist physical thought, Aristotelian proportionality as expounded in *Ethics* v.5, and even by metaphors and exempla drawn from the triumphant marketplace. Oresme could not talk about justice, even in this religious context, without bringing in Aristotle, *Ethics* v, and the central idea of the

[137] Odonis, *Ethics* v, lect. 9, 29b (1500, 105vb): "videmus enim quod quando est magna indigentia rerum tunc plus valent, sicut tempore messium sunt cariores famuli conducti, et tempore infirmitatum medici et tempore belli arma et sic de similibus. Res enim non summuntur hic in ordine nature sed necessitatis humane."

[138] For a clear fourteenth-century statement, see Guido Terrenia, *Ethics*, 45ra: "si res sit empta in loco ubi est magnum forum et porta ad locum ubi propter eius indigentia est carior, tunc potest vendi plus quam sit empta . . . ratione est quia res estimatur secundum indigentiam principaliter . . . in tempore autem ubi est minor indigentia emitur iusto pretio pro minori et in quam est maior indigentia venditur iusto pretio pro maiori." Notice Guido's use of the common expression *magnum forum* (*bon marché*) to mean both plentiful and inexpensive.

[139] Nicole Oresme, *Sacrae conciones* (*SC*), ms. BN lat. 16893, 56r.

[140] Oresme, *SC*, 9b: "mirum est et miserabile quia dolent de amissione denarium et non tristantur de amissione subditi." [141] Oresme, *SC*, 111r, 120r. [142] Oresme, *SC*, 13vb, 112r, 119r.

[143] Oresme, *SC*, 77vb, 81v–82v, 94v, 103v.

[144] Oresme, *SC*, 1a: "Vir stultus est qui hominem aut ex veste aut ex conditione estimat." In support, Oresme cites the foolish man in the story by Seneca who bought a horse because of its fancy saddle and bridle. Similar lessons occur throughout the Sermons. There is a continual association of moderate poverty with virtue and of wealth with shallowness. Oresme's French humanism is very apparent in these sermons; Seneca, Cicero, tales from the Latin histories, and Aristotle's *Ethics* are often used as authorities.

[145] Oresme, *SC*, 126rb: "Sed bona spiritualia non sic faciunt quia quanto plus dividuntur plus multiplicantur quia non habent quantitatem continuam que divisa diminuitur sed quantitatem virtutis que divisa non minuitur sed augetur." A similar point is found on 12va–b and 112rb.

proportional mean.[146] While constantly warning against the acquisition of wealth, he interprets the mediating role of the merchant in his society as that of the fabled tree in the middle of the earth whose fruit and branches provide food and shelter for all.[147] Perhaps most striking for one who would underscore the difference between the spiritual and the material was his physical conception of Christian qualities such as faith and virtue as "motions."[148] He used Aristotle's famous definition that "time is the measure of motion" to preach that it is better to run than walk toward the spiritual life, since in the same time the faster will cover a greater magnitude than the slower.[149] He associated Christian *caritas*, the "root" of all virtue, with the physical quality of heat – the hotter it was, the higher the *gradus*, the deeper it penetrated, the more perfect its work.[150] The association of Christian virtue with heat is an old topos. What was new was the physical reality granted to what had been, essentially, a metaphor, as if one could better understand virtue and faith as "motions," better understand the quality of charity through an analysis of the form and activity of heat.

Even more indicative of the power of the marketplace to define thinking in areas far removed from economic life are those passages in the sermons where spirituality is explained in terms of examples taken from the world of commerce.[151] To demonstrate that one spiritual act could have multiple effects, Oresme uses the example of a society of merchants, in which one economic loss is felt by all of its members.[152] In an elaborate allegory he links the thumb as the measure of the fingers to justice as the measure of the virtues, and both to money as the measure of all goods in exchange.[153] More striking still, when Oresme searched for the worldly and physical analog to the love of God as the scale and measure

---

[146] Oresme, *SC*, 56va: "Primo iustitia requirit equalitate medii proportionabilitatem et operis stabilitatem."

[147] Oresme, *SC*, 17vb: "ex ea vestebantur omnis caro, arbor in medio terrae sint mercatores qui tenent medium locum inter nobiles et populares."

[148] Oresme, *SC*, 98c: "motus fidei per quem anima a deo movetur"; 100a: "motus virtutis qui impellit hominem ad actum virtutis."

[149] Oresme, *SC*, 22b.

[150] Oresme, *SC*, 108ra. In this sermon Oresme also discusses charity as if it had an "altitude" expressed as a direction upward (love of God), and a "latitude" expressed as a direction sidewards toward the love of those near. For other physical representations of spiritual realities, see 9vb, 22rb, 70va, 75vb, 97rb–98va.

[151] For the history of the practice of phrasing religious ideas in economic terms and one striking example, see Courtenay, "King and Leaden Coin."

[152] Oresme, *SC*, 90va: "videmus quod isti mercatores qui habent communem societatem in mercatoribus quando unus recipit dampnum omnes de societate dolent."

[153] Oresme, *SC*, 48vb–49ra. To make this link, Oresme brings in Aristotle's discussion of commutative justice from *Ethics* v, and the invention of money as common measure to facilitate exchange from *Politics* i.

of our spiritual life, he chose as his example the merchant's scale, by means of which even the wonderfully stable golden florin must be tested in order to assess its true gold content or degree of perfection.[154]

Jean Buridan, more than any other thinker, had a vision of the marketplace as a system ordered on the basis of relative values. In a gloss on a passage from Book VII of the *Ethics*, he noted that all things can be considered relative to some other thing, and therefore all things are in some sense in relation.[155] When in Book II of the *Ethics* Buridan sought to illustrate Aristotle's point that virtue must be considered relatively *ad nos* rather than *in re*, he chose an example from the marketplace: the same amount of money does not always buy the same amount of grain, but sometimes more and sometimes less according to its abundance. He then referred his readers to Book V for a more complete discussion of relative value in the context of economic exchange.[156]

Buridan *played* with the idea of value relativity in the marketplace. He noted that although market price was a "common" price, and was generally the same for all buyers rich or poor, in relative terms the poor paid much more than the rich. For if one measured value in terms of the amount of labor involved in earning an equivalent price, then the poor man might have to work twenty times as long as the rich man to buy the same item at the same price.[157] Buridan, who in many areas of his *Ethics* commentary followed Geraldus Odonis, may have based this extension

---

[154] Oresme, *SC*, 111ra: "In omnibus rebus humanis quarum perfectio consistit in pondere et mensura, defectus earum cognosci non potest vel valor nisi quando fuerunt applicate ad bilenciam vel mensuram sicut patet de floreno aureo nam quantumcumque sciatur ipsum esse aureum . . . nunquam iudicatur bene de bonitate eius nisi quando ponitur in bilanciam et ponderatur si est boni ponderis et bonae mensurae . . . Spiritualiter loquendo . . . mensura vel statera vel bilancia [of the spiritual life] est ipsa dilectio dei . . . videatur bonus nunquam bene cognoscitur nisi quando in bilancia dei." For more on stable money as *statera* and *bilancia*, see 49ra.

[155] Buridan, *Ethics* VII.30, 683: "Prima est, quod omnis res est *ad aliquid*, quia omnis res est causa vel causatum, principium vel principiatum, idem vel diversum, et esse causam vel causatum, principium vel principiatum, idem vel diversum est esse ad aliquid . . . quod *omnis res est relatio*" (emphasis added).

[156] Buridan, *Ethics* II.16: "nec semper aequali pecunia bladum aequale emitur, nec ubique similiter: sed magis et minus, secundum nostram exigentiam vel abundantiam. Sed hoc usque ad quintum librum dimittatur." Also in this chapter, "Et non aequaliter venditur pannus in foro Brugis, et in foro Parisius."

[157] Buridan, *Ethics* V.16, 431: "Ad primum istorum dicendum est, quod indigentia istius hominis vel illius non mensurat valorem commutabilium: sed indigentia communis eorum qui inter se commutare possunt. Vel dicendum, quod pauper, quoad ea quibus abundat multo pluri precio emit ea quibus indiget quam dives, plus enim apponeret de labore corporali pro uno sextario frumenti, quam dives pro viginti: sed plus pecuniae non apponeret eo quod indiget ea, sicut frumento; universaliter enim indiget exterioribus bonis." Here both the value of the grain and the need for the grain are relatively greater for the poor man since he is willing to give a larger proportion of his labor for the same commodity than the rich man.

of market relativity on Odonis' earlier insight concerning the relativity of labor value.[158]

Buridan went still further in his analysis of the relativity of market value by recognizing that on the personal level *indigentia* varied in accordance with both objective wealth and subjective desire. The rich "needed" different things than the poor and in different degrees. He recognized that an analysis of economic value had to take into consideration the cost not only of such staples as grain and wine, but also of luxuries like jewels, fine clothes, silver, and gold. How could the high cost of luxuries be explained if the factor of need itself did not have a component of relativity, so that the rich felt they "needed" these excesses, and price duly reflected this imagined need just as it did real *indigentia humana*?[159] To this Buridan added that even the categories "rich" and "poor" might be better understood relatively, since, as Seneca suggested, poverty itself was not so much a matter of having less as of desiring more than one had.[160] By the end of his commentary on Book v, Buridan has managed to attach highly refined concepts of relativity and proportionality to every aspect of economic measurement and value.

COMMON VALUATION IN EXCHANGE

As sophisticated as was Aristotle's model of exchange with its recognition of proportional requital according to a geometric *figura*, it was limited as a description of the marketplace because it assumed that value and equalization were determined by the needs of individual exchangers. Albert and Aquinas made the most important corrections to Aristotle's analysis in this area with the recognition that market price was determined by

---

[158] Odonis, *Ethics* v, lect. 9, 30b (1500, 105b): "nec labor artificis venit ad extimationem comutationis. Plus enim laborat agricola pro una parva quantitate tritici, quam advocatus pro formatione unius libelli et tamen dabit advocato agricola multum de tritico in quo laboraverit in centuplo pro illo parvo libello." Once again, for Odonis, this complex problem of value was solved by the estimation and agreement of the parties involved. This sense of the relative worth of labor was implied in the Aristotelian text in regard to varying skill. The eleventh-century Greek commentator on the *Ethics*, Eustratius, whose commentary was translated by Grosseteste in the middle of the thirteenth century, was the first to attach the idea of relative labor value to the rarity or commonness of the task. Odonis explicitly recognizes the influence of Eustratius in his discussion of this question (1500, 105vb). For the influence of Eustratius' commentary on Buridan, see Langholm, *Wealth and Money*, 23–24.

[159] Buridan, *Ethics* v.16, 431: "non solum indigentia necessarii mensurat apud egenos commutabilia, sed etiam indigentia excessus apud divites." Buridan, *Ethics* v.17, 433: "Oportet igitur primo quod valor pecuniae indigentia humana mensuretur; licet enim forte non indigeamus ad nostras necessitates auro, vel argento, tamen divites indigent eis ad excessus suos in apparatibus vel exterioribus." For a discussion and appreciation of these insights of Buridan, see De Roover, *Pensée*, 48; Langholm, *Price and Value*, 125–26; Langholm, *Wealth and Money*, 95, 106.

[160] Buridan, *Ethics* v.16, 431.

common rather than individual estimation (*aestimatio communis*), based on common rather than individual need and utility.[161] One of the great accomplishments of the early commentators was their philosophical comprehension of this difficult yet crucial economic concept. The way was prepared for them, in part, by the conception of exchange embodied in Roman law, particularly in the judgment: "The prices of things are determined not by their value and utility to individuals, but by their value determined commonly."[162] In his gloss on *Ethics* v and in his commentary on the *Sentences*, Albert introduced the concept *secundum aestimationem fori* into scholastic discourse.[163]

Thinking on this subject continued to evolve and deepen in the commentary tradition with the recognition that the basis of value estimation, *indigentia*, was not determined, as Aristotle implied, by individual need but rather by an aggregate, common need (*indigentia communis, indigentia totius communitatis*).[164] The solidification of this understanding was essential to the expansion of Aristotle's model of exchange between individual producers into a model of the marketplace as a self-ordering, supra-personal process, functioning as a mechanism of equalization.[165] Within this new model, price was conceived as a geometric *product* based on aggregate estimations and needs.[166]

The precocious comprehension of the mechanism of common valuation was congruent with one of the dominant intellectual themes of the thirteenth and fourteenth centuries: the great value placed on the concept of the "common good." The ideology of the common good is too large a subject to be considered here, but the conception of the community as an aggregate entity with a common voice is evident in the growing assertion of the political rights of common assemblies and the legitimacy of the principle of common election. The economic analysis of every thinker so far considered, from Albertus Magnus to Nicole Oresme, was framed within the larger concept of the *bonum commune*.

The degree to which the value of the common good was extended in

---

[161] See chapter 3, 72–76.

[162] *Digest*, 35.2.3. See also Noonan, *Usury*, 81–99; Cahn, "Roots," 30–32; Baldwin, *Just Price*, 30, 54; De Roover, *Pensée*, 55–58; Gordon, *Economic Analysis*, 132.

[163] Albertus Magnus, *iv* SENT., d. XVI, art. 46 (Borgnet edn.), 638: "Justum autem pretium est, quod secundum aestimationem fori illius temporis potest valere res vendita."

[164] Aegidius Aurelius (Gilles d'Orleans), *Ethics* v, 216ra: "valor et preciositas rerum commutabilium non attenditur solus ex indigentia unius sed magis ex indigentia totius communitatis politice."

[165] For examples of later fourteenth-century statements of *communis aestimatio*, see Langholm, *Price and Value*, 107–09. See also Henry of Freimar, *Ethics* v, 187va: "indigentia hominis istius vel illius non mensurat valorem commutabilium, sed indigentia communis eorum qui inter se communicare possent in uno loco regione vel civitate."

[166] Langholm ("Scholastic Economics," 125) said of *communis aestimatio* that "it came to mean specifically the kind of estimate which the total community makes unintentionally through the impersonal working of the economy."

the fourteenth century can be seen in Remigius of Florence's treatise, *De bono communi*. Not only did he maintain that the common good was of greater importance than the welfare of any individual, but he asserted that the community was held nearer to God than the individual.[167] Remigius himself did not use his glorified concept of communal privilege to redefine the boundaries of economic liceity, but other thinkers did.[168] The scholastic acceptance of the merchant's right to profit by re-selling a thing for more than it had originally cost came only when thinkers began to appreciate the merchant's role in serving and benefiting the community.[169]

When at the very beginning of his *Tractatus* Olivi asked whether the desire to buy cheaply and sell dear was licit, he gave a practical answer to this theoretical question: yes, otherwise the whole *communitas* would be sinning since everyone desired it.[170] The controlling principle of Olivi's economic argument was the consideration of the common over the private good.[171] Aspects of the economic process that contributed or conformed to that good were deemed licit, whether or not they satisfied older, theoretical requirements for equality.[172] From the valorization of the common Olivi derived his most forward-looking economic concepts: his recognition of "probable profit" as profit commonly expected by those involved in commerce; his linkage of the just price to a common *aestimatio* that shifted relative to changing conditions; his consequent conceptualization of an estimated latitude of just price in opposition to an

---

[167] L. Minio-Paluello, "Remigio Girolami's *De bono communi*: Florence at the Time of Dante's Banishment and the Philosopher's Answer to the Crisis," *Italian Studies* 11 (1956), 56–71. Remigio realized (66) that he had to reinterpret St. Augustine in order to maintain these positions.

[168] Capitani, "Il 'De peccato usure' di Remigio de' Girolami." Remigio upheld the definition of usury as an unqualified sin against nature. He condemned most forms of *interesse* as well as the idea that there existed any fructifying potential in money, despite the obvious connection of such fructification to the power and wealth of his beloved city. For more on Remigio, see the introduction to chapter 4.

[169] Spicciani, *La mercatura*, 221. See above where Duns Scotus set the merchant's service to his community as the first condition for legitimate profit. The degree to which this reasoning was extended by mid-century is indicated by Buridan's position on moneychangers and monetary mutation, discussed on 156–57.

[170] Olivi, *Tractatus*, 51: "Et videtur quod sic. Quia aliter fere tota communitas vendentium et ementium contra iustitiam peccarent, quia forte omnes volunt care vendere et vile emere." We have seen in the previous chapter that this reasoning from common practice was not sufficient for Aquinas (*ST*, 2, 2, 77, 1, ad 2).

[171] Langholm, *Economics in the Schools*, 360: "The general rule [in Olivi's economic thought] is that the factors which influence value should not be interfered with provided that they are in accordance with the 'common estimate of the common good (*communis taxatio boni communis*)."

[172] For estimates of the centrality of the concept of the common good to Olivi's thought, see Kirshner and Lo Prete, "Olivi's Treatises," 247. Such acceptance did not extend to usury where money other than productive *capitale* was involved.

individually knowable point of equality; and his emphasis on voluntary agreement as a sufficient guarantee of exchange equality.[173]

A similar preoccupation with the common good appears in the thought of Nicole Oresme.[174] The most important glosses in his commentary on Aristotle's *Politics* expanded the powers of the community in political and economic decision making even beyond those granted by Aristotle.[175] In his treatise on money and minting, the *De moneta*, Oresme considered a wide range of topics, but its primary purpose was to convince the king that he had no right to alter or debase the coinage of the realm, despite the fact that from the last decades of the thirteenth century such debasement had been continually practiced by French kings to meet the multiplying fiscal demands of royal government. Oresme's opposition to monetary debasement was based on two lines of argument. In the first, mutation of the currency was a sin against the very nature of money as instrument and as measurement. A just measure was a fixed and certain measure.[176] Oresme's second argument turned on the valuation of the *bonum commune* above the *bonum privatum* of the prince.[177] In a chapter entitled "Who owns the money?," Oresme lectured the prince that, although it was his duty to put his stamp on money, he did so on behalf of the community, and he could claim neither lordship nor ownership of it. For money was a "balancing instrument" invented by the community to facilitate the exchange of their natural wealth, and was therefore the

[173] Olivi, *Tractatus*, 51: "Item secundum ordinem iuris et iustitie et caritatis commune bonum prefertur et preferri debet bono privato, sed communi saluti hominum post lapsum expedit quod taxatio pretii rerum venalium non sit punctualis nec secundum valorem absolutum rerum, sed potius ex communi consensu utriusque partis vendentium, scilicet, et ementium libere pretaxetur." Market values are determined, therefore (53), "non punctualiter, seu sub ratione seu mensura indivisibili in plus et minus, sed sub aliqua latitudine competenti, circa quam etiam diversa capita hominum et iuditia differunt in extimando."

[174] Buridan's influence on the ideas found in Oresme's justly famous treatise, the *De moneta*, has been debated since the beginning of this century. R. Kaulla, "Der Lehrer des Oresmius," in *Zeitschrift für die gesamte Staatswissenschaft* 60 (1904), 433–62, found precursors to Oresme's most important formulations on the nature of money in Buridan's commentaries on both the *Ethics* and the *Politics* of Aristotle. Bridrey, *Théorie de la monnaie*, chapter 4, disagreed, and, establishing the date of the *De moneta* as 1355, stated that it must have been written before Buridan's *Politics*, which he dated after 1360. Marie-Odile Piquet-Marchal, "Doctrines monétaires et conjonture aux XIVe et XVe siècles," *Revue internationale d'histoire de la banque* 4 (1971), 327–405, accepted many of Bridrey's conclusions, including the fact of Buridan's posteriority (342). Langholm's *Wealth and Money* deals extensively with the thought of both Buridan and Oresme, and his analysis (18–20, 99–103) does justice to the complexity of their relationship.

[175] Susan M. Babbit, *Oresme's "Livre de Politiques" and the France of Charles V* (Philadelphia, 1985).

[176] Oresme, *De moneta*, 20. Oresme cited the biblical injunctions against diverse weights and measures, Proverbs 15:10 and Deut. 25:16, and quoted Cassiodorus: "For what is so criminal as to permit oppressors to sin against the very nature of the balance, so that the very symbol of justice is notoriously destroyed by fraud?"

[177] Oresme pointedly called the prince the "most public person" (*persona magis publica*): *De moneta*, 10.

community's common property.[178] The central point of this treatise was that money was the possession of the community, naturally directed toward the common good, and therefore properly ordered and controlled by community decision.[179]

As if in concrete demonstration of his belief in the rights and responsibilities of the community, Oresme took the step, unusual for its time, of translating the Latin *De moneta* into the vernacular French so that it might be read and appreciated by the wider audience of the royal court.[180] Later in life, his continuing concern with matters touching the common good led him to undertake (at the direction and expense of his king, Charles V) the translation (with brief commentary) of the Latin versions of Aristotle's *Ethics* and *Politics* into French (1370–72).[181] Oresme's commentaries, copied into sumptuous manuscripts for the pleasure of the court, represented the first translations of complete works of Aristotle into any vernacular language.[182]

Jean Buridan went as far as any thinker in defining economic liceity in terms of service to the community. In regard to the mutation of money he took Oresme's position that if done for the *bonum privatum* of the prince it was utterly illicit, but the same act done on behalf of the common good could in many cases be deemed licit.[183] Concerning the work of the moneychanger, despised by both Aristotle and many contemporaries, he again wrote that if the changer performed his work solely for his private profit he was sinning against the nature of money as *medium*. If, on the other hand, he took only moderate profit and ordered his work toward the good of the whole community, then his work was licit and he could licitly charge for his labor and expenses.[184] John

[178] Oresme, *De moneta*, ch. 6, entitled "Cuius sit ipsa moneta" (10–11): "Quamvis pro utilitate communi princeps habeat signare nummisma, non tamen ipse est dominus seu proprietarius monete currentis in suo principatu. Moneta siquidem est instrumentum equivalens permutandi divicias naturales, ut patet ex primo capitulo. Ipsa igitur est eorum possessio, quorum sunt huiusmodi divicie . . . Est igitur pecunia communitatis et singularium personarum et ita dicit Aristotiles septimo Politice, et Tullius circa finem veteris Rethorice."

[179] Oresme, *De moneta*, 8: "pro utilitate communi, racione cuius moneta inventa est et ad quam naturaliter ordinatur."

[180] *Traictie*. The relationship between the Latin and French version is discussed in Gillard, "Nicole Oresme, sujet théorique, objet historique."

[181] The letter authorizing payment from the king is published in *Mandements et actes divers de Charles V (1364–1380)*, L. Delisle (ed.) (Paris, 1874), item 889.

[182] For more on this undertaking, see Menut in his introduction to Oresme's *Ethiques*.

[183] Buridan, *Politics* I.11, 54: "In nullo casu propter bonum privatum, nulla mutatio monetae est licita: sive apud principem: sive apud quemcumque alium . . . Propter commune bonum in multis casibus licita est mutatio monetae." Furthermore, the *bonum commune* was specifically linked to the facilitation of commerce (53): "Alio modo propter utilitatem reipublicae, i.e., ad dandum stipendia mercatoribus, et illa est licita."

[184] Buridan, *Politics* I.15: "si campsor utendo pecunia ordinat pecuniam in pecuniam ad propriam utilitatem, talis usus est illicitus. probatur. Nam talis intendit bonum privatum. Secundo quia

Noonan, the economic historian and historian of canon law, noted that Buridan's relativism in his judgment on moneychangers represented a sharp departure from traditional (particularly Thomistic) reasoning by justifying an unjust act by its ends.[185] But Buridan's position here is not all that different from the position developed by many thinkers over the course of the thirteenth century justifying the profit and activity of the merchant if it served the *bonum publicum* rather than merely the *bonum privatum* of the merchant himself. A generation before Buridan, Odonis constructed the following syllogism: that which benefits the *civitas* is just; geometric *contrapassum* governing economic exchange benefits the *civitas*; geometric *contrapassum* (despite its destabilizing aspects) is just.[186] By the late thirteenth century, the value placed on the common good had grown sufficiently great to change the designation of particular occupations and particular economic acts from evil to good within scholastic discourse.[187] In this way, the exaltation of the *bonum commune* fed into the growing power of the economic order to expand traditional definitions of justice and right order.[188]

The valorization of the common as a second point of ethical and economic reference gave added weight to the relativistic thinking so important to proto-scientific innovation in the fourteenth century. It was linked in addition to complex and scientifically fertile conceptions such as probability, estimated ranges, the notion of an aggregate product, and, above all, the model of a self-ordering, supra-personal system. The idea of linking value in the marketplace to common estimation developed alongside the problematic recognition that personal decisions based on personal needs often led to *distorted* and unjust value determinations in exchange. The position that developed within economic thought, which

ordinat pecuniam contra naturam suam. Secundo conclusio, si campsor pecuniam ad vite necessitatem ordinat et ad utilitatem et bonum totius communitatis talis campsor est licitus." Buridan in other decisions in his commentary on the *Politics* (II.2) justified the sins of both monopoly and monetary debasement if they are ordered *ad bonum commune*.

[185] Noonan writes (*Usury*, 185): "With this surprising thought, Buridan makes a radical departure from the scholastic tradition whose normative rule was that objective inequality and injustice could not be rectified by the end served."

[186] Odonis, *Ethics* v, lect. 9, 27b (1500, 105rb): "Illud est iustum in quo vel per quem civitas commanet et cives inter se commanent. Sed per proportionale contrapassum facere commanent et civitas et cives quare istud contrapassum est iustum."

[187] Buridan made his relativist judgments in an economic context despite being cognizant of the scholastic rule against rendering the illicit licit through circumstance. E.g., *Politics* 1.13, 68, "nihil de se malum ex iuramento facto est licitum. Probatur, quia malum de se nunquam potest bene fieri." Buridan's use of the common good to justify moneychanging was preceded as well by Guido Terrenia's similar solution to the question of usury. On this subject, see Giorgio Marcuzzi, "Una soluzione teologico-giuridica al problema dell'usura in una questione *De quolibet* inedita de Guido Terreni (1260–1342)," *Salesianum* 41 (1979), 647–84.

[188] Langholm, "Scholastic Economics," 133–34, has been particularly sensitive to this important form of redefinition.

held that the common view was a more trustworthy guide to value than the personal view (*pretia rerum non ex affectione nec utilitate singulorum, sed communiter funguntur*),[189] favored the association of truth with common judgment – an association that proved crucial to the future development of scientific thought.

THE SOCIAL GEOMETRY OF MONETIZED SOCIETY

The image of money as a common instrument touching all, touched by all, measuring diverse needs, goods, and services, and bringing all into relation was the scholastic descendant of Aristotle's *civitas* "held together" through geometrically proportional exchange.[190] What began to emerge in the fourteenth century was a radically new image of the world: from a static world of points and perfections to a world of *lines* in constant expansion and contraction. The new conceptual world of lines comprehended both nature and society, but there is strong reason to believe that the roots of this reconceptualization lay in the experience and comprehension of monetized society. A number of economic concepts previously discussed in this work support the association of monetized exchange with a new social geometry of intersecting lines. Among them are: the standard description of money as the connecting *medium* of exchange; money as the numbered *mensura* of the intensible quality of *indigentia* and so of a constantly expanding and contracting market value;[191] money's description as a common continuum facilitating exchange by permitting "toutes choses estre mesurées ensembles";[192] the Aristotelian notion that economic gain and loss should be conceived of as a line to be bisected and equalized;[193] Buridan's remarkable addition that each person carried around his own potential line of *lucrum/damnum*, measurable by money and conceptually extended in every act of exchange; the association of money as instrument with the crossed diagonals in Aristotle's geometric *figura* of exchange; the development of the legal doctrine of *laesio enormis*,

---

[189] *Digest*, 35.2.63.

[190] Aristotle [1132b32–35]. Albertus Magnus was directly within this line of descent with his precocious concept of the "flux and reflux" of proportional services (*Ethics*, Borgnet edn., 356a). An early fourteenth-century example: Burley, *Ethics* v, 83rb: "Sed per hoc quod cives proportionaliter compatiuntur: commanent cives sibi invicem in civitate. ergo iustum commutativum continet contrapassum secundum proportionem."

[191] From *HG* VI.22, 208: "medium sive regula sive pretium sive mensura"; Langholm, *Wealth and Money*, 69. Money was specifically conceived to measure the "excess and defect" of things compared mutually, underscoring its ability to expand and contract. Oresme, *Ethiques* v, 294: "car par elle mesure l'en toutes teles choses et la superhabundance et la deffaute."

[192] Oresme, *Ethiques* v, 297.

[193] Odonis, *Ethics* v.10, 16b (1500, 102rb): "sciendum quod bonum et malum restringi possent ad lucrum et damnum."

whereby just value was recognized as a range within numbered limits defined along a line of price;[194] and the acceptance by theologians like Olivi and Duns Scotus of the "just price" not as an arithmetic point, but as a continuous range or monetarily expressed and graded *latitudo*.

The expanding place of the monetized market in the life of the *civitas* acted to focus attention away from idealized points and toward the dynamics and mathematics of the continuum. Philosophers reformulated the definition and requirements of equality from exactitude to approximation, from the knowable to the estimated. Economic equilibrium came to be recognized as the product of a self-ordering, aggregate process, rather than the result of individuals consciously maintaining balance and order. But where in this picture was Aquinas' and Henry of Ghent's exchanger as judge? Where was the exchanger who based his economic decisions on an interiorized *ideal* of equality and consciously ordered each transaction toward that ideal? When Aquinas wrote his gloss on Aristotle's statement that the *civitas* was held together through proportional exchange, he was sure to emphasize the individual's responsibility to return an equivalence proportional to that which he had received.[195] For Godfrey, Olivi, Odonis, and Buridan the voluntary decision to exchange became in itself the guarantee of equality. Even while equality remained the ideal end, theorists came to recognize that profit and not equality motivated most exchangers, and that exchanges motivated by the desire for profit could nevertheless be just. Olivi and Buridan observed that the decision to exchange was based not on the striving for equality, but on the desire by each party to get *more* out of the exchange than the other.[196] Rather than condemning this situation, they recognized it as a sufficient basis for just agreement, provided that all parties were in a position to make *rational* estimations and decisions concerning their own advantage. In the emerging conception of market exchange, imbalance (desire for profit) was routinely transformed through a geometric process of equalization into the balance of exchange agreement.[197] Aristotle had been left far behind.

Money as common measure facilitated the rational calculation of

---

[194] For the recognition of *laesio* within fourteenth-century scholastic economic thought, see Buridan, *Politics* I.16; Oresme, *Ethiques* IX, ch. 1.

[195] Aquinas, *Ethics* (1969), 291a: "per hoc commanent cives sibi invicem in civitate quod sibi invicem proportionaliter contrafaciunt, prout scilicet si unus pro alio facit aliquid, *alius studet proportionabiliter facere pro eodem*" (emphasis added).

[196] Buridan, *Ethics* v.10, 409: "His ergo observatis, lucrum vel damnum si accidant, non sunt contra justitiam, immo sic in commutationibus est vera mercatura, secundum quam communiter utraque pars accipit *quod magis est utile sibi*" (emphasis added). Compare this to Aquinas' opinion, 96–99.

[197] Olivi, *Tractatus*, 61: "Nam ex utriusque partis libero et pleno consensu incohatur et rectificatur. Ita quod emptor vult sibi plus rem emptam quam pretium eius, et venditor econtrario."

self-interest now seen as the basis of exchange equality. In Buridan's economic analysis even the smallest losses and gains, the seemingly most subjective services and qualities, could be measured, quantified, and proportionally equalized by monetary price. Thomistic *judgment* toward the end of an ideal equality had been replaced by *calculation* in the direction of a self-interest minutely observed. In a number of passages in the *De moneta*, Nicole Oresme showed that he, like Buridan, made the connection between money as measure and the most delicate calculations based on self-interest. Oresme talked of merchants and bankers as if they were elements in a geometric equation, responding with the greatest sensitivity to changes in the quantity and quality of the money supply. His observation that merchant activity was determined by the measurement of self-interest even when such interest violated the law and interests of the *civitas* underlined the distinction between the new calculation and older scholastic requirements of individual judgment and responsibility. To illustrate the dangers of the king's monetary policy he wrote:

Again, such alterations and debasements diminish the amount of gold and silver in the realm, since these metals, despite any embargo, are carried abroad, where they command a higher value. For men try to take their money to the places where they believe it to be worth the most. And this reduces the material for money in the realm.[198]

Oresme made this statement without condemning the merchants who so acted.[199] Instead he criticized the king for creating economic conditions that would, almost inexorably, lead merchants to export gold and silver from the realm. For Oresme, activity based on the close calculation of gain and loss was a primary element in the social geometry of commercial society. The power and efficacy of minute calculation was proved every day by the merchants, bankers, speculators, and moneychangers who kept the *civitas* healthy and supplied, while amassing riches and social capital for themselves.[200]

The extent to which thinkers came to recognize rationalized models of economic behavior is shown in another passage from the *De moneta*. Here

---

[198] Oresme, *De moneta*, 32: "Rursum aurum et argentum propter tales mutaciones et impeioraciones minorantur in regno, quia non obstante custodia deferuntur ad extra, ubi carius allocantur. Homines enim libencius conantur suam monetam portare ad loca, ubi eam credunt magis valere. Ex hoc igitur sequitur diminucio materie monetarum in regno."

[199] Rather than judging here, Oresme was simply recording the existence of a sensitive international market in precious metals that had been in place from the late thirteenth century. The royal *Ordonnances*, however, were less understanding. For the frequent harsh condemnation of metal exporters, see *Ordonnances*, vol. I, 324, 423, 770, 792 (1294, 1304, 1322, 1326); vol. II, 255, 390 (1346, 1350), and chapter 1.

[200] Ibanès, *Doctrine*, 78; Cazelles, *Histoire de Paris*, 95–118.

the merchant is described as if he were responding almost mechanically to the shifting pressures of the marketplace:

Again, because of these alterations, good merchandise or natural riches cease to be brought into a kingdom in which money is so changed since merchants, other things being equal [*ceteris paribus*], prefer to pass over to those places in which they receive sound and good money.[201]

By the fourteenth century it was well understood: the merchant calculated in terms of his profit by translating values into monetary terms. So too did all exchangers. So too did Buridan, Oresme, Burley, Odonis, Bradwardine, and Heytesbury in their private purchases and their public administrative services. Though it went against some of the basic principles of their philosophical and theological heritage, these same thinkers were brought to recognize that money and the calculation of self-interest lay at the basis of exchange equality and therefore at the connecting center of the *civitas*. For scholastic philosophers working within an intellectual framework that encouraged the transference of insights between disciplines and "sciences," the recognition of a social geometry constructed around money as connecting and measuring *medium* had intellectual repercussions far beyond the economic sphere. Nicole Oresme, who observed his fellow Parisians calculating their actions in terms of minute movements along numbered lines of price, also invented a remarkable "geometry of qualities" to quantify and represent continuous qualitative changes by means of geometric lines and figures.[202] His was only one manifestation of the new passion to devise systems of measurement and relation within fourteenth-century natural philosophy constructed around the geometric line or latitude.

Within natural philosophy, qualitative change was broadly reconceived as were the methods of measuring and comparing this change. How does something become whiter or hotter? How does the human quality of charity increase and decrease? How does human charity relate to Christ's charity? In answering these and similar questions, the search for knowable points of perfection gave way to the passion to measure changes in qualitative intensities, newly conceived as motions along gradable continua. Geometry replaced arithmetic as the mathematical key to nature. *Calculatores*, natural philosophers, mathematicians, and visual artists began to analyze a world now seemingly composed of lines

---

[201] Oresme, *De moneta*, 33: "Item propter istas mutaciones bona mercimonia seu divicie naturales de extraneis regnis cessant ad illud afferri in quo moneta sic mutatur, quoniam mercatores ceteris paribus prediligunt ad ea loca transire, in quibus reperiunt monetam certam et bonam."

[202] Oresme's plan to quantify qualities through geometric representation is discussed at many points in chapter 7.

in expansion and contraction, creating complex mathematical and logical languages to facilitate their measurement and relation. Within this new geometric world of lines, natural philosophy evolved into science.

Chapter 6

LINKING THE SCHOLASTIC MODEL OF MONEY AS MEASURE TO PROTO-SCIENTIFIC INNOVATIONS IN FOURTEENTH-CENTURY NATURAL PHILOSOPHY

A complex and dynamic vision of the natural world emerged within fourteenth-century natural philosophy and, with it, a cluster of logico-mathematical languages capable of describing and bringing order to this vision. Over the course of the century, the traditional image of a world of hierarchically fixed and absolute values began to dissolve, gradually replaced by a shifting, relational world in which values were in constant flux. As philosophers turned from the investigation of perfections and essences to the investigation of motion and change, questions of measurement and gradation came to dominate scholastic discourse. Since the pioneering work of Pierre Duhem in the early years of this century, historians of science have come to recognize the critical place this shift in philosophical concern played in the development of modern scientific thought.[1] The concluding chapters of this book examine the many ways in which the intellectual process of scientific innovation was tied to the social experience of monetization and market development.

The suggestion that fourteenth-century philosophical speculation was influenced by social developments taking place in the society beyond the schools raises a number of questions. If natural philosophers were deeply involved in the world beyond the classroom and sensitive to its changes, why is the phenomenal world – including the world of social experience – so absent from their speculation? Why in the study of nature do

---

[1] Pierre Duhem, especially his monumental *Le système du monde: histoire des doctrines cosmologiques de Platon à Copernic*, 10 vols. (Paris, 1913–59); Anneliese Maier, *Die Vorläufer Galileis*; Maier, "La doctrine d'Oresme," esp. 349–52; Marshall Clagett, "Some Novel Trends," esp. 302–03; Clagett, "Richard Swineshead and Late Medieval Physics," *Osiris* 9 (1950), 131–61, esp. 131. At the same time that historians recognize the importance of fourteenth-century science in the development of mathematical physics, they warn against the anachronistic "reading back" of later approaches into the thought of the fourteenth century. Again, Maier, *Die Vorläufer Galileis*, 1–6; Clagett, "Nicole Oresme and Medieval Scientific Thought," *Proceedings of the American Philosophical Society* 108 (1964), 298–310, esp. 298.

empirical observations play a minuscule part within arguments dominated by technical logic? Why, for example, when the subject of investigation is a quality such as whiteness, do thinkers use the same denatured examples of Socrates and Plato when examples from personal experience were easily available? Why do natural philosophers often choose to discuss the physics of an imaginary case (*secundum imagin- ationem*) rather than the physics of openly observable and measurable occurrences? Why in a natural philosophy preoccupied with questions of measurement did no one actually bother to measure anything?

The answers to these questions involve one of the primary intellectual continuities between the thirteenth and the fourteenth centuries: a shared understanding, following Aristotle, that the goal of philosophy is to establish propositions that are universally true, necessary, and certain.[2] Within the medieval university the term *scientia demonstrativa* applied not to demonstrations by experience but to a system of syllogisms and propositions satisfying the Aristotelian requirements of logical demon- stration.[3] With few exceptions, philosophers believed that the certainty required of science was to be found through the application and the test of technical logic rather than through direct observation of the contin- gent object world.[4]

The development of a logic of supposition or "positional difference," and its refinement in the philosophical thought of William of Ockham, intensified this belief.[5] In Ockhamist logic, while the knowledge of singulars is possible through intuitive cognition, and intuitive cognitions derived from direct experience provide "evident assent" with respect to contingent fact, these cognitions do not provide a basis of certainty sufficient for science. In order to attain its required necessity and univer- sality, the proper objects of demonstrative science must be propositions rather than natural phenomena.[6] In positional-difference logic, proposi- tions found verifiability in the internal fit and agreement of their parts

---

[2] John F. Bowler, "Intuitive and Abstractive Cognition," in Kretzmann, Kenny, and Pinborg, *Cambridge History of Later Medieval Philosophy*, 460–78; Eileen Serene, "Demonstrative Science," *ibid.*, 496–517.   [3] Serene, "Demonstrative Science," 496.

[4] Murdoch, "Unitary Character." This is not to say that thinkers denied the importance of experience in the acquisition of knowledge, or that all agreed that the proper immediate objects of science were propositions. See Katherine Tachau, *Vision and Certitude in the Age of Ockham: Optics, Epistemology, and the Foundation of Semantics 1250–1345* (Leiden, 1988), particularly her treatment of Peter Aureol (110) and Adam Wodeham (303–09), who criticized Ockham on these points. For Aquinas' position on the requirements of scientific truth, see *St. Thomas Aquinas: The Divisions and Methods of the Sciences – Questions v and vi of His Commentary on the "De trinitate" of Boethius*, Armand Maurer (ed. and tr.), (Toronto, 1963).

[5] John Murdoch, "William of Ockham and the Logic of Infinity and Continuity," in Norman Kretzmann (ed.), *Infinity and Continuity in Ancient and Medieval Thought* (Ithaca, N. Y., 1981), 165–206, esp. 195–204.   [6] Serene, "Demonstrative Science," 513.

rather than in their agreement to sense observation.[7] Truth or falsity was judged in isolation from the world of contingent objects. In the exposition of their insights into nature, the Oxford Calculators (less so the Paris masters) consciously sought to cleanse their language and logic of contamination from the contingent world of both society and nature. This attempt to distance and cleanse explains the seemingly paradoxical existence in the fourteenth century of what John Murdoch has called a "natural philosophy without nature."[8] Since a willful isolation from the contingent determined the form and language of philosophical presentation, it is necessary to look below the surface of formal presentation to discover both the experiential matrix and the perceptual changes that underlay proto-scientific innovation.[9]

The development of a concern for quantitative analysis was a precondition for the emergence of science. This critical development has been traced back to the late thirteenth century when philosophers became fascinated with the question of qualitative change.[10] No longer satisfied with a vague vocabulary of measurement, philosophers began to create complex logico-mathematical languages to investigate ever more intricate problems of comparison and gradation. The proto-science of *calculationes* emerged from this philosophical context in the first quarter of the fourteenth century. English scholastics, in particular those associated with Merton College, Oxford, were the first to develop and apply mathematical rules to problems of qualitative motion and gradation. These scholars, now known collectively as the Merton School or the Oxford Calculators, extended the logic and mathematics of qualitative measurement to questions of local motion or motion in space.[11] The influential Calculator

---

[7] Murdoch, "Involvement of Logic," esp. 15–18; Murdoch, "Logic of Infinity," 204; Buridan, *Jean Buridan's Logic: The Treatise on Supposition, The Treatise on Consequences*, Peter King (ed. and tr.) (Dordrecht, 1985), 35, 195. For a discussion of Ockham's use of meta-linguistic analysis in his investigation of the natural world and his influence on later natural philosophy, see Murdoch, "Analytic Character," esp. 179–83, 197–98. Here Murdoch links the use of sophisms, logic, and meta-linguistic analysis in natural philosophy to "the rapidly growing belief in the absolute contingency of the natural world and in the concomitant concerns with the certitude of knowledge and the grounding of this certainty, not in contingent *things*, but in *propositions* as the only possible bearers of the requisite universality and necessity." On this point, see also André Goddu, *The Physics of William of Ockham* (Leiden, 1984), 75–77.

[8] Murdoch chose to underscore this point by incorporating it into the title of his essay, "The Analytic Character of Late Medieval Learning: Natural Philosophy Without Nature," and see esp. 191–96. Note Murdoch's explanation (197–98) for William Heytesbury's rigidly logical, analytical, and abstract approach to physical questions in terms of his perception of the relative uncertainty of experience in comparison to logic.   [9] Murdoch, "*Mathesis*," esp. 225ff.

[10] Clagett, *Science of Mechanics*, 206; Clagett, "Swineshead," 132–50; Maier, "Das Problem der intensiven Grösse," esp. 59ff.; Weisheipl, "Developments in the Arts Curriculum," esp. 173–74.

[11] Clagett, *Science of Mechanics*, 199–219.

Thomas Bradwardine treated velocity as if it were a qualitative ratio capable of being intensified or lessened in the same manner as whiteness and heat.[12] Under this assumption he devised (c. 1328) the first *workable* (i.e., internally consistent) mathematical function describing the proportional relationships between velocities, forces, and resistances in the measurement of motion.[13] Bradwardine's function was so clearly superior to previous mathematical treatments of motion, including Aristotle's, that it was soon avidly adopted and applied to a wide range of philosophical problems involving ratio and relation.[14] In the process of refining their analysis of motion and qualitative change, the Calculators laid the foundations for the development of mathematical physics.

The mathematical potentialities of measurement pioneered by the Oxford Calculators in the area of velocities, forces, and resistances proved exceptionally attractive to theologians and arts masters at the University of Paris. Beginning in the 1330s, the Paris masters extended the subtle methods and models of the Calculators into all areas of scholastic thought.[15] Every "quality" capable of increase or decrease, whether physical or mental, came to be visualized as a divisible, continuous magnitude in the process of expansion or contraction. Theology proved especially susceptible to the new passion to quantify. Philosophers of the period did not make the distinctions between physical and mental qualities that we now make.[16] In one theological treatise after another, the intensities of qualities we today consider essentially subjective and immeasurable, such as love, grace, and charity, were treated by the Calculators as measurable and capable of relation in the same way that lines or spatial magnitudes are measurable.[17] In Oxford and Paris,

---

[12] Weisheipl, "Dumbleton," 447.

[13] The word "workable" is applied here only in a mathematical sense. Bradwardine's function was intended to be internally consistent rather than consistent with empirical observation. See Bradwardine, *De proportionibus*; Maier, "Der Funktionsbegriff." This work has been partially translated by Sargent as "The Concept of the Function in Fourteenth-Century Physics," in *On the Threshold of Exact Science: Selected Writings of Anneliese Maier on Late Medieval Natural Philosophy* (Philadelphia, 1982), 61–75.

[14] Bradwardine, *De proportionibus*, 11–18; Murdoch, "*Mathesis*," 225–32.

[15] Murdoch, "*Subtilitates Anglicanae*"; Murdoch, "Unitary Character," esp. 280–303; Murdoch "*Mathesis*," 238ff. For an example of such an application, see Coleman, "De Ripa." The attempt of de Ripa and other Parisian theologians to apply mathematical rules to theological problems is discussed further on 232–35.

[16] Rega Wood, "Calculating Grace: The Debate about Latitude of Forms According to Adam de Wodeham," in Asztalos, Murdoch, and Niiniluoto, *Knowledge and the Sciences*, vol. I, 373–91, esp. 373.

[17] Underscoring the essential connection between mental and physical qualities is the fact that the locus of the earliest debates on the question of the physical basis of qualitative change were the many commentaries on the theological question of the increase of charity from book I, d. XVII, of Peter Lombard's *Sentences*.

elaborate logical and mathematical languages were devised to describe and conceptually measure quantified qualities now conceived as divisible continua.[18]

The pervasiveness of this habit of quantification has led modern scholars to talk of a "measuring mania" in the philosophy of the period.[19] How do we explain the strength or the characteristics of this mania, and why did it appear at this time? There have been a number of recent attempts to answer this question.[20] While the best answers have recognized the influence of social factors on the Calculators' fixation with the logic of measurement, the social factors considered have been limited to an artificially isolated university culture. Edith Sylla's analysis of the culture of logical disputation within the schools, and her conclusion that the Calculators' primary goal was to do and teach logic rather than natural philosophy, explains essential elements of their use of measurement: that it remained entirely conceptual and theoretical; that problems in measurement were investigated primarily through the use of *sophismata* or imaginary cases rather than through cases based in physical reality; that terminist logic rather than physical theory was used to solve most of the questions raised.[21] But analyses rooted in the social context of the university have their limits. In themselves they cannot explain the passion to quantify qualities at the heart of this intellectual movement, nor the form and function of the conceptual instruments devised by the Calculators as an aid to their program of measurement and relation.

Taken together, the analysis of the disruptive social process of monetization and the insights and models derived by scholastic thinkers to make sense of this social process provide new perspectives from which to view the emergence of a measuring "frenzy" within philosophy. On the most basic level, the two aspects of fourteenth-century science that most strike historians are: (1) the generalized attempt to extend a formal system of measurement into every area of life and thought, and (2) the quantification of an extraordinary range of mental and physical qualities extending from heat and whiteness to love and grace. Both characterized as well the activity of money as measure in the society of the fourteenth century. When commenting on money and its uses, scholastic thinkers never failed to stress its success as an instrument of measurement and the great range of subjects and services it quantified and brought into relation: "ad

---

[18] The best general description of these measuring languages is given in Murdoch, "Unitary Character," 280ff.
[19] This is John Murdoch's phrase ("Involvement of Logic," 19).
[20] See 3–5 for a brief description of current historical thought on this question.
[21] Sylla, "Oxford Calculators"; Murdoch, "Unitary Character," 279.

hoc enim inventum est nummisma ut sit medium et mensura omnium commutabilium."[22]

The price of a good or service expresses the quantification of its quality as a commodity.[23] In effect the ever-shifting value of every article and service in the marketplace is measured and quantified by moneyprice, and every sale demonstrates the effectiveness of monetary measurement in overcoming extremely complex problems of valuation and commensuration. The range of subjects and qualities the Calculators of the fourteenth century sought to measure and quantify was matched only by the range of commodities and services being measured and related by moneyprice in the marketplace around them. As the century progressed, money measurement extended far beyond objects to accommodate the most subjective of services and qualities. Following a century of commutation, the values of labor and land, formerly fixed by tradition and tied to a complex system of social reciprocities, were increasingly commoditized and measured by a shifting market price. The quantification of quintessentially subjective qualities that characterized the science of *calculationes* was embedded in contemporary religious practice as well, seen most clearly in the developing system of "accounting" surrounding the granting of papal indulgences. By the beginning of the fourteenth century (1308), monetization and rationalized price measurement had so invaded the realm of official theology that the proportion of payment to reward in indulgence could be officially fixed by Clement V at one penny of Tours for each year of pardon conferred.[24] The repercussions of this commoditization, the sense that everything was for sale, that all ancient values and structures had a price attached to them, was profoundly experienced in every European society of this period.[25]

Fascination with the measurement and comparison of subjective qualities such as charity and grace in a philosophical and theological context had its parallels in scholastic *economic* thought and practice. From the middle of the thirteenth century, philosophers weighed the difficult question of the legitimacy of merchant profit. We have traced the emergence of a consensus (shared by thinkers as normally opposed as Godfrey of Fontaines and Henry of Ghent) around a scheme legitimizing

---

[22] Burley, *Ethics* v, 83vb.
[23] Karl Polanyi, *Primitive, Archaic, and Modern Economies: Essays of Karl Polanyi*, G. Dalton (ed.) (Garden City, N. Y., 1968), 182: "The specific characteristic in the payment use of money is quantification." For an earlier discussion of the pervasive quantification of qualities in the social realm, see Joel Kaye, "The Impact of Money on the Development of Fourteenth-Century Scientific Thought," *Journal of Medieval History* 14 (1988), 251–70, 252–57.
[24] William E. Lunt, *Papal Revenues in the Middle Ages*, 2 vols. (New York, 1965), vol. I, 111–25, vol. II, 458. See also Kaye, "Impact," 256.
[25] See chapter 1, 18–19, for examples illustrating the urgency of this theme at this time.

merchant profits but only in proportion to the expenditure of the merchant's physical labor and to his possession of quantifiable *mental* qualities such as solicitude, honesty, and professional expertise.[26] Within this consensus, mental qualities could licitly be translated into a numerable price value in the marketplace.[27] That is to say, the measurement of mental qualities by the numbered scale of price was current in society on a grand scale before the quantification of subjective qualities became current in natural philosophy. And the measurement of subjective qualities by money price was being analyzed in scholastic economic thought at the same time that the concern for quantifying mental qualities was growing in the realm of theology and natural philosophy. Jean Buridan's already noted calculation of the money value of a simple "thank you, sir," or *grates Domine*, provides perhaps the clearest indication of the degree to which price measurement had invaded formerly subjective areas and was understood to measure personal and mental qualities in the marketplace.[28]

Money measurement found its most rationalized application within the growing administrative bureaucracies of the century – city governments, royal governments, the administration of diocese, chapter, and order, the ballooning bureaucracy of the papacy at Avignon, the expanding university itself. Bureaucracies require standardization to function. The first step in bureaucratic standardization is the introduction of moneyprice as standardized measure. By the thirteenth century, money, often expressed as numbered annual income, served as the prime bureaucratic instrument in the quantification of qualitative social grades and responsibilities. Thus, bureaucratic quantification of social qualities through the use of a monetary scale was already established in society, especially educated society, before the mania to measure and quantify came to dominate philosophical speculation.[29] Furthermore, the century's most highly reputed scholars were continually employed as servants in those civil and ecclesiastical administrations most responsible for the extension of monetary measurement and gradation.

The frequent focus of fourteenth-century natural philosophy on the seemingly "abstract" intricacies of a highly technical mathematics and logic, and the dry, refined intellectuality of its formal presentation,

---

[26] *HG* I.40, 228; Olivi, *Tractatus*, 56; Duns Scotus, *Sent. iv*, 317; Langholm, *Economics in the Schools*, 55–56, 258; and see chapter 4, 103–04, and chapter 5, 139–41.

[27] See chapter 5, 141–42, for Olivi's complex discussion of payment in the labor market for the relative expenditures of skill, energy, and care in the various occupations; Olivi, *Tractatus*, 56–57. Again, even in the case of Olivi's detailed discussion, it is clear that scholastic commentators were recording only a fraction of the actual complexities of price measurement taking place around them in the marketplace.   [28] See chapter 5, 143; Buridan, *Ethics* v.14, 425.

[29] See chapter 1.

disguises the fact that these logical and mathematical investigations of nature were being carried on by men alive to the social and economic realities of their societies. Godfrey of Fontaines, Henry of Ghent, Peter John Olivi, Duns Scotus, Geraldus Odonis, Thomas Bradwardine, William Heytesbury, Richard Kilvington, Walter Burley, Jean Buridan, Nicole Oresme, Albert of Saxony – all served as administrators for Church, order, university, or king; all of them (with the exception of Bradwardine and Heytesbury) have left writings demonstrating knowledge of economic affairs and of the critical role of money as an instrument of measurement in exchange; all can be found making the most innovative and forward-looking contributions in the area of natural philosophy.

Given the pervasive context of money's rationalization of measurement within the marketplace and within the rapidly developing bureaucracies of civil, ecclesiastical, and university government, I propose that the answer to John Murdoch's question "how and why did the near frenzy to measure everything imaginable come about in the fourteenth century" must include the consideration of the social context of a monetized society.

MONEY AS *MEDIUM* AND AS MEASURE

How did money function within the monetized society of the fourteenth century, and how is the scholastic model of money as measure reflected in *particular* technical developments within fourteenth-century natural philosophy?

Monetization, it needs to be said, is a social context defined by the increased importance and use of a specific tool, the instrument of money. The growing consciousness of money as a physical instrument culminated in its analysis as an "instrument of equalization" (*instrumentum equivalens, l'egal instrument*) in the *De moneta* of Nicole Oresme.[30] Building on this idea, Oresme maintained that it must be well designed in order to function as an *instrumentum aptum* of measurement and relation.[31] This led him to list the necessary properties of money as tool, a list with an important place in the history of economic thought.[32] First on this list was the very physical requirement that money be "easy to handle and to feel with the hands," and second, that it be light to carry.[33] Given the

---

[30] Oresme, *De moneta*, 10.

[31] Oresme, *De moneta*, 5: "Et quoniam moneta est instrumentum permutandi divicias naturales, ut patet ex capitulo precedenti, consequens fuit quod ad hoc tale instrumentum esset aptum."

[32] For the history of this list of money's ideal properties from the "averroist" manuscripts of the late thirteenth century through Buridan and Oresme, see Langholm, *Wealth and Money*, 80–88.

[33] Oresme, *De moneta*, 5: "quod fit, si sit faciliter manibus attractabile seu palpabile, leviter portabile, et quod pro modica ipsius porcione habeantur divicie naturales in quantitate maiori."

culture-wide acceptance and everyday use of this money-instrument and the daily necessity on the part of everyone, from peasant to scholar, to calculate with it and to use it carefully and knowingly and well, I have suggested that monetization represented not only a major social advance but an important, perhaps epochal, *technological* advance.[34] What effect did the continual use and reckoning with this measuring tool have on the consciousness of its users? In particular, what effect did the technological impact of money have on the development of models of measurement and relation by philosophers and mathematicians of the fourteenth century? To answer these questions the technological form of money must be analyzed in greater detail.

Unlike most instruments, the shape of money does not indicate its range of uses. Its technological form is, in a sense, hidden. When people of the fourteenth century looked at the coined money in their hand, they saw a round, discrete object – in rough geometrical terms, a point. However, as writers on money from Aristotle through the scholastic theorists to the twentieth century have realized, money in fact functions as a *line*, a connecting *medium* (to use the word attached to it by Aristotle and the scholastics), a measuring scale composed of a divisible and expandable *continuum* of value.[35] All goods, all commodities, all services (as Jean Buridan noted, even highly subjective services that at first glance seem immeasurable) have their values expressed in terms of their price, i.e., their position on this common, numbered, measuring line or continuum.[36]

Guided by the rich Aristotelian and scholastic discussions of money as measure, I have defined the technological form of money as the following: an extendable, divisible, graded, and numbered continuum used as a common measuring scale, capable of expressing constantly shifting and diverse values in common numerical terms, and thus facilitating relation between seemingly incommensurable goods and services in exchange.[37] I call this model *the technological form of money in exchange*. It describes the function of money as experienced by the great majority of the population in the great majority of economic transactions. As students and masters living in commercial centers, as procurers and financial officers within their universities, natural philosophers of the fourteenth century manipulated the instrument of money following this model in countless transac-

---

[34] For an earlier statement of this point, see Kaye, "Impact," esp. 259–60.

[35] A list of the many ways in which money was treated as a line in the monetary thought of scholastic thinkers is given in the conclusion to chapter 5, 193–94. The continuation of the concept of money as *medium* in the economic thought of the twentieth century can be seen in L. von Mises, *The Theory of Money and Credit*, H. E. Bateson (tr.) (New Haven, 1953), 48.

[36] For Buridan's discussion, see *Ethics* v.14, 425.

[37] For an earlier formulation, see Kaye, "Impact," 260.

tions. Their writings on money reveal their awareness of the elements that comprise it. But for the majority of leading natural philosophers – those who exercised important administrative functions beginning in their university days and on through their careers – this model is insufficient. We must imagine for them a second technological model in which the primary use of money is as a line instrument of legal and administrative gradation.

From the late twelfth century civic and Church officials began to standardize the collection of dues and services by pegging them to monetary income. By the thirteenth century, the laws and administrative ordinances of England and France contain numerous examples of the use of money as a grading continuum, with very specific taxes and duties required of those who fell above or below designated points on the numbered line of income. To take one of many examples, the French *Ordonnances* of 1302–03 require those who have *in bonis mobilibus* 100 pounds of Paris, or *in bonis mobilibus et immobilibus* 200 pounds of Tours, to present themselves to fight in Flanders. Those with a lesser income are excused from fighting, but they owe financial support for the army, again graded according to income: for each 100 pounds of Tours in income, 20 pounds of Tours is required, except for those with an income above 500 pounds who are taxed at a rate of 25 pounds per 100. A second *Ordonnance* follows requiring the sending of "un Gentilhomme bien armé" for each 500 pounds in income from land. The gentleman must be mounted on a horse worth at least 50 pounds of Tours, with armor of iron. For townsmen with smaller holdings, six foot-sergeants are required for each one hundred households. The "poorer sort living under the king's law" are required to pool their resources and send four foot-soldiers for every 100 pounds in income.[38]

As complicated as they were, such public laws were simple affairs compared to the actual tasks of practical administration. An *Ordonnance* of 1338 listed the different daily wage rates, down to the penny, of eight grades of soldier, from a simple footman, to lesser knights with and without a horse, to chevaliers who receive different rates according to the value of the horse they have brought with them to battle.[39] Even more detailed than the war levies were the many *Ordonnances* concerned with economic life. Again to take but one example, a French law of 1350 listed the maximum prices allowable for twenty-six separate named breads, determined by the type and weight of flour, to the ounce, in each. Reading these and similar governmental directives from the fourteenth century, one senses the great flexibility of the instrument of money and

---

[38] *Ordonnances*, vol. I, 370 (1302); 383ff. (1303). [39] *Ordonnances*, vol. II, 121 (1338).

the complex problems in gradation and relation it was regularly called upon to solve.

A French *Ordonnance* of 1268 "against those who blaspheme against God, the Virgin, and the Saints" provides an example of how money was used as an administrative and legal tool to grade not only concrete duties but highly subjective qualities and acts.[40] According to this law, anyone above the age of fourteen who speaks a "horrible" blasphemy must pay between 20 and 40 pounds in amends; the exact payment was to be determined according to the enormity of the blasphemy and the social condition of the blasphemer, i.e., his ability to pay.[41] One whose blasphemy is "less horrible" is to pay between 20 sous and 10 pounds, according to the same formula.[42] And if the blasphemy is even less severe, the payment required is between 5 sous and 40 sous *"according to the quality of the misdeed*, and the condition of the person [emphasis added]."[43] Here we have an attempt to measure and impose the sliding scale of price on highly subjective qualities within a thoroughly relational framework – an attempt that would later be paralleled in the use of the "latitude" (*latitudo qualitatum*) as a conceptual measuring instrument in the realm of theology and natural philosophy.

Similar administrative uses of money as a measuring and relating scale were familiar to the large number of scholars who participated in the bureaucracy of the university. Within the medieval university, every examination the student took, every grade passed and degree earned, *determinatio, licentia, inceptio*, had a price attached to it – a price most often assessed, collected, and registered, by the university masters themselves.[44] Each fee charged was determined relative to the ability of the student to pay. On entry into a Nation at the University of Paris, each student declared under oath the resources of his family and how much he could afford to pay per week.[45] An inquest undertaken by the teaching masters into the student's financial standing often preceded the official oath taking.[46] The amount sworn to was the student's *bursa*, and on the basis of

---

[40] *Ordonnances*, vol. i, 99 (1268).
[41] *Ibid.*: "Se aucune persone de l'aage de quatorze ans, ou de plus aura proferé quelque horrible blaspheme payera quarante livres d'amende, ou au moins vingt livres, selon la condition de la personne, et l'enormité du blaspheme."
[42] *Ibid.*: "Celuy qui a cet age aura proferé quelque blaspheme moins horrible, payera dix livres, ou au moins vingt sols, selon l'enormité du blaspheme."
[43] *Ibid.*: "Et si le blaspheme est encore moins grand que le precedant, l'amende sera de quarante sols, et au moins de cinq sols, suivant la qualité du meffait, et la condition de la personne." The formula "selon la qualité du meffait" is found often in French legal writings of the period.
[44] See chapter 1, 30–36, for the administrative experience of the leading French and English natural philosophers. For a general statement about the engagement of the fourteenth-century scholar in his world, see W. A. Pantin, *The English Church in the Fourteenth Century* (Notre Dame, Ind., 1962), 133.   [45] Léo Moulin, *La vie des étudiants au moyen âge* (Paris, 1991), 56–59.
[46] Moulin, *La vie*, 59.

this sum his monetary dues *pro scholis* were determined proportionally.[47] The *bursae* were critical to the financial functioning of the university and were recorded with care. Each monthly entry in the one surviving Parisian *Liber procuratorum* of the fourteenth century – that of the English Nation – contains a statement of funds on hand, an accounting of numerous receipts and expenses, and a list of students ranked by their numbered *bursae*. The responsibility for figuring and keeping these accounts was given through election to a different teaching master every month.[48]

The model of money as measure in administration lacked the rich Aristotelian textual reinforcement found with the model of money in exchange. Nevertheless, it played such an important role in the professional lives of our scholars that the model of money as instrument in both administration and exchange must be taken together when considering the impact of monetization on the development of conceptual measuring "instruments" within scholastic natural philosophy. In seeking to define the technological model of money in administration, many of the elements of the model of money in exchange are applicable. In both cases money serves as an extendable, divisible, graded, and numbered continuum used as a common scale of measurement. But there are important differences. Money in exchange facilitates relation by expressing the diverse and shifting values of all commodities in common numerical terms. Money in administration also facilitates relation but it does so along a single axis. It provides a numbered scale superimposed on pre-existing relationships within a unified field. Within each field it acts as an instrument of proportionalization. Where money in exchange serves as a *medium* for *connecting* diverse goods and services by making them commensurable, money in administration serves as a *medium* for *dividing* values along a single continuous axis. In exchange, the primary function of money is to facilitate the finding of an equalizing point or range (latitude). In administration, equality remains the ideal, but the goal is to determine points (or latitudes) of proportionally equal division along a continuum of value. In exchange, money was recognized by the fourteenth century as having replaced the intervening judge as an impersonal "instrument of equalization" within a dynamic process.[49] In administrative measurement, subjective judgment remains necessary in deciding where and how to divide the measuring continuum. Taking similarities and differences into account, I define the *technological form of money in administration* as a

---

[47] For example, in 1350 the student with a *bursa* of 2–3 sous paid 20 sous *pro scholis*, while a *bursa* of 8–9 sous paid 50 sous *pro scholis*. See Moulin, *La vie*, 56.

[48] Courtenay, "Registers of the University of Paris." Courtenay estimates (33) that most masters could expect to be elected proctor at least once every two years.   [49] See chapter 5, 132–33.

continuous numbered scale superimposed by the administrator on a given problem in measurement and gradation, to the end of finding ranges of equalization and points of proportional division.

We know that philosophers of the fourteenth century were fascinated by the subject of the measurement and quantification of qualities. With the technological forms of money in exchange and in administration in mind, we turn to consider in greater detail how they conceived of qualities and what models or conceptual instruments they devised for their measurement and relation. This is a complex question, since the process of developing a measuring system for qualities was pursued by many thinkers and took many paths. However, general descriptions of the historical process can be and have been made. These historical descriptions have in common the recognition that, from the late thirteenth century, philosophical analysis took the direction of conceptualizing qualities as if they existed as *continua* in a state of constant expansion (intension) or contraction (remission). Qualitative change came to be conceived of as real motion within a real, continuous "qualitative space."[50] The path-breaking historian of medieval science, Anneliese Maier, was particularly sensitive to this new mode of understanding qualities. She went so far as to say that fourteenth-century philosophers constructed a whole new "image of the world" based on the perception of continuous and continuously changing qualitative magnitudes.[51]

Since the work of Maier, historians have increasingly come to focus on the central place of questions concerning the continuum and the analysis of change along a continuum in the logical and measuring languages of the fourteenth century. At the same time, they have continued to explain the great elaboration of these questions from within the intellectual tradition.[52] The evidence for intellectual and particularly Aristotelian influences here is strong. Aristotle considers logico-mathematical problems relating to the continuum at many points in his writings, particularly in the *Physics*. For Aristotle, understanding the logic of the continuum is a

[50] Edith Sylla, "Medieval Concepts of the Latitude of Forms: The Oxford Calculators," *Archives d'historie doctrinale et littéraire du moyen âge* 40 (1973), 223–83, esp. 234, 263; Coleman, "De Ripa," 155; Wilson, *Heytesbury*, 21; Clagett, "Swineshead," 139; John Murdoch, "Thomas Bradwardine: Mathematics and Continuity in the Fourteenth Century," in Edward Grant and John Murdoch (eds.), *Mathematics and Its Applications to Science and Natural Philosophy in the Middle Ages: Essays in Honor of Marshall Clagett* (Cambridge, 1987), 103–37, 110.

[51] Maier, "La doctrine d'Oresme," 337: "de saisir les phénomenes commes des grandeurs intensives et de construire avec elles l'image du monde."

[52] For example, Murdoch, "Naissance et développement de l'atomisme au bas moyen âge latin," in *La science de la nature: théories et pratiques* (Montreal, 1974), 11–32, 19, traced the fourteenth-century fascination with the continuum from the writings of the Arab thinker Algazel, through the works of Duns Scotus and from there to the "prodigious elaboration and multiplication" it found in the fourteenth century.

key to understanding crucial questions about motion, space, and time.[53] Any discussion of motion within an Aristotelian framework, including the motion of the increase and decrease of a quality, could involve the discussion of continua. The doctrine of contraries at the center of Aristotle's conception of qualitative change also lent itself to being understood as involving motion along a connecting continuum.[54] But Aristotle himself never held it to be so. On the contrary, Aristotle held that qualities were not continuous magnitudes and so could *not* be conceived or treated as continua.[55] He insisted on the categorical separation of "quality" and "quantity." According to his definition, quantities were composed of parts and were divisible. Qualities, though admitting of variation in degree, were not in themselves composed of parts and therefore not divisible, either conceptually or actually.[56] Qualities could not, therefore, by definition, be continua.[57] Aristotle did not share the scholastic conception that qualitative change involved motion over real "qualitative space" – a conception central to the measuring schemata of the fourteenth century. The conceptualization of qualities as divisible, continuous magnitudes in expansion and contraction was a scholastic invention. It was not only non-Aristotelian, it was anti-Aristotelian.[58]

There was, for Aristotle, only one indivisible form for each quality: one form of whiteness, defined by its perfection or essence, one form of heat. Thomas Aquinas' support of Aristotle's categorical distinction between quantity and quality was typical, and this position was maintained until the end of the thirteenth century. Of course Aquinas and others who supported this distinction recognized that subjects were variably white or hot and that human qualities such as charity or love were possessed in varying degrees. But they explained these phenomena by saying that apparent variations resulted from the varying acceptance by subjects of uniform and indivisible qualities.[59] The potential of

---

[53] For the central place of the continuum in Aristotelian physical thought, see especially his *Physics*, bks. III–VI; Murdoch, "Analytic Character," 177; A. G. Molland, "The Geometrical Background to the 'Merton School,'" *British Journal for the History of Science* 4 (1968), 108–25, esp. 121.

[54] See for example the diagram of qualitative change through the contrary drawn by Shapiro, "Walter Burley and the Intension and Remission of Forms," *Speculum* 34 (1959), 413–27, esp. 414. Aristotle recognized that those who talk of contraries often use spatial metaphors, though they do so imprecisely. See Aristotle, *Categories* VI [6a15–19].   [55] Aristotle, *Categories* VI.8.

[56] *Ibid.* Duhem, *Système*, vol. VII, 480–81. For a discussion of this distinction in Aristotle's thought and its importance to the quantifying program of the fourteenth century, see Shapiro, "Burley," 415–18 and n. 6.

[57] Aristotle's categorical separation of quantity and quality is discussed in Sylla, "Medieval Quantifications of Qualities," esp. 9.

[58] Maier, "Das Problem der intensiven Grösse," esp. 6–7; Sylla, "Medieval Quantifications of Qualities," 9.

[59] Aquinas *ST*, 1, 2, 52, 1. See Clagett, "Swineshead," 132–35. For a different interpretation of Aquinas' position on qualitative variation, see Shapiro, "Burley," 423.

the subject for the expression of the quality varied, not the quality itself.[60] By the late thirteenth century, however, philosophers began to abandon this explanation of qualitative change. The Aristotelian explanation no longer accommodated the constant flux of qualities and values they perceived in their world. They searched for new models of measurement to make sense of their new perceptions. Of the many competing solutions to the question of qualitative change that existed at the end of the thirteenth century, the one most frequently adopted, the "additionist" theory, was the one that went most directly against Aristotelian authority.

Those who came to accept the new additionist theory conceived of qualities as continuous magnitudes divisible into parts.[61] Increase (*intensio*) and decrease (*remissio*) of qualities occurred through the addition or subtraction of these numerable and quantifiable parts of the quality.[62] The anti-Aristotelian conclusion this solution represented – that qualities could be quantified – opened the door to new realms of mathematical and scientific investigation.[63] Once the quantification of qualities had been theoretically justified, vague notions of increase and decrease were no longer sufficient within the structure of scholastic debate. Looking for added precision, philosophers used the tool of logic to sharpen, grade, and calibrate notions of increase and decrease. In doing so they prepared the study of qualitative change for the application of logical, numerical, and mathematical analysis.

One of the most striking aspects of the new passion to measure and quantify based on the logic of the continuum was the development of what John Murdoch has identified as "limit languages": *De primo et ultimo instanti* (First and last Instants); *Incipit et desinit* (Beginning and ceasing); *De maximo et minimo* (Maximum and minimum). Questions raised in Book VI of Aristotle's *Physics* concerning the infinite divisibility of the continuum provided the textual point of departure for the fourteenth-century Calculators.[64] Given that motion is necessarily continuous, and that the continuum is infinitely divisible, is there a "primary when" at the

---

[60] Duhem, *Système*, vol. VII, 482–85; Clagett, "Swineshead," 131.

[61] For the history of this process, see Maier, "Das Problem der intensiven Grösse"; Duhem, *Système*, vol. VII, 480–533; Clagett, "Swineshead," 132–38; Wilson, *Heytesbury*, 18–22; Sylla, "Medieval Quantifications of Qualities."

[62] Sylla, "Medieval Quantifications of Qualities," 11–15 and n. 17. For the metaphysical problems involved in the additionist position, see Shapiro, "Burley," 423–26.

[63] Sylla ("Medieval Quantifications of Qualities," 27–28) makes a distinction between the additionist quantification of the "intensity" of a quality, which became common in the fourteenth century, and the true quantification of qualities, found in the work of Nicole Oresme, which required considering the varying intensities of qualities in relation to their varying extension in the subject. Oresme's measuring scheme is considered in chapter 7.

[64] Aristotle, *Physics* VI.5 [232b32–236b27].

beginning of a motion, an indivisible point at which a motion can be said to have begun, and is there a last indivisible instant at which the motion can be said to have ceased?[65] Aristotle's solution is to deny that there is such a point for the beginning of a motion but to accept that there is one for its end.[66] With these statements of Aristotle as their point of departure, scholastic philosophers took the logical investigation of limits to extraordinary lengths. The Calculators Walter Burley, Thomas Bradwardine, Richard Kilvington, and William Heytesbury all wrote treatises centered on this investigation. Using ever more complex imaginary cases and sophismata designed to bring abstract issues into sharper focus, they elaborated increasingly technical logical techniques to resolve limit problems.[67]

From the continuum of time, the area of investigation extended to limits involving continuous powers and capacities.[68] Here the point was to define the boundary between the maximum that could be done (*maxima quod sic*) and the minimum that could not be done (*minima quod non*) or the minimum capacity sufficient for a task and the maximum that remained insufficient. William Heytesbury dedicated a chapter in his treatise *Rules for Solving Sophismata* (*Regulae solvendi sophismata*) to the problem of maxima and minima.[69] In the words of John Longeway, the editor of this question, "Heytesbury's rules are put in the form of a 'division' of the limits of capacities into several sorts, with instructions on how to select the sort appropriate to a given case."[70] Heytesbury lists the following types of questions that can be approached under one of the several categories he treats:

For instance, if it is asked whether there is a maximum Socrates suffices to carry or a minimum he does not; or a coldest thing the hot thing A suffices to assimilate to itself, or a least cold it does not; or a maximum part of the difform *medium* B that Plato suffices to divide, or a minimum he does not; or a maximum period of time for which Socrates can endure or a minimum for which he cannot; or a first instant in which Antichrist can be or a last in which he cannot; or a maximum velocity at which Socrates, for a given period of time, can move weight A, or a

---

65 Murdoch, "Analytic Character," 181–98; Murdoch, "Logic of Infinity," 170–91; Murdoch, "Involvement of Logic," 19. The phrase "primary when" appears in R. P. Hardie's translation of the *Physics* as well as in Murdoch's analysis of Aristotle's position.

66 Aristotle, *Physics* [236a4]: "That in which the completion of change has been effected must be indivisible."

67 *Sophismata of Richard Kilvington: Introduction, Translation, and Commentary*, xii–xiii; Alain de Libera, "Le développement de nouveaux instruments conceptuels et leur utilisation dans la philosophie de la nature au XIVe siècle," in Asztalos, Murdoch, and Niiniluoto, *Knowledge and the Sciences*, vol. I, 158–97, esp. 189.

68 Murdoch, "Logic of Infinity," 187–90, notes another point of "textual take-off" for the application of limits in *De caelo* I.11 [281–a1–27].   69 Heytesbury, *Maxima and Minima*.

70 Longeway, "Introduction," in Heytesbury, *Maxima and Minima*, 3.

minimum which he cannot; or a most remiss degree that is more intense than the middle degree of the whole heat or a most intense that is not; or a least visible thing a lynx can see unaided, or a most visible it cannot; and so on for an infinite number of similar cases.[71]

The mere listing of questions here cannot do justice to the complexities that ensue when Heytesbury applies the necessary logical rules to their solution or tests them through the use of sophismata. Clearly the elaboration of the logic of limits exhibited by Heytesbury and the Calculators cannot be explained without reference to the central place of logic in the Oxford curriculum in the fourteenth century – and the central role of disputations in the teaching and practice of logic. But recourse to this context is less successful in explaining why the logicians focused so tenaciously on this particular class of problem. John Murdoch suggests that these developments *might* be explainable as the natural working out of the "twists and turns" in Aristotle's treatment of continua in the *Physics*. But his conclusion on this point is characteristically honest and open: "It is presently difficult to say just why such a way of approaching and resolving problems increased so markedly in this century."[72] I would like to suggest the technological model of money as a contributing explanation, and in particular the model of money as measure in administration.

In both exchange and administration money functions as an extendable, divisible, graded, and numbered continuum used as a common scale of measurement. In the particular case of administration I have defined its function as a "numbered continuous scale superimposed by the administrator on a given problem in measurement and gradation, toward the end of finding ranges of equalization and points of proportional division."[73] The questions continually before the administrator are: "How do I convert qualitative or subjective grades into monetary terms for the purpose of measurement? How and where do I divide the resulting continuum of money to achieve just proportionality? How and where and with what rationale do I set limits along the continuum of value? How do I treat those cases that fall close to the points of division on the continuum?" The areas around these points become exceedingly problematic, given that small changes in position to either side of the dividing points can have large repercussions in terms of duties owed. These and similar administrative questions are reflected (though never stated) in the Calculators' language of limits. Within the literature of limits there is not one mention of taxes or prices; not one practical administrative question is asked. As Heytesbury's list of questions reveals, the questions the

---

[71] Heytesbury, *Maxima and Minima*, 12.   [72] Murdoch, "Analytic Character," 197.
[73] See 174–75.

Calculators did pose were in every case imaginary and purposely denatured. They invariably chose to consider the whiteness, or carrying
capacity, or speed, or endurance of Socrates, even though real cases of
actually measurable limits in these areas were readily available to them. In
doing so they believed they were insulating their logic from the contamination of contingent experience, thereby providing their arguments with
the necessity and universality required by a proper science of nature.

Due to this shared concern, when focusing on the highly evolved
technical logic of the Calculators it is easy to overlook how central the
practical goals of career training were to the culture of English universities
in the fourteenth century.[74] Most students in this period (as in our own)
attended university for the purpose of receiving professional credentials
and professional training. The majority who would not go on to be
masters themselves valued logic because of its potential application to
their future careers in law and administration. A student trained by
Heytesbury or Kilvington in the "pure" logic of limits would be well
prepared to deal with the administrative questions posed above. But it
was the *practice* of administration as well as its preparation that can be seen
in the elaboration of the limit languages. The authors of the treatises
themselves had extensive administrative experience. The complex task of
running and financing Merton College was performed throughout the
century by its teaching masters.[75] In 1338–39, soon after the writing of the
Regulae, Heytesbury had moved up the administrative ladder to be
elected bursar of Merton, responsible for determining dues, auditing
accounts, and collecting revenues.[76] Burley, Bradwardine, and Kilvington also performed important services for their university and later took
high offices in Church and kingdom.[77] In addition, all three were attached to the wealthy London household of Richard de Bury, the bishop
of Durham.[78] De Bury had risen through the ranks to become the most
powerful official in England: clerk in the exchequer of Chester, constable
of Bordeaux, treasurer of the king's wardrobe, keeper of the privy seal,
lord chancellor, lord treasurer, and finally bishop of Durham, the weal-
thiest diocese in England.[79] On de Bury's frequent sojourns in London,
Burley, Bradwardine, and Kilvington joined other Oxford theologians

---

[74] Courtenay, *Schools and Scholars*, 118–46.

[75] Aston, "Administration of Merton," 311–12, 331–44, 364.

[76] Details of Heytesbury's administrative service are discussed in Aston, "Administration of Merton," 334–38, and chapter 1, 34–35. In later life (1353–54, 1371–72) Heytesbury served as
chancellor of Oxford.   [77] See chapter 1, 33–34.

[78] See Courtenay, *Schools and Scholars*, 133–37, 244; Courtenay, "London *Studia*."

[79] N. Denholm-Young, "Richard de Bury (1287–1345)," in *Collected Papers on Mediaeval Subjects*
(Oxford, 1946), 1–25; Kretzmann and Kretzmann, "Introduction," *Sophismata of Richard Kilvington: Introduction, Translation, and Commentary*, xxvii.

and philosophers – Robert Holcot, John Maudit, and Thomas Fitz-Ralph among them – to share daily readings and disputations after dinner around de Bury's table.[80] The mixing of administrative and intellectual tasks and social circles characterized English university culture of the period, both for the students and for the masters. Seen in this light, it is not surprising that this mixture characterized the intellectual products of the university as well.

The clearest way to study how the Calculators approached the quantification and measurement of qualities is to investigate the origin of their prime conceptual instrument of measurement, the "latitude of qualities" (*latitudo qualitatum*) or "latitude of forms" (*latitudo formarum*), so named because qualities were conceived of as forms inhering in subjects. I know of no current historical interpretation that suggests the influence of models from outside the philosophical-textual tradition on the origin of the "latitude" and its use in measuring qualities. But there are striking similarities between the early philosophical application of the latitude to set limits of variation within qualities in natural philosophy and the use of the latitude within scholastic economic thought to denote the legitimate numbered range of a commodity's just price.[81]

Historians of science have traced the development of the concept of the latitude from thinker to thinker and manuscript to manuscript, beginning with Galen in the second century A. D. and spanning centuries, cultures, and disciplines to arrive at the fourteenth century.[82] Since, however, this intellectual trail includes thinkers such as Peter John Olivi and Duns Scotus, whose application of the concept of "latitude" to the question of just price derived from both economic experience and legal discussions of economic exchange, it is clear that the conceptual development of the latitude was open to many sources of influence coming from outside the philosophical-medical textual tradition.[83]

The case of Arnald of Villanova is instructive here.[84] In Arnald's

---

[80] Denholm-Young, "De Bury," 20.

[81] For thirteenth-century uses of the concept of legitimate range in just price theory, see chapter 4, 99–100. For late thirteenth- and fourteenth-century developments in this concept, see chapter 5, 124–26.

[82] Sylla, for example ("Latitude of Forms," 226–29), discusses a possible chain of textual transmission from the descriptive use of the latitude in the medical writings of Galen to its use in the work of Avicenna and Averroes, to the medical writings of Arnald of Villanova and the theological works of Aquinas, Henry of Ghent, and Duns Scotus, and from these sources to the Oxford Calculators.

[83] See for example the place of Godfrey of Fontaines, Peter Olivi, and Duns Scotus in Maier's discussion of the early history of the conceptualization of qualities as intensive magnitudes, in "Das Problem der intensiven Grösse," 36–43, 46–55.

[84] The connection between Arnald's mathematical treatment of pharmacological theory and the mathematics of the Merton School was first suggested in Michael McVaugh, "Arnald of Villanova and Bradwardine's Law," *Isis* 58 (1967), 56–64, esp. 59.

medical writings on pharmaceutical theory, historians have found important new elements of quantification attached to the concept of the latitude of qualities. Specifically, in his *Aphorismi de gradibus* Arnald measured the components of medicines in terms of their qualitative intensity, and he conceived of this intensity as a line situated in space (*diametrum altitudinis*).[85] Furthermore, Arnald described a mathematical method of measuring and comparing qualitative intensities in degrees (*gradus*).[86] The use of the latitude in Arnald's medical writing is currently considered an important early fourteenth-century link in its textual development.[87]

Arnald was in intellectual contact with an older medical tradition stretching from Galen through Avicenna's *Canon* in which the latitude was used as a broad limit-measure of both the strength of medicines and the health of subjects.[88] But he was also familiar with contemporary, non-medical philosophical discussions of qualitative change. Michael McVaugh, the editor of Arnald's *Aphorismi de gradibus*, has suggested that Peter John Olivi was the source of Arnald's familiarity with contemporary philosophical debate, surmising that Arnald was a student of Olivi's at Montpellier in the 1290s. McVaugh maintains that Olivi provided the "philosophical foundation" for Arnald's mathematical and quantitative theory of compound medicines.[89] However, since McVaugh was unaware of Olivi's use of the *latitudo* in his economic writings, he was not in a position to recognize any connections between Arnald's use of the measuring latitude and Olivi's technical use of this concept in his analysis of market value and the just price. If, as McVaugh and others maintain, Arnald's writings later influenced the use of the latitude by the Oxford Calculators, and if Arnald's use of the latitude was shaped in part by Olivi's conception, then, even within a history concerned with demonstrating connections between writers and texts, social influences intrude.[90]

---

[85] Arnald of Villanova, *Aphorismi de gradibus*, Michael McVaugh (ed.), vol. II in *Arnaldi de Villanova opera medica omnia* (Granada, 1975), 36, 39, 54ff. McVaugh (93) posits the medical *Canon* of Avicenna as the textual origin of Arnald's use of the concept of the latitude.

[86] It is Arnald's exponential relation of degree to intensity that is given most attention because of its similarity in form to Bradwardine's mathematical function relating forces and resistances exponentially. The textual similarities between Arnald's formulation and Bradwardine's law led Sylla ("Latitude of Forms," 228) to suggest that his medical latitude had a similar influence on the Calculators' concept of the latitude.   [87] Arnald of Villanova, *Aphorismi de gradibus*, 90–96.

[88] Arnald of Villanova, *Aphorismi de gradibus*, 92–93.

[89] Arnald of Villanova, *Aphorismi de gradibus*, 96.

[90] Although there is no conclusive proof of a direct textual link between Arnald and the Calculators, and although there are considerable differences between the form and function of Arnald's latitude and the Calculators' latitude, textual similarities alone seem to some to be a sufficiently strong indication of textual transmission. See McVaugh, "Arnald of Villanova," 64, 59 n. 8; Clagett, "Some Novel Trends," 284; Sylla, "Latitude of Forms," 228.

In its earliest use in the second half of the thirteenth century, the latitude of a quality denoted a range within which a quality or form could vary and still remain identified as that quality or form.[91] Thomas Aquinas is one of the first scholastic philosophers to have used the word "latitude" in this sense.[92] Following Aristotle, Aquinas defined virtue as the active seeking of the virtuous mean (*medium*) by particular subjects in particular situations, rather than as an abstract ideal. The specific concept and word "latitude" entered Aquinas' ethical discussion when he noted that the mean of virtue was not to be understood as some absolute, indivisible point (*non est omnino indivisible*), but rather as a legitimate range or "latitude" (*habet aliquam latitudinem*) on either side of this point. The requirements of virtue were satisfied if the subject's actions and thoughts fell within this latitude in attempting to approach an ideal mean that changed according to changing circumstances.[93] The resemblance of Aquinas' "latitude" of virtue to his influential solution to the problem of just price in the *Summa theologiae* is striking.[94]

There, in his discussion of just exchange, Aquinas maintained that the "just price" of a commodity could not be limited to one single point of equality (*non est punctualiter determinatum*). Instead, just equality was satisfied by the necessarily approximate estimations made by the exchangers in the direction of that point of equality. Therefore, he concluded, just price and the requirement of economic equalization were satisfied within a range of values extending from a little above to a little below (*modica additio vel minutio*) an ideal mid-point of equality.[95] It is impossible to say whether the economic solution of a legitimate range of just price influenced the philosophical understanding of a *latitude* of qualities, or vice versa, or whether (as I believe) the influence went in both directions. But what can be said is that, while the philosophic conceptualization of an ethical quality in terms of a legitimate range or latitude along a continu-

---

[91] *HG* IV. 5: "quia natura et essentia rei non consistet in terminio definito seu determinatio magnitudinis sibi debite, sed in latitudine quodam." Again: "natura forme non consistit in simplici et determinato sed in latitudine quadam indeterminata in qua secundum gradus diversos potest salvari formae ipsius speciei."   [92] Sylla, "Latitude of Forms," 228.

[93] Thomas Aquinas, *Quaestiones disputatae. De virtutibus in communi*, q. I, art. XI (cited in Sylla, "Latitude of Forms," 228 n. 17): "quod ratio virtutis non consistit in indivisibili secundum se, sed ratione sue subjecti, inquantum quaerit medium: ad quod quaerendum potest aliquis diversimode se habere, vel pejus vel melius. Et tamen *ipsum medium non est omnino indivisibile : habet enim aliquam latitudinem:* sufficit enim ad virtutem quod appropinquit ad medium, ut dicitur 2 Ethicorum" (emphasis added). For an earlier discussion of Aristotle's discussion of a range in virtue from the *Ethics* II.16, and its applicability to the later scientific use of the latitude, see chapter 5. Sylla cites a second use of this concept by Aquinas in his *Sentence Commentary*, bk. III, d. III, q. 5, art. 2.

[94] See the discussion of Aquinas' position in chapter 4, 99–100.

[95] Aquinas, *ST*, 2, 2, 77, 1: "quia justum pretium rerum quandoque non est punctualiter determinatum, sed majus in quadam aestimatione consistit; ita quod modica additio vel minutio non videtur tollere aequalitatem justitiae."

um was new to Aquinas, the similar solution of a latitude of just price was already firmly established and much discussed within both medieval legal theory and scholastic just price theory, as the Roman law doctrine of *laesio enormis*.[96]

In a more general sense, Aquinas' recognition of a legitimate latitude of qualities was born out of a new *acceptance of the approximate*. As we have seen, the concept of just equalization, indeed the whole direction of economic thinking in the thirteenth century, was leading toward the acceptance of economic decisions based on estimation and approximation, and *against* a strong philosophical and theological preference for exact and knowable points of equality.[97] Again, the idea of a legitimate, approximate range of shifting values conceived as a latitude seems to have occurred first in the economic perception of fluctuating prices within legal ranges, and only later in the philosophical concept of acceptable ranges in qualitative change.[98]

Though Aquinas did not yet use the word "latitude" to describe the legitimate range of just price, at the end of the thirteenth century, Peter John Olivi and Duns Scotus did.[99] As far as I am aware, historians of science, in their many discussions of the history of the measuring latitude, have missed both Olivi's and Duns Scotus' early use of this concept in their "latitude" of just price.[100] At the least, Olivi's use of the word "latitude" in this economic context suggests the connection between the conceptualization of the measurement of qualitative variation in scholastic thought, and the everyday existence and measurement of price variations in the marketplace.[101]

---

[96] See *Codex*, 4.44.2; 4.44.8; Baldwin, *Just Price*, 18–23; Cahn, "Roots," esp. 12–22.

[97] See the conclusion to chapter 4.

[98] The influence of practical, sensual experience on Aquinas' concept of the latitude is apparent in another of his uses of term: "Ad tertium dicendum quod quantitas determinata alicuius speciei non est determinata secundum aliquid indivisibile; sed habet aliquam latitudinem: quia in species humana invenitur major et minor quantitas, et in eodem individuo secundum diversa tempora, et in diversis, ut ad sensum patet" (cited by Sylla, "Latitude of Forms," 228 n. 17, from Aquinas, *Sentence Commentary*, bk. iii, d. iii, q. 5, art. 2).

[99] See Olivi, *Tractatus*, 53: "pensatio valoris rerum usualium vix aut numquam potest fieri a nobis nisi per coniecturalem seu probabilem opinionem, et hoc non punctualiter, seu sub ratione seu mensura indivisibili in plus et minus, *sed sub aliqua latitudine competenti*, circa quam etiam diversa capita hominum et iuditia differunt in extimando. Et ideo *varios gradus* et paucam certitudinem, multumque ambiguitatis iuxta modum opinabilium" (emphasis added).

[100] When writing my earlier paper linking the scientific concept of the measuring latitude to the use of money (Kaye, "Impact," 1989), I also was unaware that scholastic thinkers had actually used the term "latitude" in their discussion of price measurement.

[101] Compare, for example, Duns Scotus' concept (*Sent. iv*, d. xv: see chapter 5) of a *magna latitudo* involved in all commutative exchange, in which monetary values within this latitude were represented as *gradus* or "degrees," and his concept of a qualitative latitude in *Sent. ii*, d. xv (1893, vol. viii, 7b–8a): "subiectum enim sive elementum habet talem qualitatem in quadam latitudine sed non praecise sub tali vel tali gradu."

Olivi's choice of the word and concept "latitude" in relation to just price was not accidental. It was central to his understanding of market exchange as a system of dynamic equalization based on estimation and approximation, and it was defined with a technical precision unusual for the period. No more the "little above and a little below" of Aquinas' range of price – Olivi's *latitudo iusti pretii* had numerical price limits defined by the Roman law rule of one half above and one half below the normal market price.[102] If we stop to picture Olivi's *latitudo*, we see that it described a range of market values along a line of price, composed not only of numbered outer limits, but of numbered uniform degrees (*gradus*), represented by coins, within these limits. The latitude of just price for some given commodity would range, for example, from 5 pence to 15 pence around a normal market price of 10 pence.[103] It was understood that the fair market value of any commodity could vary continually up and down along this numbered and graded latitude.[104] In effect, every commodity and every service finding exchange in the market was conceived of as if it were carrying its own proper latitude, within which its value could and did legally vary. This potent image of an economic world in which values were continually fluctuating within specific latitudes along the continuum of money was in place from the late thirteenth century. In just price theory, the latitude of accepted price variation for any given commodity was itself understood to move along money's numbered continuum as changing market factors caused the estimated market price, and thus the latitude's mid-point, to shift. This dynamic image redefined the model of value determination from one based on points and perfections to one based on lines and approximations, reinforcing the sense of a "world of lines" generated within the monetized marketplace.

In its earliest thirteenth-century uses within natural philosophy the latitude of qualities resembled in a simplified way the latitude of just price, functioning first as a range within which a quality could vary and still remain defined as that quality, and somewhat later as a range in which the intensity of the quality could vary up to a certain terminus of perfection – perfect whiteness or perfect heat. With the first generation of Calculators in the fourteenth century, the latitude moved closer to its eventual identification as a measuring continuum. Walter Burley (whose conception of money as measure was discussed in chapter 5) used the latitude as a

---

[102] See chapter 4 for the way canon lawyers and theologians of the thirteenth century interpreted this rule derived from the doctrine of *laesio enormis*.
[103] This is the example often given in earlier legal discussions of the problem. See Baldwin, *Just Price*, 45; Cahn, "Roots," 29 n. 55; Hostiensis, *Summa aurea*, bk. III, cols. 943–44.
[104] Olivi, *Tractatus*, 53; Duns Scotus, *Sent. IV*, 282b–284a.

way of ordering and representing the indivisible degrees a quality obtained when moving between its contraries through its mid-point, as, for example, moving from perfect whiteness to perfect blackness.[105] Burley cited Book II of Aristotle's *Ethics* (on the *medium* of virtue) to support his position that the *medium* between two contraries is difficult to determine with precision. This is the same locus where Aquinas introduced the term "latitude," and it shows the continuing connection between ethical thought and the physical concept of an approximating range.[106] Each indivisible degree within the latitude joining the contraries represented an actual form the quality was capable of becoming.[107] In Janet Coleman's words, Burley's latitude provided "a static basis, a grid, for possible alteration in intensity within a particular species while the species remains the same."[108] Composed as it was of indivisible degrees (potential forms the quality might assume), Burley's latitude could not yet be defined as a continuum, and thus could not yet be said to quantify qualitative intensity.[109]

When Burley was writing, a number of competing and mutually inconsistent conceptions of the latitude of qualities were held by philosophers, each chosen because of its fit within different perceptual schemes and different sets of philosophical definitions. The question of qualitative change was of such importance, and the progression of positions was so marked to contemporaries, that outlines of the historical development of this concept began to appear within contemporary philosophical debates.[110] More surprising than the simultaneous existence of contrary conceptions is that an essentially anti-Aristotelian "additionist" theory of qualitative change had, by the late 1330s, been adopted by the majority of the Calculators.[111] As the model for conceptualizing qualities and qualitative intensity approached ever closer to a linearly ordered quantity, the *latitudo qualitatum* moved ever closer in form to an instrument capable of translating qualitative values into quantitative terms.

---

[105] Shapiro, "Burley," 414.

[106] Burley, *Tractatus primus,* a partial edition of which is included in Sylla, *The Oxford Calculators and the Mathematics of Motion 1320–1350: Physics and Measurement by Latitudes* (New York, 1991), appendix 1, 471–520, 516–17.   [107] Sylla, *Oxford Calculators,* 339.

[108] Coleman, "De Ripa," 143.

[109] Sylla, *Oxford Calculators,* 344, 343: "Whereas the later Calculators can say that a latitude of quality is identical to its maximum degree, both containing all the lesser degrees, for Burley the maximum degree of a latitude corresponds only to the end point of the latitude, and the latitude itself is only a conceptual way of ordering these indivisible degrees."

[110] See for example Duns Scotus' summary of earlier positions in his *Sentences,* bk. I, d. XVII, q. 4. A list of opinions taken from the *Summa logicae* of the English Calculator John Dumbleton can be found translated in Coleman, "De Ripa," 153, and in fuller form in Sylla's *Oxford Calculators,* 565ff.   [111] Clagett, "Swineshead," 138–39.

The additionist theory of qualitative change proposed in its earliest form by Duns Scotus and adopted by William of Ockham held that qualitative intensity was divisible into parts or degrees, and that increases and decreases in qualitative intensity occurred through the addition and subtraction of these parts.[112] Within this new theory, later adopted by those whom Edith Sylla has called the "core Oxford Calculators" (Thomas Bradwardine, Richard Swineshead, William Heytesbury, and John Dumbleton), the latitude became the prime conceptual instrument of an influential and self-confident movement toward the quantification of qualities. In recent decades, historians of science have isolated a series of stages in the conceptual development of the measuring latitude, and they have discerned an overall direction to this development.[113]

For Walter Burley in the early stages of the debate, the latitude had only a conceptual, not a real existence. It provided a means of conceptually ordering qualitative motion along a series of steps or degrees, but it did not measure those steps. The degrees themselves, as potentially existing forms of the quality, had more reality than did the conceptual latitude. By the second quarter of the fourteenth century, in the writings of the core Calculators and in the work of certain Parisian natural philosophers, the latitude moved from being a description of an abstract range of values to having real existence as a measuring continuum *physically identified* as the intensity of a quality at a given point.[114] It had come to be imagined (in Edith Sylla's words) on the model of the "geometric line" as an "homogeneous continuum," composed of qualitative parts.[115] Degrees (*gradus*) of intensity were imagined as intervals ranged along the measuring continuum.[116] The increase or decrease in a quality such as heat was thus imagined as a motion through degrees along a linear "qualitative distance." Such an imagining of the latitude as an homogeneous continuum was predicated on the radical redefinition of qualitative intensity as a kind of quantity, linearly ordered and additive.[117]

---

[112] Maier, "Das Problem der intensiven Grösse," 59ff.; Clagett, "Swineshead," 136ff. See Sylla, "Latitude of Forms," 230–31 and n. 21 for a precise description of the early additionist theory of Duns Scotus and William of Ockham. Here, the quality increases through the addition of forms of the quality, each preceding form preserved in the new form, so that the lower degree of the quality is contained within the higher degree.

[113] The work of Edith Sylla is particularly valuable in this area. She notes, for example (*Oxford Calculators*, 345), that from Henry of Ghent to Nicole Oresme, "there is a fairly direct transition from the idea of a latitude as an abstract or conceptual range to the idea of a latitude as a particular measure of the intensity of a quality at a given point."

[114] Sylla, "Latitude of Forms," 239, 252; Sylla, *Oxford Calculators*, 345, 378.

[115] For the actual physical representation of latitudes by geometric lines in the margins of Dumbleton's *Summa*, see Sylla, "Latitude of Forms," 264–66.

[116] Sylla, "Latitude of Forms," 255–56; Wilson, *Heytesbury*, 144.

[117] Wilson, *Heytesbury*, 21; Clagett, "Swineshead," 139–40.

These developments appear most clearly in the writings of the Oxford Calculator John Dumbleton.[118]

An important step in the evolution of the latitude into an instrument of measurement was its conceptual division into homogeneous and infinitely divisible parts, on the model of the division of the line into points.[119] Equal intervals on Dumbleton's measuring latitude represented equal changes in qualitative intensity.[120] Once the Calculators had established the latitude as an homogeneous measuring continuum representing intensities, they could then compare different qualitative intensities within the same species by comparing the actual lengths of their respective latitudes, just as geometrical lines or distances can be compared and measured together.[121] In Dumbleton's words: "Every quality in the same species having the same latitude is of equal intension."[122] A greater intensity (of heat, for example) would be represented by a longer latitude, a lesser intensity of heat by a shorter latitude, making immediately clear the relation of the two intensities.[123] Moreover, since qualitative increase and decrease (intension and remission) involved motion along the latitude of quality, qualitative *change* could now be measured (and represented) in terms of distances traversed along this continuum.[124]

As the latitude came ever more to resemble a common measuring line, it gained the capacity to act as an instrument of *relation* and *representation*. On the most basic level, Dumbleton used the *latitudo qualitatum* to relate qualitative intensities to qualitative distances, and then to relate qualitative distances to quantitative distances.[125] The latitude became for him the

---

[118] Sylla, *Oxford Calculators*, 385–86: "Dumbleton's conception of latitudes and degrees, therefore, is significantly different from the notions of earlier theorists. No longer is a latitude imagined as similar to a series of numbers, where each number in itself is higher than previous numbers as well as occurring higher in the scale, and no longer are degrees imagined as single numbers. Instead a latitude is imagined as similar to a geometric line, where any part of the line is similar to any other part and where a degree corresponds to part of the line beginning at its origin and ending after a greater or lesser interval depending on the intensity of the degree."

[119] Sylla, *Oxford Calculators*, 385–87. Dumbleton, *Summa logicae et philosophiae naturalis*, partially edited in appendix 1, pt. 4, by Sylla, *Oxford Calculators*, 565–625, 572: "Primo tamen intelligendum est quod omnis qualitas intensibilis est infinitum divisibilis intensive, sic videlicet quod infinitum est aliquis gradus minus intensus alio gradu eiusdem qualitatis."

[120] Dumbleton, *Summa logicae*, in Sylla, *Oxford Calculators*, 571: "in omni distantia qualitativa gradus non est aliud nisi distantia illa ex partibus suis composita."

[121] Sylla, "Latitude of Forms," 252, 263.

[122] Dumbleton, *Summa logicae*, in Sylla, *Oxford Calculators*, 571: "Ex hiis patet hec conclusio: quod omnes qualitates in eadem specie eandem latitudinem habentes equales existere intensive."

[123] Sylla, *Oxford Calculators*, 400.

[124] Sylla, "Latitude of Forms," 252–56; Wilson, *Heytesbury*, 144; Dumbleton, *Summa logicae*, in Sylla, *Oxford Calculators*, 571: "Causa precisa et essentialis intensionis in qualitate intensibili et remissibili est distantia sive latitudo in qua fit motus verus in qualitate."

[125] Dumbleton, *Summa logicae*, in Sylla, *Oxford Calculators*, 571: "Pro istis et omnibus consimilibus supponendum est quod in omni distantia qualitativa gradus non est aliud nisi distantia illa ex partibus suis composita."

common measure linking motion of alteration over "qualitative space" with local motion over "quantitative distance."[126] Once this was established, he and other Calculators could use Euclidean principles to claim that equal latitudes measured equal qualitative intensities within the same species of quality.[127] Dumbleton writes: "all qualities in the same species, whether uniform or non-uniform, containing equal latitudes (i.e., equal qualitative distances) are necessarily equally intense in their natures."[128]

The next step was to correlate latitudes in order to investigate the mathematics of motion. Dumbleton sought to represent Bradwardine's important function relating forces, resistances, and velocities by juxtaposing a latitude of proportions against a latitude of velocity.[129] In his scheme, equal parts on the latitude of velocity corresponded to equal differences in velocity, and equal parts on the latitude of proportions corresponded to equal proportions.[130] In the margins of Dumbleton's *Summa logicae*, Bradwardine's function is actually illustrated by the drawing of two parallel lines representing the two latitudes.[131] Sylla's observation on the use of these lines is important here:

It seems very likely that the drawing of parallel lines and labeling of them to represent Bradwardine's function was not only an illustration but a practical tool. One could manipulate the variables and determine the results using the parallel lines more easily than by calculation. Thus the parallel scales or latitudes could function like a log table or slide rule in simplifying mathematical operations.[132]

There were no existing slide rules or logarithmic scales to serve as models for Dumbleton's practical calculating instrument. Although Dumbleton's latitude of proportions is divided on principles different from those of the linear scale of money, there were practical instruments associated with money and exchange in common use in Dumbleton's

[126] Dumbleton, *Summa logicae*, cited by Sylla, *Oxford Calculators*, 397: "Dum diversi generis sunt inter se distantia qualitativa et quantitativa, nec unum continuum ex hiis fieri potest, quare propter extensionem intensive nihil accrescit qualitati, sicut nec quantitati accidit maioritas propter qualitatem. Unde si nominaremus spatium in quo est motus alterationis 'spatium qualitativum' sicut dicimus spatium in quo est motus localis 'distantiam quantitativam,' non magis diceremus distantiam nec latitudinem in qualitate habere medietates inequales in quantitate intensive quam dicimus quantitatem finitam habere medietates inequales in quantitate."
[127] Euclid, *Elements*, Book X, proposition 12: "Magnitudes commensurable with the same magnitude are commensurable with one another also."
[128] Dumbleton, *Summa logicae*, in Sylla, *Oxford Calculators*, 570: "Tertia dicit quod omnes qualitates in eadem specie sive fuerint uniformes sive difformes equales latitudines continentes, id est distantias qualitativas, equeintensas necessario in suis naturis consistere."
[129] Sylla, "Latitude of Forms," 264–67; Sylla, *Oxford Calculators*, 402–07.
[130] Sylla, *Oxford Calculators*, 402–06. Sylla notes that the latitude of proportion is illustrated in some margins of the *Summa logicae* as a drawn line on which the proportions 1:1, 2:1, 4:1, 8:1, and 16:1 are equal distances apart.
[131] Sylla, *Oxford Calculators*, 403–07; Sylla, "Latitude of Forms," 265–66.
[132] Sylla, *Oxford Calculators*, 407; Sylla, "Latitude of Forms," 267.

time that could well have served as the model for his calculating lines. The merchant's abacus or lined counting table, like the calculator's latitude, simplifies mathematical operations by translating values into linear terms and by calculating values through the relation of points on juxtaposed lines. The abacus was arguably the most ubiquitous calculating instrument of the time – an essential tool of administration as well as of commerce.[133]

In addition to the abacus, the calculation of values through the comparison of degrees on two related continua also characterized the common use of "money of account" in the commerce of the fourteenth century. Money of account functioned as an idealized monetary scale of artificially fixed ratios of named coins that were often no longer in circulation, against which the actual value of the coin in circulation was measured.[134] As Fernand Braudel described it: "A money of account is thus a scale, a measure. It makes possible the classification of prices and creates a continuous accounting procedure."[135] By the fourteenth century this "imaginary scale" was used in everyday exchange and record keeping to measure monetary values, necessitating that both traders and their customers become adept in its use. The manipulation and comparison of monetary scales to determine values was thus an essential component of both the models of money in exchange and money in administration. The determination of economic values through the comparison of continua was commonplace in the marketplace of the fourteenth century.

This is apparent in Jean Buridan's attempt to reformulate Aristotle's mathematical model of *iustitia directiva*, discussed in chapter 5.[136] For Aristotle's single line of *lucrum/damnum* bisected by the judge, Buridan substituted a model involving comparison and ratio between multiple lines. In his view, each party to an exchange carried his own line of *lucrum/damnum*. The proper assessment of economic gain and loss necessitated that it be determined relative to the differing potential for each person to gain and lose from the transaction.[137] To illustrate how a

---

[133] For the use of the abacus in commercial culture and its influence on mathematical innovation, see Frank J. Swetz, *Capitalism and Arithmetic: The New Math of the Fifteenth Century* (La Salle, Ill., 1987), 29–32, 177.

[134] For a discussion of money of account, see Peter Spufford, "Coinage and Currency," in *The Cambridge Economic History of Europe*, vol. III (Cambridge, 1965), 593–95; Carlo Cipolla, *Money, Prices, and Civilization in the Mediterranean World* (New York, 1967), 38–50; Kaye, "Impact," 259.

[135] Fernand Braudel, "Prices in Europe from 1450 to 1750," in *The Cambridge Economic History of Europe*, vol. IV (Cambridge, 1967), 374–486, 379.   [136] 144–45.

[137] Buridan, *Ethics* v.10, 410: "quod plus debito et minus debito sunt duo iniusta, hoc quidem ex parte facientis, illud autem ex parte patientis. Et ita in operatione iusta sunt duo iusta secundum unum ex parte facientis et alterum ex parte patientis . . . propter quod oportet in operatione iusta duo esse iusta et duo esse iniusta modo predicto."

medium could be established between multiple continua, Buridan chose the example of the comparative measurement of qualitative intensities – comparing the whiteness/darkness continuum of Socrates to the whiteness/darkness continuum of Plato in order to properly measure the equalizing mean between whiteness and blackness proper to each. This example, appearing in Buridan's discussion of money as a measure in exchange in Book v of the *Ethics*, was central to contemporary philosophical debates concerning qualitative measurement by latitudes.[138]

As the conceptual development of the *latitudo qualitatum* continued, its form and function diverged ever farther from traditional (Aristotelian) philosophical definitions. Central to Aristotle's description of the physical universe was the idea that qualities existed as pairs of contraries, and that the increase of a quality such as whiteness or heat within a subject was tied to the decrease of its contrary (blackness or cold) within the same subject.[139] As the century progressed, the Calculators moved toward eliminating the Aristotelian doctrine of contraries from their measurement of qualitative intensities. Philosophers continued to believe in Aristotle's contraries, but they ignored them in their qualitative measurements. In doing so, the "core" Calculators freed themselves from having to explain observable qualities in terms of movements toward or away from contrary perfections (absolute heat or absolute cold) that were not in themselves perceptible.[140] This not only simplified their measuring schemata, but it also served to locate measurement within the realm of practical observation rather than of philosophical definition.

A similar movement in the direction of practical measurement can be seen in the growing consensus among natural philosophers to measure the intensity of qualities from the zero degree. Before the work of John Dumbleton and Richard Swineshead in the 1340s and 1350s, all measurements of qualities had been taken from the maximum (perfect) degree down, following the universally accepted metaphysical definition of

---

[138] *Ibid.*: "quod nullum est inconveniens duo fusca ejusdem gradus inveniri in duobus subiectis, et tamen utrumque esset medium albi et nigri. Et etiam forte dici posset, quia illa duo fusca non essent media albi et nigri secundum eandem particularem rationem albi et nigri, sed secundum hoc subiectum esset innatum esse album aut nigrum fuscum in eo esset medium albi et nigri . . . Fuscum igitur in Socrate non esset medium albi et nigri, nisi secundum quod Socrates est innatus esse albus et niger, et non secundum quod Plato potest esse albus aut niger."

[139] For the doctrine of contraries in fourteenth-century discussions of qualitative change, see Shapiro, "Burley," 413–14. Burley, writing in the early part of the century, and not considered a representative of the "core" Calculators, remained attached to this position. For his description of color as a latitude stretching from perfection (whiteness) to absolute imperfection (blackness), see Sylla, *Oxford Calculators*, 341 n 53, and in the appendix IV.N.I. Compare Burley's conception of qualitative measurement to that of Henry of Frimaria (145, in chapter 5) who directly links money as a measure of *indigentia* to whiteness as the scale against which color is measured.

[140] Coleman, "De Ripa," 151–53.

qualities in terms of their forms of perfection.[141] The "core" Calculators for the first time recognized the advantages of standardization in qualitative measurement – of applying the measuring latitude to every quality in the same way, beginning at the zero intensity of each quality, and extending it to the intensity's maximum existent degree.[142] In chapter 1 of his treatise entitled *Calculationes*, under the heading *De intensione et remissione*, Richard Swineshead asks from what point the intensity of a quality should be measured. There are three possibilities: (1) intension can be measured from the perfection of the quality (*summum gradum* or *gradum intensissimum*), while remission is measured by its "distance" from the least intense degree;[143] (2) intension can be measured by its "distance" from the zero degree (*non gradu*), while remission is measured from the most intense degree;[144] (3) both intension and remission can be measured from the zero degree.[145] Like Dumbleton, Swineshead rejects the first two and accepts the third, essentially because measurement from zero degree renders it easier to compare intensities by comparing the length of their respective latitudes.[146]

Dumbleton and Swineshead made the move toward measurement from the zero degree despite the position taken by earlier Calculators that serious philosophical problems attached to such measurement. Heytesbury had argued that the first "step" in augmentation from zero to any degree would involve a "mutation," i.e., an instantaneous, discontinuous motion, which was an impossibility. Furthermore, comparing the endpoint of the measuring latitude to the zero degree would involve an infinite product – another impossibility.[147] Nevertheless, the later Calculators persisted, primarily because the new zero-based schema provided an additive and more easily quantifiable basis for measurement. The zero degree (*non gradus*) of a quality was more easily determined than some ideal or definitional maximum degree.[148] Outside the context of scholas-

---

[141] Sylla, "Latitude of Forms," 272.

[142] Clagett, "Swineshead," 140–61; Sylla, "Latitude of Forms, 262, 272–73; Bradwardine, *De proportionibus*, 50; Coleman, "De Ripa, 153: "[For Dumbleton] movement is real in qualitative change and it is additive, successive, divisibly acquired and measured by the attainment or the loss of qualitative distance from the zero degree up to some intensified state."

[143] Richard Swineshead, *Subtillissimi Ricardi Suiseth Anglici Calculationes noviter emendate atque revise* (Venice, 1520), 2ra: "Prima positio ponit quod intensio cuiuslibet qualitatis attenditur penes appropinquantionem ad summum seu ad gradum intensissimum illius latitudinis, et remissio penes distantiam a gradu intensissimo."

[144] *Ibid.*: "Secunda positio ponit quod intensio habet attendi penes distantiam a non gradu; et remissio penes distantiam a gradu intensissimo."

[145] *Ibid.*: "Tertia positio dicit quod intensio attenditur penes distantiam a non gradu; et remissio penes appropinquationem ad non gradum."   [146] Clagett, "Swineshead," 139–40.

[147] Sylla, *Oxford Calculators*, 735–36.

[148] Sylla, "Latitude of Forms," 273; Sylla, *Oxford Calculators*, 398–99: "He [Dumbleton] rejects the measurement of remission from the maximum degree, which Roger Swineshead had favored . . .

tic philosophy, however, the Calculators can hardly be said to have "invented" either the zero-based scheme or the neglect of the contrary within this scheme. This is, of course, the way that cloth or grain or the duration and severity of penance, or the size of benefices, or the price of goods and services in exchange were continually being measured in the world around them.[149]

Thus, although the product of more than a half-century of intense philosophical debate, and although couched in the most subtle and complex logic, and although lacking all traces of contact with contingent experience, the measuring model of the Calculators (if not their language) in many respects was moving away from philosophical abstraction, away from metaphysical definitions, *against* the weight of intellectual authority, and toward more "practical" and concrete solutions to the problems of measurement. This situation, rather than being paradoxical, adds weight to the assertion that the form of the Calculators' measuring model was strongly influenced by practical and successful models of measurement in constant use in the world around them.[150]

Of all these models, from the simplest daily measurements by volume and length, to the complex measurements of land area, seed-to-yield ratios, and profits required by agricultural estate managers, to the mechanical scales and abacuses of the merchants, to the mechanical clock whose appearance in the first half of the fourteenth century coincided with the development of the measuring latitude, the model of money as measure had arguably the most profound influence on the conceptual instruments the Calculators devised to facilitate their "mania" to measure. The strongest indication of this influence is the constant reflection of money in the form and function of the *latitudo qualitatum*.[151]

A theoretical system built only on distance from zero could deal with all the same cases as the system built on both the distance from zero degree and the distance from the maximum degree, and avoid some of the anomalies which the double system of measurement seemed to lead to."

[149] Sylla ("Latitude of Forms," 273) explained the movement toward zero-based measurement not in terms of practical influences, but in terms of a general reframing of the question from metaphysical into mathematical terms.

[150] The historian of science Pierre Souffrin, a mathematician and astronomer in his own right, is especially sensitive to the influence of practical measurement on the "abstract" measuring schemes of the fourteenth century. He writes of the scholastic attempt to measure and quantify qualities: "Ces notions trouvent leurs sources dans les efforts des philosophes pour assimiler théoriquement, pour intégrer à la philosophie, des pratiques de quantification et de mesure qu'ils voyaient fonctionner avec efficacité dans tous les aspects de la vie sociale" (Souffrin, "La quantification," 66).

[151] For example, compare the technological form of money as outlined above to Wilson's characterization of William Heytesbury's definition and use of the latitude (*Heytesbury*, 144): "That differences in intensity of a given quality are quantitative; i.e., that intensities may be arranged according to a scale of degrees, of such character that it will be physically meaningful to speak of equal numerical differences between degrees as representing the same qualitative distance or intensive magnitude."

Among these reflections are: the seeking of one single instrument of conceptual measurement which could be universally applied to all qualities and motions; the redefinition of qualitative increase as additive; the conception of the latitude as a homogeneous continuum or geometric line; the gradation and division of this line into numbered degrees; using this graded conceptual line to measure expanding and contracting values by translating value change into distances on the measuring continuum; using the latitude as an instrument of relation to compare qualitative values; abandoning contraries in the measurement of qualitative intensity; adopting a standardized application of the latitude from the zero degree to the existent degree. As we have seen, all these innovations were present in the "core" Calculators' latitude, all went against Aristotelian definition, and all can be found in the already defined technological model of money as measure.

The objection can be raised that money could not have served as the model of the Calculators' latitude because their latitude was, by definition, a true continuum. The Calculators were particularly sensitive to the question of the composition of the continuum, and the majority agreed with Aristotle in denying the real existence of indivisibles within it.[152] They would therefore not have seen money as a continuum, since the scale of money is composed of indivisibles represented by the smallest coin.

There are a number of reasons why this objection is questionable. In a practical sense, it is the everyday *use* of money that determines how it will be experienced, and it is this experience that is reflected in the consciousness of its users. In the fourteenth century, the value of the widely circulating "black" or "petite" money was so small that what struck scholastic observers, and what they chose to emphasize in their commentaries, was the divisibility and flexibility of money – its ability to measure and bring together the most seemingly diverse commodities and incommensurable values – and not its ultimate indivisibility. Again on the practical level, the many imperfections in the coinage of the day (traceable primarily to changes in the precious metal content within a single currency and to the proliferation of competing currencies) led to the everyday use of "money of account."[153] The values of the coins in circulation were continually measured against the fixed, because ideal, scale of the money of account.[154] Thus, the value of the "indivisible" individual coin was understood to float within a latitude or range of value, and was itself measured against a continuum defined by monetary ratios.

---

[152] Murdoch and Synan, "Two Questions"; Sylla, "Latitude of Forms," 260–62.
[153] For a discussion of "money of account" in a slightly different context, see Kaye, "Impact," 259.
[154] Spufford, "Coinage and Currency."

The fact that monetary values are expressed in terms of unit coins assures that money will be experienced as a *numbered* line. This is one of the factors that gave money such extraordinary influence as a measuring instrument. The numbers in the continuum of money are a prime model for the philosophic attempt to conceptualize *gradus*, or numbered degrees along the Calculators' measuring *latitudo*. The numbering of the monetary scale did not imply, within the philosophical understanding of the fourteenth century, that money could not at the same time be seen as a true continuum. Within scholastic philosophical discourse it was generally understood that measuring continua could be divided into smaller measuring units without losing their character as continua. Jean Buridan, for example, considered this question while commenting on Aristotle's discussion of measurement in Book x of the *Metaphysics* (so important to Albertus Magnus and Thomas Aquinas in their discussions of money as measure).[155] He noted that, although time was continuous, the year was divisible into and measured by months, the months by days, and so on, each larger part being measured by a smaller unit part.[156] Similarly, the continuous space of a constellation could be divided into degrees and those degrees into minutes, without compromising the continuity of space.[157] Buridan concluded that, although no continuum is properly said to be composed of or measured by indivisibles (which have no real existence), nevertheless in daily life we commonly measure larger continuous magnitudes by their smaller composite units.[158] Philosophers of the fourteenth century understood that numbers could stand both for discrete units and for numbered parts of a continuous magnitude.[159]

Every commentator on Aristotle's *Ethics*, including many who were in the forefront of proto-scientific thought, fastened onto the fact that

[155] For Albert and Aquinas, see chapter 3, 59, 67.

[156] Jean Buridan, *In metaphysicen Aristotelis quaestiones* (Paris, 1588; reprint Frankfurt a/M, 1964) x.1, 61ra: "et sic annus mensuraretur per menses: menses per dies etc. omnino apparet quod omne quod sic mensuratur mensuratur per suas partes." See Aristotle, *Metaphysics* x.1 [1052b32–1053a14]. Buridan returns to the same question in his commentary on the *Physics* iii.7, 60vb: "signum enim in celo non est divisibile in signa sed bene in gradus nec gradus in gradus sed bene in minuta etc."

[157] For Buridan's attitude toward this question of continua measured by units, and his reliance on Aristotelian philosophical definitions in its solution, see J. M. Thijssen, "Buridan on Mathematics," *Vivarium* 23 (1985), 55–78, 67.

[158] Buridan, *Metaphysics* x.1, 61va: "Ad aliam concedo quod nullum continuum mensuratur per indivisibile simpliciter sed continuum mensuratur per suas partes que secundum aliquam numeralem discretionem reputantur indivisibiles."

[159] Buridan, *Physics* iii.7, 60vb: "Uno modo quod solum loquamur de rebus ab invicem separatim existentibus ita quod nec una sit pars alterius . . . Alio modo possumus numerum accipere pro rebus numerabilibus non ab invicem separatis per discontinuationem, sed tamen ab invicem diversis, sicut partes magnitudinis continue." See Thijssen, "Buridan on Mathematics," 64: "'Number' stands for counted or countable things being part of one and the same continuum."

money was remarkable for its ability to measure and commensurate all goods and services in exchange. But does that mean that any thinker ever thought of the Aristotelian analysis of money when speculating on the philosophical problem of relating diverse species through the measuring continuum? Direct evidence of such connection within scholastic natural philosophy is thin – understandably so given the prevailing attempt to distance *scientia demonstrativa* from the taint of contingent experience. There is, however, an important exception found in a commentary on Aristotle's *Physics* by the influential natural philosopher and analyst of economic exchange, Jean Buridan.

In question 6 of Book VII Buridan asks whether in order for things to be compared or measured against each other it is required that they be of the same species (*specialissime univoca*).[160] I have already discussed the medieval hesitation to make comparisons and measurements across species based on a misreading of Euclid. Adding to this was the often-repeated definition from Book X of the *Metaphysics* that the measure is always of the same species as the measured. Yet there are, noted Buridan, many examples where measurement and comparison across species both occurs and is justified. His first point is that God is clearly recognized to be superior to all of His creatures, and substance is clearly understood to be superior to accident, but these recognitions rely on comparisons being made between things of different species. His second point involves money as an instrument of measurement and relation: "all goods in exchange are mutually comparable as is clear from Book V of the *Ethics*, yet they are of greatly varying natures and species [*sunt valde diversarum rationum et specierum*]."[161] His third point is that Aristotle himself maintains that "time is the measure of motion" even though they are neither of the same species nor the same genus (*tamen nec sunt eiusdem speciei nec eiusdem generis*); similarly circular and straight motion can be compared though of different species; similarly the area of a square and a triangle can be compared.[162] The setting of these comments demonstrates that, when weighing the question of measurement across species in a physical-philosophical context, Buridan considered the economic case of monetary commensuration as naturally as he did the theological comparison of God to his creatures or the scientific comparison of time and motion as continua.

Given the existence of these common cases of intra-species comparison, two of which are drawn from the writings of Aristotle himself,

---

[160] Buridan, *Physics* VII.6, 106va: "Utrum quod aliqua sint comparabilia requiritur et sufficit quod ipsa sint specialissime univoca."

[161] *Ibid.*: "Item omnia venalia sunt adinvicem comparabilia ut pate quinto ethicorum et tamen sunt valde diversarum rationum et specierum ut equi, quadrige, vinum, etc."   [162] *Ibid.*

Buridan asks what Aristotle intended when he repeatedly denied the possibility of such comparison in Book VII of the *Physics*.[163] He reaches two conclusions concerning this question. The first is to say that Aristotle clearly recognized that the commensuration of two things of different species *is* possible, but only when measured in relation to some third quality possessed in the same manner by both. Thus we can compare a dog and a horse in respect to whiteness or size or number.[164] So too can we compare different species of motions *secundum velocitatem*.[165]

Buridan's second conclusion is that the cases cited at the beginning of his question are "common" comparisons but not "proper" comparisons. "Common" comparison is when two things are compared and one exceeds the other but the excess has no determined or determinable proportion. "Proper" comparison involves a measurement in which the excess of one to another is in some determinable proportion.[166] When we compare God to any of his creatures we do it in a "common" rather than a "proper" sense because he is clearly superior but we cannot say exactly how much (*in quanto*).[167]

Having established his conclusions, Buridan applies both of them to the question of money as common measure. He writes: "All goods are commensurated [*comparantur adinvicem*] in exchange, but only through their common relation to human need and not 'properly' in their natures according to a determinable proportion."[168] Buridan has added nothing new here. He has repeated what Aristotle and every commentator on the *Ethics* had to say about measurement and commensuration in exchange. But he has done it in the context of a philosophical discussion concerning the measurement and relation of motions and qualities.

Having given two reasons why money can be said to measure and relate all goods in exchange only in an improper or common sense,

---

[163] Chapter 4 of Book VII [248a10–249b26] touches on this question at many points. A noteworthy outgrowth of Aristotle's attempt to limit comparisons to within a single species is his denial that an alteration and a local motion can be commensurated, for otherwise "an affection will be equal to a length which is impossible" [248a15]. As we have seen, this Aristotelian "impossibility" lay at the core of the Calculators' measuring program and formed the basis of Dumbleton's *latitudo qualitatum*.

[164] Buridan, *Physics* VII.6, 106va–vb: "Tunc ergo est prima conclusio quod ad proprie dictam comparationem non requiritur quod termini supponentes pro illis rebus que comparantur sint adinvicem eiusdem speciei specialissime quia equus et canis vere possunt adinvicem comparari secundum albedinem aut secundum magnitudinem."   [165] Buridan, *Physics* VII.6, 106vb.

[166] *Ibid.*: "secundum determinatam proportionem ut duplam, triplam vel sexqui alteram etc."

[167] *Ibid.*: "Sed non potest dici in quanto sit melior nec potest aliquod capi in deo quod sit equalis vel similis bonitatis cum asino . . sic etiam dicimus deum esse sapientiorem homine sed non potest dici in quanto."

[168] Buridan, *Physics*, 107rb: "Ad aliam dicitur quod omnia venalia secundum relationem ad humanam indigentiam comparantur adinvicem sed non proprie scilicet secundum certam proportionem ex natura rei."

Buridan then adds a third. Where the first two objections were shared by the other common examples he gave, the third was directed specifically at money as measure. The measurement of value in exchange is never precise because the selling price is reached through a process of free bargaining in which deception is common: the seller often claims that his goods are worth more money than the buyer offers, and the buyer claims the contrary.[169] It is not only that money cannot measure things in their nature, but that buyer and seller actively seek to subvert the measure in their search for advantage. Accurate measurement, the "proper" measurement that science requires, is not even attempted in the process of exchange.

In this question Buridan reveals two attitudes that are particularly relevant to the role of money as a model for philosophical measurement. The first is that money's capacity to bring the most diverse goods and services into relation in the marketplace was present in Buridan's mind when he considered the mathematical-philosophical-theological question of the measurement and comparison of diverse species – a question that was central to the proto-scientific program of fourteenth-century natural philosophy. The second is that money can never be consciously claimed as a model for scientific measurement. In Buridan's view, the measurement of *indigentia* by price involves not only a relative and shifting component but one that is also intentionally distorting. Given the hyperbolic association of science with unchanging truth, it is doubtful that any philosopher could claim the form and function of money as a model for his philosophical measurement. Money could be recognized as a remarkable instrument of measurement and relation speaking *communiter*, but never speaking *proprie*. This would be true not only for the Oxford Calculators, who consciously cleansed their writing of reference to contingent experience, but for thinkers like Buridan and Oresme in whose work one sees glimpses of nature, personal observation, and the record of social experience.

Nummisma est medium et mensura omnium commutabilium.[170] So impressed were scholastic thinkers by the measuring successes of money and by the multiplicity of its uses that they credited ancient philosophers with having invented it: "O inventa prudencium! O provisa maiorum!"[171] Within the

---

[169] *Ibid.* (continuing from n. 168): "Immo oportet voluntarie concordare in quantitate proportionis unde venditor dicit sepe rem suam in valore excedere pecuniam emptoris et emptor dicit contrarium et non possunt concordare aliter aut veniens qui plus indiget illa re concordat ita cum venditore et comparat illam rem sue pecunie vel alteri rei sue de hoc magis debet videri in speciali quinto ethicorum."   [170] Burley, *Ethics* v, 83v.

[171] Oresme, *De moneta*, 17, citing the words of Cassiodorus. For Brunetto Latini's paean, see chapter 5, 137.

same intellectual culture that was preoccupied with devising conceptual instruments to measure, relate, and commensurate qualitative "magnitudes," money was recognized as a remarkably effective instrument of measurement, relation, and equalization. Just as the Calculators' *latitudo qualitatum* was specific to specific qualities, so the graded continuum of money was conceived as a kind of measuring latitude specific to the single intensible quality of *indigentia* or need common to all goods in exchange.[172] The philosophers involved in "inventing" the measuring *latitudo* actively participated, in private and public capacities, in highly money-conscious societies in which the impact of the "technological form" of money was magnified through the necessity of its constant and careful use. Aristotle's discussion in the *Ethics* focused their understanding of the form of money as a geometric line and the form of exchange as a geometric system of equalization. Here the confluence of the practical and the theoretical, the actual and the textual, provided a rich environment for intellectual influence and transference.

[172] See chapter 5, 139, for a discussion of this point in an economic context. The description of money as *une mesure commune* is from Oresme, *Ethics* IX.1, 452.

Chapter 7

LINKING SCHOLASTIC MODELS OF MONETIZED EXCHANGE TO INNOVATIONS IN FOURTEENTH-CENTURY MATHEMATICS AND NATURAL PHILOSOPHY

By the second half of the thirteenth century, the social conception of a world connected by commerce and held together by a monetarily measured and regulated "flux and reflux of services" was being articulated in the *Ethics* commentaries of Albertus Magnus and Thomas Aquinas.[1] As the social process of monetization gathered speed, scholastic thinkers expanded Aristotle's geometric model of exchange from its base in personal relationships to a supra-personal system of relations in which aggregate needs and decisions determined value. The great claims Aristotle made for the success of money as a measuring and commensurating *medium* were expanded as well. In the writings of the most sophisticated economic thinkers of the fourteenth century, the use of money as an instrument of equalization was understood to permit exchangers of unequal social status and occupation, at crossed purposes (each wanting to buy cheap and sell dear), with unequal needs, exchanging unequal goods of unequal value, to arrive, nevertheless, at an approximate equality in their economic transactions – an equality, moreover, sufficient to be named "just." The monetized marketplace was seen to bind all producers and consumers into a geometrically conceived, self-equalizing system, regulated through a shifting market price determined by common need and estimation.

In order to show that the experience and comprehension of this social context influenced the development of scientific thought, I have been following six clusters of insights, organized into category headings, that in my view characterize the most innovative aspects of both economic thought and natural philosophy in the fourteenth century. For the most part these insights appeared first in scholastic economic thought as thinkers sought to comprehend the workings of their monetized society.

[1] Aquinas, *Ethics* (1969), 291a; Albertus Magnus, *Ethics* (Borgnet edn.), 355b, 356a; and see chapter 3, 71–72.

At times contemporaneously, but most often within a generation or two, reflections of these economic insights then appeared in scholastic natural philosophy applied to the conception and comprehension of nature. In the previous chapter I explored how the insights developed within one of these socioeconomic categories, "Money as *medium* and as measure," were reflected in specific, proto-scientific innovations within fourteenth-century natural philosophy. In this concluding chapter I do the same for the five remaining categories comprising the new geometric model of economic exchange: the social geometry of monetized society; mathematics and the geometry of exchange; equality, the mean, and equalization in exchange; common valuation in exchange; and relation and the relativity of value in exchange. Within each category I focus on one proto-scientific development, chosen primarily because of its recognized importance to the history of science. Where the last chapter was centered on the writings of the Oxford Calculators, here I concentrate on the work of the two great Parisian masters, Jean Buridan and Nicole Oresme. They are natural subjects for this study: their writing is "friendlier" to the modern reader than that of the Calculators because closer to personal observation and experience; their speculation in the area of natural philosophy is marked by boldness and acuity; their contributions to the history of science are broadly recognized; and their writings reveal them to have been among the most sophisticated economic observers of their day.

THE SOCIAL GEOMETRY OF MONETIZED SOCIETY

The historian of science Anneliese Maier brought attention to the creation of a new *image du monde* in the natural philosophy of the fourteenth century in which all phenomena and processes were conceived of as continuous magnitudes in constant expansion and contraction.[2] When Maier made this comment, she had in mind not only the mathematics of the continuum developed by the Oxford Calculators but also the proto-scientific speculation of the masters at Paris who first adopted and later extended the logico-mathematical insights of the Calculators.[3] Perhaps the clearest image of a world composed of continuous magnitudes can be found in the opening words of Nicole Oresme's *Tractatus de configurationibus qualitatum et motuum* or *Treatise on the Configurations [Geometry] of Qualities and Motions*:

---

[2] Maier, "La doctrine d'Oresme," 337.
[3] On the intellectual influence of the Oxford Calculators on Paris masters, see Murdoch, "*Subtilitates Anglicanae*." For intellectual connections between Paris and Oxford on questions of perception and epistemology, see Tachau, *Vision and Certitude*, esp. 278ff.

Every measurable thing except numbers is imagined in the manner of continuous quantity. Therefore, for the mensuration of such a thing, it is necessary that points, lines, and surfaces, or their properties, be imagined. For in them (i.e., the geometrical entities), as the Philosopher [Aristotle] has it, measure or ratio is initially found, while in other things it is recognized by similarity as they are being referred by the intellect to them (i.e., to geometrical entities). Although indivisible points, or lines, are non-existent, still it is necessary to feign them mathematically for the measures of things and for the understanding of their ratios.[4]

In this treatise, written in the same half-decade as the *De moneta*, his work on money and minting, Oresme showed both his affiliation with the Oxford Calculators and his willingness to expand upon their mathematical analysis of qualities and motions. His system of configurations, essentially a graphing system devised for the measurement and relation of qualities, has long been recognized as one of the extraordinary accomplishments of scholastic mathematics and natural philosophy. Oresme was a highly accomplished mathematician who saw himself as a geometer as much as a natural philosopher. In constructing his system he consistently framed philosophical innovations in the area of measurement, including the Calculators' measuring latitude of qualities (*latitudo qualitatum*), in geometrical terms. Within his novel and ambitious geometry of qualities, Oresme brought the use of the measuring line into still closer connection to the model of money as measuring continuum and connecting *medium*.

Oresme believed along with the Merton Calculators that any quality open to change in intensity was equally measurable, whether that quality was physical or mental, whether one considered whiteness or heat, joy or friendship, the power of the intellect or the depth of religious vision. Like the Calculators, Oresme based his quantification of such measurement on the "addition theory" of intension and remission, where increases or decreases in qualitative intensity resulted from the addition or subtraction of parts of the quality, and where qualitative motion was conceived to occur over a qualitative "distance" open to quantification. Because of the linear implications of the addition theory, Oresme believed that "there is no more fitting way for [qualitative intensity] to be imagined than by that species of a continuum which is initially divisible and only in one way,

[4] Nicole Oresme, *Nicole Oresme and the Medieval Geometry of Qualities and Motions: A Treatise on the Uniformity and Difformity of Intensities Known as "Tractatus de configurationibus qualitatum et motuum"*, Marshall Clagett (ed. and tr.) (Madison, 1968), I.1, 164–65: "Omnis res mensurabilis exceptis numeris ymaginatur ad modum quantitatis continue. Ideo oportet pro eius mensuratione ymaginari puncta, lineas et superficies, aut istorum proprietates, in quibus, ut vult Philosophus, mensura seu proportio per prius reperitur. In aliis autem cognoscitur in similitudine dum per intellectum referuntur ad ista. Etsi nichil sunt puncta indivisibilia aut linee, tamen oportet ea mathematice fingere pro rerum mensuris et earum proportionibus cognoscendis."

namely by a line."[5] Thus, "the measure of intensities can be fittingly imagined as the measure of lines . . . equal intensities are designated by equal lines, a double intensity by a double line, and always in the same way if one proceeds proportionally."[6]

On one level, as Oresme himself points out in his opening statement, he derived the idea of the line as universal measuring reference from Aristotle.[7] But the uses to which Oresme put the line were far from anything conceived by Aristotle. The modern economist Ludwig von Mises has described money as a "medium for expressing values."[8] That is what the geometric line and geometric figures became for Oresme. He not only perceived a world composed of quantifiable continuous magnitudes, but he also saw these magnitudes as part of a dynamic, interrelated system in constant change, with quantities and qualities in perpetual intension and remission. He chose the geometric line as the reference for measurement not only because it could measure all intensities, but because it could also measure the continual *changes* in intensities that defined the world around him. He wrote: "the measure of intensities can be fittingly imagined as the measure of lines, since an intensity could be imagined as being infinitely decreased or infinitely increased in the same way as a line."[9]

With the single continuum of the Calculators' latitude, one could measure and compare the intensities of heat in two different subjects, e.g., two containers of water, or a container of water and a piece of metal, but one had no way of representing the subjects themselves in any way.[10] If the heat in two containers of water of vastly different size were compared, the length of the latitude measuring the intensity of heat in each would not and could not reflect the difference in their size. Thus, while the

---

[5] Oresme, *De configurationibus* I.1, 166–67: "non potest convenientius ymaginare quam per illam speciem continui que est primo divisibilis et uno modo tantum, scilicet per lineam."

[6] *Ibid.*: "Ergo mensura intensionum potest ymaginari congrue sicut linearum mensura . . . Ideoque intensiones equales per equales lineas designantur et dupla intensio per duplam lineam et sic semper proportionaliter procedendo."

[7] In an earlier work, the *Questiones super geometriam Euclidis*, Oresme associated his general use of geometric figures as the common measure for all continuous magnitudes with Aristotle's definition of time as the measure of motion. This work was first edited by H. L. L. Busard (Leiden, 1961). Clagett later made his own partial edition covering questions 10–15 and included it as appendix 1 of the *De configurationibus*, 526–75. See q. 11 (Clagett edn.), 536–37.

[8] Von Mises' definition (*Theory of Money*, 48) indicates the continuity of the Aristotelian understanding of money as *medium et mensura*.

[9] Oresme, *De configurationibus* I.1, 166–67: "Ergo mensura intensionum potest ymaginari congrue sicut linearum mensura, cum etiam intensio possit eodem modo sicut linea in infinitum diminui et quantum est ex se in infinitum augeri."

[10] For the exception to the Calculators' use of a unitary scale, see the discussion of Richard Swineshead's "multitude of forms" in Clagett, *Science of Mechanics*, 335–36; Sylla, "Medieval Quantification of Qualities," 27–29. Sylla stresses, however, the originality of Oresme's full recognition of extension and intension as components of qualitative quantity.

latitude could theoretically quantify qualitative intensities, it could not quantify the amount of a given quality in a given subject.

Oresme recognized this deficiency. His new approach, which he outlined clearly in the first part of his *De configurationibus*, was to construct a dual system of coordinates capable of representing at the same time both the intensity of a quality and the extension of the subject in which the quality inhered. In Oresme's scheme, the extension of a given subject in space or time was measured by a base line, and the intensities of the quality or motion in that subject were represented by perpendicular lines erected on the base line.[11] Greater or lesser intensities at various points in the subject were represented by proportionally longer or shorter lines erected on the base line at these points. When drawn, these measuring lines along two coordinates formed two-dimensional surfaces of varying geometrical configurations and sizes.

Oresme maintained that the quantity of any given quality could be represented by the surface area of these drawn figures. Two quantities (not just intensities) of the same quality could be compared by comparing both the shape and the size of the figures representing them. The varying shapes of the configurations could be further used to indicate both how the quality was distributed at different points within the subject, and how the intensity of the quality increased or decreased over time. For example, all uniform qualities (qualities of stable, equal intensity in all parts of the subject) were imagined as rectangles formed by equal lines of intensity erected at two or more points upon the base line of the subject.[12] All qualities figured by a right triangle were "uniformly difform," i.e., they increased in intensity or decreased in intensity uniformly.[13] The relation of the quantities of any two uniformly difform qualities of the same species could then be easily determined by comparing the shape and area of the triangles representing the qualities.[14] The manuscripts of the *De configurationibus* contain dozens of drawn figures illustrating the possi-bilities of representation and relation within Oresme's new graphing system.[15]

Oresme intended his figures to represent motions in precisely the same way that they represented qualities. Thus, a right triangle represented a

---

[11] Clagett, "Some Novel Trends," 287–89; Sylla, "Medieval Quantification of Qualities," 10; Oresme, *De configurationibus* 1.4, 172–75: "The quantity of any linear quality is to be imagined by a surface whose length or base is a line protracted in a subject of this kind . . . and whose breadth or altitude is designated by a line erected perpendicularly on the aforesaid base."

[12] Oresme, *De configurationibus* 1.11, 190–95.   [13] Oresme, *De configurationibus* 1.8, 184–87.

[14] Clagett discusses this important mathematical development and its influence on future scientific thought in "Some Novel Trends," 293ff.

[15] Clagett has reproduced more than a dozen figures accompanying the first explanatory chapters of the *De configurationibus*.

"uniformly difform" motion, i.e., a uniformly accelerating motion. A rectangle represented a motion of uniform speed. By comparing the figures representing the two species of motion, uniform and difform, and by demonstrating that the areas included within the two figures were equal when the height of the rectangle coincided with the mid-point of the hypotenuse of the triangle, Oresme arrived at an ingenious geometrical proof for what has come to be known as the "mean speed theorem."[16] The Oxford Calculators of the 1330s first formulated the mathematical rule that a uniformly accelerating velocity could be measured by its mean speed. Oresme's geometric demonstration of this theorem proved exceptionally fruitful as a tool for analyzing motion in the centuries that followed.[17] At the same time, by equating uniform with uniformly difform motion, Oresme demonstrated, against traditional objections, that although of different natures and species these motions could nevertheless be related and equated through the *medium* of his configurations.[18]

Oresme's two-dimensional geometric configurations represented and quantified qualities in ways that had been impossible using the Merton Calculators' single measure of intensity alone. Oresme's enterprising innovations in this area indicate that he *expected* more from a system of measurement and quantitative representation than was provided by the Calculators' latitude of qualities. Where does such expectation come from? The question is an important one because expectation is often at the root of invention. There are, I believe, strong indications that Oresme's expectations were informed by the existence of practical systems of measurement functioning in the world around him – systems that were considerably more successful in performing their measuring functions than was the conceptual latitude in current use by philosophers.[19]

Scholastic thinkers considered money a relative measure, incapable of

---

[16] Oresme, *De configurationibus* III.7, 408–11. This geometrical proof was first offered for uniformly difform qualities and then extended to uniformly difform motions (410–11): "De velocitate vero omnino dicendum est sicut de qualitate lineari, dum tamen loco puncti medii capiatur instans medium temporis velocitatem huiusmodi mensurantis."

[17] For its influence on Galileo's analysis of motion, see Clagett's introduction to Oresme, *De configurationibus*, 46–47, 71–73, 103–06; Clagett, *Science of Mechanics*, 409–16.

[18] Before it appeared in the *De configurationibus*, the outline of Oresme's geometrical representation of the mean speed theorem occurred in questions 10 and 13 of his *Questiones super geometriam Euclidis* in the context of traditional objections against the relation of unlike species and natures. Q. 11 (Clagett edn.), 536–37: "Arguitur quod non, quia ista sunt diversarum rationum, igitur inter illa non est comparatio." Oresme uses the equation of the rectangle of uniform velocity with the triangle of uniformly difform velocity in part to show in what ways two different species of motion can be related (and equated) through the common *medium* of geometric figures.

[19] Souffrin has speculated that everyday practices and instruments of measurement "functioning with efficacy" in society served as an important "source" of philosophical conceptions about measurement and quantification, including Oresme's configurational system. On this subject, see Souffrin, "La quantification," 66.

quantifying the true, inner nature of the measured commodity. But in many cases price does measure the qualities possessed by commodities. When, for example, two horses are sold in the same market, their market price, reflecting the common estimation and demand for them, measures them not only as generic horse, but also measures their age, training, health, temperament, etc. As Olivi's concept of *complacabilitas*, and Buridan's concept of relative *indigentia* reveal, price often says something about the quality of the subject itself.[20] Furthermore, price directly represents differences in "extension" within the same species. In the marketplace the price of bread was regulated according to the weight of flour it contained, measured to the ounce. A two-pound loaf could be sold at twice the price of a one-pound loaf, even though the "intensity" of *indigentia* attached to both might be equal. Such measurement by price of "extension" was carried to great lengths in the fourteenth century. In every city the regulation of both quality and quantity in products for sale, and their minute measurement by price, was routine.[21]

The Calculators' latitude was incapable of measuring in this fashion since it considered only the intensity of a given quality at a given point and not the quality's full value within a subject. Oresme, by considering both intensity and extension, and by using two-dimensional figures as the common measuring referent, was thus the first to invent a workable schema of quantifying and representing qualities in their totality.[22] This represented more than an advance in figuration. Within Oresme's system quantity (extension) and quality (intension) were bound together in a dynamic proportional relationship. Thus, if the quantity of a quality remained constant while the extension of the subject decreased, the line of qualitative intensity would increase, approaching infinity as the space approached a point.[23]

Oresme was quite conscious of the advantages in measurement his system provided. He made a point of asserting its superiority as a practical system, which, in the context of fourteenth-century natural philosophy, was unusual. From the opening of his treatise he puts forward the claim that ease, quickness, simplicity, and *commonality of reference* are primary virtues of his measuring system – and indeed they are. In his view, geometric representation can be used as a practical aid in the processes of

[20] See chapter 5, 148, 152.
[21] See for example the forty-seven separately listed prices of bread by weight of flour (pegged to the market price of wheat) under the heading "De modo et forma faciendi panem et pro quanto precio vendi debeat," in *Statuti di Verona del 1327*, Silvana Anna Bianchi and Rosalba Granuzzo (eds.), 2 vols. (Rome, 1992), vol. II, 557–60.
[22] Sylla, "Medieval Quantification of Qualities," 27.
[23] Sylla, "Medieval Quantification of Qualities," 30–34.

measurement and relation because it is tied directly to the senses and common experience.[24] He chose to represent qualitative intensities by geometric lines not only because of the line's mathematical properties, but also because "it is better known and is more readily conceived by us."[25]

Oresme was not the first scholastic to recognize the heuristic advantages of visual geometric forms. In their writings on mathematical perspective, Robert Grosseteste and Roger Bacon, the great admirer of practical knowledge and invention, preceded him in this.[26] It is likely that he was also influenced in this direction by his close reading of Witelo's *Perspectiva* – a late thirteenth-century Latin compendium of optical theory that used geometric figures (particularly the pyramid) to represent varying intensities of light.[27] Whatever his debts to intellectual precursors, Oresme's position on the representational use of figures is considerably more developed than any before him. In a noteworthy passage in the *De configurationibus*, he demonstrates his concern to widen the community capable of comprehending discussions of qualitative measurement, arguing for the heuristic superiority of geometric figures on the basis that the information they contain is more immediately apparent and available to a larger audience:

But it is apparent that we ought to imagine a quality in this way in order to recognize its disposition more easily, for its uniformity and its difformity are

[24] For the development of geometry as a system of common representation, see Funkenstein, *Theology and the Scientific Imagination*, esp. 310.

[25] Oresme, *De configurationibus* I.1, 166–67: "linearum quantitas sive proportio notior est et facilius a nobis concipitur."

[26] For statements by both Grosseteste and Bacon on the "usefulness" of geometric figures for conveying information to the senses, see David Lindberg, "Roger Bacon and the Origins of *Perspectiva* in the West," in Grant and Murdoch, *Mathematics and Its Applications*, 249–67, esp. 254–55. For a discussion of Bacon's argument in the *Opus maius*, that philosophical truths are more easily understood when presented in the visual terms of lines and geometrical forms, see Samuel Edgerton, "From Mental Matrix to *Mappamundi* to Christian Empire: The Heritage of Ptolemaic Cartography in the Renaissance," in David Y. Woodward (ed.), *Art and Cartography: Six Historical Essays* (Chicago, 1987), 10–50, esp. 30.

[27] Oresme was strongly influenced by writings on optics and particularly impressed by Witelo's *Perspectiva*. Much in Witelo's work was inspired if not derived from Ibn al-Haitham's (Alhazen) work on optics, circulating in the thirteenth century under the Latin title *Perspectiva*. In question 11 of his *Questiones super geometriam Euclidis* (Clagett edn., 536–37), Oresme cites Witelo's (and Grosseteste's) use of figures to represent light intensity as an argument in favor of his use of figures to represent qualities. Oresme's discussion of visual perception in the *De causis mirabilium* also relies heavily on Witelo. Stefano Caroti discusses Witelo's considerable influence on Oresme's conception of motion in "La perception du mouvement selon Nicole Oresme (*Questiones super physicam* III.1)," in *Comprendre et maîtriser la nature au moyen âge: mélanges d'histoire de science offerts à Guy Beaujouan* (Geneva, 1994), 83–99, esp. 88–92. See also Sabatai Unguru, "Mathematics and Experiment in Witelo's *Perspectiva*," in Grant and Murdoch, *Mathematics and Its Applications*, 269–97, esp. 279–80.

examined more quickly, more easily, and more clearly when something similar to it is described in a sensible figure. [This is true] because something is quickly and perfectly understood when it is explained by a visible example. Thus it seems quite difficult for certain people to understand the nature of a quality that is uniformly difform. But what is easier to understand than that the altitude of a right triangle is uniformly difform? For this is surely apparent to the senses.[28]

In like manner, when Oresme discusses the remarkable "invention" of money in the *De moneta*, he focuses on the capacity of the images stamped on coins to facilitate value recognition through the use of common signs. He writes:

it was wisely ordained by the sages of that time that pieces of money should be made of a given metal and of definite weight and that they should be stamped with a design, known to everybody, to indicate the quality and true weight of the coin, so that suspicion should be averted and the value readily recognised.[29]

Both money and geometrical configurations impressed Oresme by their capacity to facilitate value recognition through their action as common *media of representation*.[30] When he discussed the qualities necessary for money to act as an "apt instrument," he listed ease of use, ease of recognition, and commonality of value representation – the same qualities he claimed for his configurations.[31] Further underscoring this connection, his most important statements in the *De configurationibus* concerning the practical advantages of geometrical figures in representing and relating values appear in a chapter he entitled *De quantitate qualitatum* (On the quantity of qualities). This title succinctly describes the measuring program of Oresme, in which the Aristotelian position that qualities cannot be quantified has been abandoned.[32] It also describes in a phrase the essential role of money in economic life. The great range and versatility of money in quantifying and representing value was everywhere apparent in Oresme's society, as social status, power, civic respon-

---

[28] Oresme, *De configurationibus*, 174–75: "Sed quod per hoc debeamus ymaginari qualitatem ut eius dispositio levius cognoscatur apparet quia eius uniformitas atque difformitas citius, facilius, et clarius perpenduntur quando in figura sensibili aliquod simile describitur quod ab ymaginatione velociter et perfecte capitur et quando in exemplo visibili declaratur. Satis enim difficile videtur quibusdam intelligere que sit qualitas uniformiter difformis. Sed quid facilius quam quod trianguli rectanguli altitudo est uniformiter difformis? Certe hoc apparet ad sensum."

[29] Oresme, *De moneta*, 8: "ideo per sapientes illius temporis prudenter provisum est, quod porciones monete fierent de certa materia et determinati ponderis, quodque in eis imprimeretur figura que cunctis notoria significaret qualitatem materie nummismatis et ponderis veritatem, ut amota suspicione posset valor monete sine labore cognosci."

[30] Aristotle had made the same point in Book V of the *Ethics*, associating value representation and recognition with the common images stamped on coins.

[31] See chapter 5, 138, and chapter 6, 170–71.

[32] Oresme, *De configurationibus* 1.4, 172–73. For Aristotle's position on the categorical distinction between quality and quantity, and against the quantification of qualities, see chapter 6, 175–77.

sibility, land, and labor – indeed, human qualities in general – were translated and located on the numbered line of price.[33]

In his economic writings and in his sermons, Oresme acknowledges the capacity of money to quantify qualities, to bring into relation the most diverse goods and services, to serve as a common measure and as a common *medium* of value representation. Why then did he never specifically refer to it as a successful, functioning model on which philosophical insights into measurement could be based? One answer is that the image of money as instrument carried with it strong and wildly contradictory connotations. Alongside the great intellectual appreciation for money as an invention, there remained throughout this period a profound religious and philosophical distrust of its powers of multiplication and corruption. This intense duality characterized Oresme's attitude just as it did Aristotle's.[34] We have seen expressions of this contradictory attitude in Oresme's surviving sermons.[35] The same man who was famous for his treatise emphasizing the importance of money and monetary policy to the life of the French *civitas* was famous for preaching a blunt Christmas sermon before Pope Urban V, denouncing the pecuniary excesses and corruption of the Avignonese court.[36]

In a rare personal observation appearing in his quodlibetal questions on the marvels of nature, Oresme revealed his awareness that the continual use of money could produce unconscious effects – on others if not on himself. He writes:

Now I have seen many who have known right away how to tell me about a florin, "This one is false, that one is good," and yet in weight and figure and colour they were so similar that I did not know how to perceive any difference, even though I would consider it intently with due purpose; in fact, not even he knew how to tell me how he knew, except that it appeared so to him.[37]

Despite his recognition of the perceptual acuity that follows from

---

[33] Included here is the constant use of graded, numbered income to define responsibilities and tax duties in the civic and religious bureaucracies of the fourteenth century. See chapter 6, 172–75, for examples from English and French royal ordinances. For the transference of such monetary gradation to common parlance, see Murray, *Reason and Society*, 186; Yunck, *Lady Meed*, 197.

[34] For Aristotle's negative assessment, so important to the scholastic treatment of money, see chapter 2, 52–55. Within a single passage in the *De moneta* (4–5) Oresme acknowledges both Aristotle's positive judgment of money in the *Ethics* and the negative assessment of Ovid: "From earth we mine a source of future ill, / First iron and then gold, more deadly still."

[35] See chapter 5, 149–50.

[36] Nicole Oresme, "Sermo coram Papa Urbano V," in Flaccius Illyricus (ed.), *Tomus secundus scriptorum veterum* (London, 1690).

[37] *Nicole Oresme and the Marvels of Nature: A Study of His "De causis mirabilium" with Critical Edition, Translation, and Commentary*, Bert Hansen (ed. and tr.) (Toronto, 1985), 308–10: "Vidi autem multos qui statim de uno floreno sciverunt michi dicere 'Iste est falsus, ille est bonus,' et tamen in pondere et figura et colore erant ita similes quod nullam differentiam scivi percipere, quamvis ex intentione propria fortiter considerarem, ymo ipsemet nesciebat michi dicere quomodo cognosceret nisi quia sic apparebat ei et cetera."

habitual use, it is likely that, rather than withholding his acknowledgment, Oresme (and other thinkers of the fourteenth century) never consciously appreciated the instrument of money as a model for philosophical measurement. The negative connotations of money, and particularly the warnings against its corrupting influence, would have rendered inappropriate its open acknowledgment as a model in philosophical discourse, even within the more relaxed formal bounds of Oresme's work. The position Jean Buridan took in his *Physics* commentary is relevant here. He expressed great admiration for money as a measure and instrument of relation speaking *communiter*, but he insisted with equal strength that it can never be considered a "scientific" measure *proprie* because its use was tainted by self-interest and willful deception.[38] Although the strong distinction between *communiter* and *proprie* rendered it unlikely that any philosopher would claim money as a model in philosophical speculation, it is precisely on the level of common, everyday experience that it had its strongest influence on perception and thought.

All that can be said for certain is that as a philosopher and scientist Oresme attempted to bring intellectual order into a complex world of expanding and contracting qualities conceived of as continuous magnitudes. He sought to do this by introducing geometric lines and figures as a *medium* of quantification, value representation, and relation, claiming for them the practical advantages of ease and speed of value recognition. And he did this with the scholastic model of money vividly present in his intellect and his writings.

MATHEMATICS AND THE GEOMETRY OF EXCHANGE

Nicole Oresme's specific geometrical solutions in the *De configurationibus* are of less importance to the history of science than the monumental intellectual reformulation of the structure of the world from which these solutions derived.[39] In a later work, Oresme demonstrated that he was perfectly aware of the magnitude of his reformulation, and with remarkable prescience he located its essence in the shift from an arithmetical to a

---

[38] Buridan, *Physics*, 107rb, discussed in chapter 6, 196–99.

[39] Anneliese Maier made this general point about the whole of fourteenth-century natural philosophy: "its achievements consist less in transforming the content of the traditional view of nature than in discovering new ways of conceptualizing and comprehending nature." See Maier, "The Achievements of Late Scholastic Natural Philosophy," in *On the Threshold of Exact Science*, 146.

geometrical worldview.[40] We have seen that, as the monetized market-place came to assume an ever-larger place in social and political life, a similar shift from arithmetical to geometrical models of equalization occurred in the scholastic analysis of economic exchange.[41]

On the most basic level money is linked to number, and one of the significant effects of monetization is the expansion of the place of numerical calculation in everyday life.[42] Isidore of Seville underlined the link between money and number in the *Etymologies*, when in chapter 3 under the heading "What number is" he wrote: "*Nummus* (coin) gave its name to *numerus* (number) and, from being frequently used, originated the word."[43] A similar association of number and coins can be found in the work of Islamic mathematicians where the common unit was called *dirham* after the name of a common coin.[44] The connection between the increased use of money as a tool and the development of both numeracy and an "arithmetical mentality" in the medieval West has been recognized and investigated in previous historical studies.[45] But what is important for the understanding of developments in fourteenth–century natural philosophy is not so much the numerical thinking associated with the use of money as the geometrical thinking associated with its use within a now fully monetized society, not so much money as numbered unit as money as numbered continuum or line.

As the marketplace multiplied in size and complexity, the instrument of money moved from being identified with points, units, and numbers to being identified with the expanding and contracting line. Aristotle's description of money as *medium* gave textual support to this move.[46] The

---

[40] *Nicole Oresme and the Kinematics of Circular Motion: "Tractatus de commensurabilitate vel incommensurabilitate motuum celi"*, Edward Grant (ed. and tr.) (Madison, 1971), pt. III, 284–323.

[41] This shift is the primary subject of chapter 4.

[42] See chapter 3, n. 18.

[43] *Isidori hispalensis episcopi etymologiarum sive originum*, W. M. Lindsay (ed.) (Oxford, 1911), bk. III, ch. 3: "Numero numus nomen dedit, et a sui frequentatione vocabulum indidit": cited in Edward Grant, *A Source Book in Medieval Science* (Cambridge, Mass., 1974), 4.

[44] The influential Al-Khwarizmi called the numbered unit *dirham*. See Michael Mahoney, "Mathematics," in Lindberg, *Science in the Middle Ages*, 145–78, 157.

[45] See for example Murray, *Reason and Society*, 193: "as money pierced its way further and further into human relationships, it brought its own innate arithmetic into everyone's lives, not just traders' . . . Because money involved sums, and sums (as Leonardo [Fibonacci] said) created mental habits, the common transactions of an increasingly liquid economy stimulated numerical thinking generally."

[46] The word *medium* found in the Latin translations of Aristotle to describe money [*Ethics*, 1133a20] was both a mathematical and an ethical term. In medieval Latin it could be used either to denote a mid-point between two extremes or to denote a continuous magnitude connecting two extremes. Rudolph Goclenius, *Lexicon philosophicum* (Frankfurt, 1613), 675: "Medium est, quod interiacet inter extrema"; J. G. Bougerol, *Lexique saint Bonaventure* (Paris, 1969), 97: "Medium namque dicit communicantium cum extremis."

change from numbered point to continuous line occurred as thinkers came to recognize that the value of goods and services fluctuated constantly in proportion to fluctuating need and supply. Prices came to be understood as *approximate* products of varying *estimations* falling within numbered *latitudes* defining the range of just price. These latitudes of just price for various commodities were themselves understood to shift along a continuum of value in relation to changes in common need and scarcity.[47]

All geometric thinking, from the *Elements* of Euclid to the work of the Calculators and Oresme, is based on the line. In the fourteenth century, thinkers sought to construct both a geometry and a logic of the continuum capable of analyzing and ordering a world now perceived as composed of graded lines in expansion and contraction. In this connection it is revealing that the geometry of Oresme and the Calculators is not what is called "pure" geometry. They were not concerned with the invention of new geometric theorems. Their concern was always to *use* geometry to investigate and explain the structure and function of the world in which they lived.[48] Aristotle's employment of geometrical *figurae* to represent economic exchange and monetary proportionality in Book v of the *Ethics* (rendered as labeled rectangles with crossed diagonals in the manuscript margins) provided an influential intellectual model for such use.

The radical intellectual implications of life in the new "world of lines" is nowhere more clearly revealed than in Nicole Oresme's new attitude to the ancient question of mathematical incommensurability. Oresme was so fascinated by the question of commensurability that it formed the central subject of two complete treatises, the *Ad pauca respicientes* and the *Tractatus de commensurabilitate vel incommensurabilitate motuum celi*, and a good part of a third, the *De proportionibus proportionum*.[49]

The term "incommensurable" means "lacking a common measure." Early in the history of Greek mathematics it was discovered that the diagonal of the square is incommensurable with its side; that is, their relationship cannot be expressed as a ratio of whole numbers. The resultant irrational number has no finality and thus no form. For this

---

[47] For Aristotelian influences on the conception of money as line, see chapter 2, 42–44. For the many economic developments of the thirteenth and fourteenth centuries leading to the conception of money as line, see the conclusion to chapter 5, 158–59. For Oresme's consciousness of the role of approximation and estimation in exchange, see *Ethics* IX (454); V (293): "[things] doivent estre proporcionnees en valeur selon juste estimacion tout considere."

[48] This characteristic scholastic use of geometry is discussed in Molland, "Geometrical Background," esp. 114.

[49] Oresme, *"De proportionibus proportionum" and "Ad pauca respicientes"*; Oresme, *De commensurabilitate*.

reason, the existence of irrational numbers created a serious problem for those who held a worldview (associated with Pythagoras) in which number and the relationship between numbers were thought to be the primary elements in the harmonic structure of the cosmos.[50] Nevertheless, such a number-centered worldview persisted and retained strong appeal to medieval philosophers, in part because it meshed so well with the powerful influence of medieval Platonism. Under this influence, thinkers looked for intelligible, eternal, and universal forms behind the sense-perceived world of change, and they identified number as an important class of such universal forms.[51]

It was only when the Pythagoreans turned from purely numerical relationships to the relationship of magnitudes that the problem of incommensurability arose. Although the number one is present as a multiple in all numbers and so acts as a measure of all numbers, there is no common measure expressible in units for all magnitudes; there is no numbered ratio that can express the relationship between the side of a square and its diagonal.[52] Oresme was well aware of this fact, as were all medieval mathematicians, but he was unusual in the way he flaunted it and *used* it to underscore the deficiencies of a Pythagorean or arithmetical worldview centered on numbers and their relationship.

Rather than treat numerical incommensurability as a special and limited case, Oresme went to great lengths to prove that it was everywhere – that any two unknown ratios of continuous quantities were more probably incommensurable than commensurable.[53] He applied this principle in particular to the study of motion, establishing the probable

<hr />

[50] Legend has it that the idea of incommensurability was so disturbing to the Pythagoreans that they murdered the unlucky mathematician who discovered the irrational relationship between the diagonal of a square and its side.

[51] A. C. Crombie, *Medieval and Early Modern Science*, 2 vols. (Garden City, N. Y., 1959), vol. I, 14.

[52] Thomas Heath, *A History of Greek Mathematics*, 2 vols. (Oxford, 1921), vol. I, 154. Book x of Euclid's *Elements* is devoted to the question of incommensurability.

[53] Oresme, *De proportionibus*, 246: "Decima conclusio. Propositis duabus proportionibus ignotis verisimile est eas incommensurabiles esse." He then goes even further (*De proportionibus*, 254–63) to prove mathematically that there are many more irrational than rational ratios. It is important to add here that Oresme's definition of commensurability involved geometrical or exponential relationships – what Oresme called the "proportions of proportions." Murdoch has defined Oresme's general proposition: "It is probable that given A to the n = B, the exponent n will be irrational." This move to the exponential is directly related to the success of Bradwardine's exponential function relating forces and resistances to velocity. See Clagett, "Some Novel Trends," 285 and n. 23; Oresme, *De configurationibus*, 10; Murdoch, "The Medieval Language of Proportions: Elements of the Interaction with Greek Foundations and the Development of New Mathematical Techniques," in A. C. Crombie (ed.), *Scientific Change: Historical Studies in the Intellectual, Social, and Technical Conditions for Scientific Discovery and Technical Invention, from Antiquity to the Present* (New York, 1963), 237–71, 265–66; Molland, "Geometrical Background," 117.

incommensurability of any two unknown distances, times, or velocities.[54] Then, accepting as a given that our senses are inadequate to ever measure or know celestial motions with precision (as they lack the power to know with precision even the things close by them), he extended the idea of probable incommensurability to cover any two (unknowable) celestial distances, times, or velocities. In doing so he extended the (to some) disconcerting notion of incommensurability into the very fabric of God's perfect heavens.[55] His purpose here was not (as it often was in his "imaginations") to assert the *possibility* of incommensurability given God's absolute power (*potentia dei absoluta*), but to insist that irrational ratios exist as real parts of the structure of the universe. Oresme made frequent and varied use of this idea, from attacking astrological determinism to constructing a theoretical model that enabled him to refute the Aristotelian tenet of the necessary eternity of the world – a great thorn in the side of Christian natural philosophers.[56]

Against the intellectual background of his day, Oresme's assertion of celestial incommensurability was shocking. He recognized that his novel conclusions would distress many of his contemporaries, even though he himself was not at all bothered by them. Rather, he boasted that he had derived scores of new and, as he called them, "wonderful" and "beautiful" insights into the workings of the world from his new perspective. Why was Oresme no longer afraid of the idea of incommensurability, and how could he delight in its possibilities? The answer lies in one of the major themes of his work: his conscious championing of geometry, geometrical entities (lines and figures), geometrical functions (exponents), and what he clearly conceived of as a *geometrical worldview* against the "ancient" and severely limited arithmetical worldview constructed around discrete units and numbers.

Oresme illustrates in wonderful fashion the clash of these two principles in part III of his *Tractatus de commensurabilitate vel incommmensurabili-*

---

[54] Oresme, *De proportionibus*, 268–305. In investigating the relationships of ratios Oresme was still in effect exploring the mathematics of the continuum. He extended his habit of conceptualizing entities as continuous magnitudes to ratios themselves, and he used the Calculators' technique of the insertion of means to demonstrate his propositions concerning proportional incommensurability.

[55] Oresme, *Ad pauca*, 386–87; Oresme, *De proportionibus*, 304–05. In still another work, his *Quaestiones super geometriam Euclidis* (q. 7, Busard edn., 95–96), Oresme asserted the probable incommensurability of the motions of the sun and the moon as a corollary to Euclid's definition: "Magnitudes measured by a common measure are commensurable."

[56] See the discussion of Oresme's attitude toward astrology and the "great year" in *De commensurabilitate* (Grant edn.), 103ff.; Oresme, *De proportionibus* (Grant edn.), 247. Edward Grant, "Scientific Thought in Fourteenth-Century Paris: Jean Buridan and Nicole Oresme," in Cosman and Chandler, *Machaut's World*, 105–24, 112–13. For Oresme's use of incommensurability in a proof against the eternity of the world, see *Nicole Oresme: "Le Livre du ciel et du monde,"* Albert D. Menut and Alexander J. Denomy (eds. and trs.) (Madison, 1968), I.29, 194–203.

tate motuum celi through the device of an epic debate between personified "Arithmetic" and "Geometry" before the judgment of Apollo. This dialogue reveals, literally, a world of opposition between Oresme and the older arithmetical idea of a fixed, precise, and ordered world based on number. Oresme characterized Arithmetic as holding to the belief that the world actually conformed to a fixed system of numbers; that the human mind, human definition, and human ideals of perfection could be projected onto nature and the heavens. From this (false) standpoint Arithmetic attacked Geometry's assertion of the probable incommensurability of heavenly motions:

the opposite conclusion of our sister, Geometry, seems to deprive us of divine goodness, diminish the perfection of the world, destroy the beauty of the heavens, bring harm to mankind, cause ignorance, and detract from the beauty of the whole universe of beings. For every ratio of incommensurables is quite distinct from a measurable ratio of numbers, and, because of its very nature, is called irrational and absurd. It seems unworthy and unreasonable that the divine mind should connect the celestial motions, which organise and regulate the other corporeal motions, in such a haphazard relationship, when, indeed, it ought to arrange them rationally and according to rule.[57]

Arithmetic viewed the concept of celestial incommensurability and the uncertainty and approximation that accompanied it with a distaste bordering on horror. It violated her ancient sense that things should be ordered according to mind. It violated her sense of a harmony based on precise mathematical intervals and ratios. As incommensurability was unsuitable to the human mind, so, she believed, it must be to the divine, creating mind.[58] Geometry's position (speaking for Oresme) was to accept the complexity of the actual over the simplicity of the definitional, and to assume that the actual was ordered, even if that order transcended human understanding.[59] Geometry mocked Arithmetic for her shortsighted concern that God conform to her vision of perfection, and went

---

[57] Oresme, *De commensurabilitate*, 288–91: "Profecto nostre sororis, Geometrie, contraria nobis conclusio divine detrahit bonitati mundi perfectionem diminuit celi tollit decorem infert nobis iniuriam hominibus affert ignorantiam et totius universitatis entium pulchritudini derogare videtur. Omnis namque incommensurabilium proportio seclusa est a ratione mensurabili numerorum et inde a proprietate rei irrationalis nuncupatur et surda. Indignum ergo videtur et irrationabile ut divina ratio motus celestes tali indiscreta habitudine connexisset per quos ceteri motus corporei debite ordinantur ac rationabiliter et regulariter fuerit." All English translations taken from the *De commensurabilitate* are by Edward Grant.

[58] Oresme, *De commensurabilitate*, 292–93: [Arithmetic speaking] "Sic ergo intellectibus non convenit nec est eis proportionalis irrationalis proportio propter quod antiqui dixerunt animam constare quadam numerali et armonica ratione."

[59] For Oresme's championing of Geometry in this debate, see Georges Molland, "The Oresmian Style: Semi-Mathematical, Semi-Holistic," in Souffrin and Segonds, *Nicolas Oresme: Tradition et innovation*, 13–30, esp. 19.

so far as to hold up the fact of incommensurability as a positive addition to the beauty of the universe. Geometry speaks:

When the measure is indeterminable, we call it irrational and incommensurable. It happens frequently that a subtle man perceives the beauty in much diversity, while an ignorant man, who fails to consider the whole, thinks that the sequence [of events] in this diversity is confused, just as he does not realize that what we call an irrational ratio is part of our order and plan. And yet the infinite plan of God distinctly recognises this diversity which, put in its proper place, is pleasing to the divine sight and makes the celestial revolutions more beautiful.[60]

There is a strong historical component to this dialogue. Arithmetic is characterized by Oresme not so much as foolish as old-fashioned. Her position is identified with philosophical positions of the past, including at times those of Aristotle.[61] With Geometry as his spokeswoman, Oresme replaces an image of the world that was relatively fixed, balanced, compartmentalized, and composed of discrete parts (numbers) with one remarkably complex, fluid, and diverse. "What song," *Geometria* asks, "would please that is frequently or oft repeated? Would not such uniformity produce disgust? It surely would, for novelty is more delightful."[62]

In Oresme's new world of continually expanding and contracting magnitudes, the numbered unit is simply too static, too rigid, and too gross to serve as the underlying principle of reality. Geometry recognizes this fact, but rather than despairing as does Arithmetic and her followers, she lives happily in the conceptual briarpatch of numerical uncertainty and approximation.[63] Why in the middle of the fourteenth century was

---

[60] Oresme, *De commensurabilitate*, 312–13: "Et dum eam comprehendere non possumus ipsam irrationalem et incommensurabilem appellamus. Solet si quidem sepe contingere ut homo subtilis in multa variatione pulchritudinem percipit cuius diversitatis ordinem homo rudis non advertens totum estimat fore confusum, sicut irrationalem proportionem vocamus quam nostra ratio capere nequit. Et ipsam tamen distincte cognoscit dei ratio infinita et divino conspectui loco suo posita placet celestesque circuitus efficit pulchriores." [61] Oresme, *De commensurabilitate*, 306–07.

[62] Oresme, *De commensurabilitate*, 316–17: "Que est ista cantilena que placeret sepe aut multotiens repetita? Nonne talis uniformitas gignit fastidium? Ymo certe, et novitas plus delectat." The appreciation of diversity is an old theme, but Oresme here applies it in a new way to the question of incommensurability.

[63] While it is true that Oresme often admits to the limitations of human knowledge (e.g., "I know that the judgment of the senses cannot attain exactness": *De commensurabilitate*, 287), his work taken as a whole bespeaks an optimism about human knowledge – even a nascent sense of scientific discovery as an historical process. In the dialogue within the *De commensurabilitate*, while the certainties of Arithmetic are ridiculed, Geometry seems well satisfied with gradually, although admittedly never perfectly, extending her knowledge and understanding of the world through the use of the measuring line. As she says (320–21): "As far as such excellent things are concerned, it would be better that something should always be known about them, while, at the same time, something should always remain unknown, so that it may be investigated further." Cf. James J.

Oresme's Geometry suddenly no longer afraid of the implications of numerical incommensurability? Because she was confident that she had found a better measuring tool than number in her own geometric lines. Very conscious of her superiority she boasts:

We say that in numbers there is no measure and no ratio that is not included within our magnitudes [i.e., geometric lines]; but along with these there can be discovered in continuous quantities an infinite number of other [ratios and measures], none of which is found among numbers. Therefore, we have what it has, and much more.[64]

In his mathematical treatise on proportions Oresme stated the same truth in the form of basic propositions: "The first is: every ratio, both rational and irrational, can be found in continuous quantities. The second: no irrational ratio is found in numbers."[65] So, numbers and ratios that are irrational and incommensurable can nevertheless find their place "within our magnitudes" and can be measured by the infinitely divisible and extendable line or latitude. Oresme was stating nothing new in these mathematical propositions. Euclid had implied the same in his definitions of incommensurability in the *Elements*.[66] Oresme's innovation was in the use he made of them and the meaning he drew from them.

Starting from the assertion of numerical incommensurability, Oresme derived a geometric system capable of bringing all magnitudes into relation through the use of the line as the common instrument of measure. Paradoxically, he used the assertion of incommensurability to break down the fear of it. Since he found arithmetical incommensurability everywhere, as a probability between any two unknown ratios and at the heart of the perfect heavens, he integrated it as an idea into the natural order of the world, and he transformed it from a frightening possibility into a simple fact. Looking to explain Oresme's fascination and innovation in terms of textual traditions makes little sense here. As far as has been determined by modern scholarship, there were only the slimmest textual treat-

Walsh, "Is Buridan a Sceptic About Free Will?," *Vivarium* 2 (1964), 50–61, esp. 51; Tachau, *Vision and Certitude*, 75, 251.

[64] Oresme, *De commensurabilitate*, 312–13: "respondemus quod nullam mensuram nullam proportionem habet in numeris quam non habeamus in nostris magnitudinibus et cum hoc infinite alie reperiuntur in continuis quarum nulla invenitur in numeris. Habemus igitur quicquid habet et multo plus."

[65] Oresme, *De proportionibus*, 156–57: "Prima est omnis proportio, tam rationalis quam irrationalis, in quantitatibus continuis reperitur. Secunda, nulla proportio irrationalis in numeris invenitur."

[66] Euclid, *Elements* x, propositions 5–8. Campanus of Novarra recognized this "advantage" in his commentary on *Elements* v.3. Oresme specifically cited Campanus' position on this point in his *Quaestiones super geometriam Euclidis*, q. 9 (Busard edn.), 101. Clagett discusses Oresme's use of Campanus in his introduction to *De configurationibus*, 52–53 and n. 6.

ments of the question of probable incommensurability before Oresme.[67] On the other hand, the form and content of Oresme's surprising treatment of the question of commensurability show many parallels to the previously outlined changes in economic actuality and economic thought that accompanied the monetization of European society.[68]

On the most basic level, money was represented in Aristotle's discussion of exchange as the great instrument of commensuration, breaking down formidable barriers of incommensurability and permitting goods and services incommensurable in themselves to find relation. In Aristotle's scheme, money overcomes not only value incommensurability, but the incommensurability of the status and skills of exchangers.[69] Every scholastic commentator from Albert to Oresme recognized this. What is more, they recognized that value incommensurability was routinely overcome in the marketplace through money acting as a common measuring continuum.[70] In his commentary on the *Ethics* Jean Buridan underlined the function of money as a commensurating line by linking the *geometric* proportionality governing exchange to the use of the line in representing and bringing into relation numerically incommensurable values.[71]

As Oresme overcame the limitations of numerical incommensurability through the use of the geometric line as a common measure, the problem

---

[67] See the thorough treatment by Edward Grant in his introduction to Oresme's *De commensurabilitate*, 78–124. Again we find Duns Scotus as a possible precursor, but Grant's general conclusion (124) is that Oresme's treatment of the question bears as much resemblance to earlier treatments "as the oak tree bears a physical resemblance to the acorn." And Grant points to Oresme's own words in the Prologue (174–75): "But if another has set out the more fundamental principles [or elements found in this book], I have yet to see them."

[68] Funkenstein, *Theology and the Scientific Imagination*, considers the new attitude toward commensurability in fourteenth-century natural philosophy of great importance to the future of scientific thought. He is concerned with commensurability primarily in terms of a new willingness to break down injunctions against the mixing of genera and categories, and of a new freedom to exchange analogies and methods between *scientiae*. He specifically considers (309–10) Oresme's use of geometry as a commonly applicable system of "representation" in this light.

[69] See Aristotle's famous statement about exchange not involving two doctors, but a doctor and a farmer, *sed hos oportet equari: Ethics* v [1133a17–22].

[70] See chapters 2, 3, and 5, under the heading "Money as *medium* and as measure." Aristotle's discussion is so concise [1133a5–1133b17] that it bears repeating here (with ellipses): "Now proportionate return is secured by cross-conjunction. Let A be a builder, B a shoemaker, G a house, D a shoe . . . there is nothing to prevent the work of the one being better than that of the other; they must therefore be equated . . . For it is not two doctors that associate for exchange, but a doctor and a farmer, or in general people who are different and unequal; but these must be equated. This is why all things that are exchanged must somehow be comparable. It is for this end that money has been introduced, and it becomes in a sense an intermediate; for it measures all things, and therefore the excess and defect . . . Money, then, acting as a measure, makes goods commensurate and equates them" (tr. W. D. Ross).

[71] Buridan, *Ethics*, 414: "Dicitur [iustitia distributiva] autem medium secundum geometricam proportionem aequalitas proportionis aliquorum duorum vel plurium; et dicitur *geometrica*, quia potest reperiri in continuis secundum nullum numerum commensurabilibus." For a more detailed treatment of this question, see chapter 5, 136–37.

of value incommensurability was being continually overcome through the use of money as a common continuum and relational *medium* in the marketplace of the fourteenth century. In the marketplace, not only were doctors brought into relation with farmers, or the shoemaker's shoes commensurated through money with the carpenter's house (as in Aristotle's examples), but armies were bought, labor and land commoditized, subjective services (including indulgences) commuted into cash, social position and duty increasingly graded by numbered income, and personal qualities measured by and translated into a numbered money price. Oresme lived in a monetized world in which the traditional inconveniences of value incommensurability, and the traditional mistrust of the approximate, were being overcome. A static world of definitionally separated categories and species was giving way to a fluidly relational world in which money commensurated the most seemingly incommensurable goods and services. Oresme's new position on incommensurability illustrates how the experience and image of monetized society could be transferred into physical and mathematical thought and projected onto the order of nature.

The closest textual parallels I have found to Oresme's debate between Geometry and Arithmetic over the structure of the natural world were the late thirteenth-century *economic* debates over the proper mathematical model governing economic exchange. We have already examined in detail one of these, and we have seen that in essence the fierce quodlibetal debate between Henry of Ghent and Godfrey of Fontaines was about whether economic equality was to be defined arithmetically or geometrically.[72] Speaking for an old ideal of arithmetical equality which he knowingly supported *against* current economic practice, Henry of Ghent sought to maintain the verities of numerical exactitude in a system based on knowable, measurable points of equality. Like Arithmetic he spoke for an order and equilibrium that conformed to the mind's ideal. Like Arithmetic he associated the approximate with the "irrational." Like Arithmetic he could not conceive the rationality of a system constructed upon a base of merely estimated rather than exact values. How uncomfortable he would have been in Oresme's universe in which "exactness transcends the human mind"![73]

Arguing against Henry and speaking for the new geometrical model of exchange, Godfrey of Fontaines located market value squarely in the *aestimatio communis*. Like Oresme's Geometry Godfrey was more intent on examining how the system of exchange actually functioned than how

---

[72] See the concluding section to chapter 4.
[73] "Nosce precisas humanum transcendit ingenium." These are the words that Apollo speaks to Oresme in his dream, *De commensurabilitate*, pt. III, 284–85.

it ought to function. Rather than demanding exact values, he accepted the fact of changing, relative values. Rather than defining human approximation as subrational, he accepted it as the basis for rational economic choice. Rather than demanding arithmetically equivalent requital, he accepted proportional requital. Rather than constructing the image of an orderly, definitionally proper system of exchange based on numbers and points, he looked closely around him and saw the dynamism of a system based on equalizing ranges and lines. In Godfrey's geometric model of market exchange, one can see the germ of Oresme's geometric model of nature. In both cases, the new geometric model involved strongly discomforting ideas – ideas that violated the established, authoritative vision of natural order.[74]

The existence of money as a numbered continuum may help to explain still other cases of the idiosyncratic treatment of the line in fourteenth-century geometric thought. Historians of mathematics have noted the peculiar fact that in the work of Oresme and the Calculators the rigid Greek distinction between the continuous and the discrete was not maintained. The geometric line was often treated as if it were a numbered line.[75] Such a treatment of number as an element of the geometric continuum had no precedent in either Euclid's *Elements* or medieval commentaries on the *Elements*. Rather than looking to the intellectual tradition for the influences on this particular development, once again the influence of practical models of measurement coming from the society beyond the schools is suggested.

EQUALITY, THE MEAN, AND EQUALIZATION IN EXCHANGE

Money as an instrument of measurement and relation was only one of a number of measuring systems, ranging from the mechanical to the conceptual, in practical use in the fourteenth century. On the conceptual end, the idea of measurement and gradation which characterized the science of *calculationes* characterized as well the influential discipline of law, which, in the university culture of Paris and Oxford, intersected at many points with the study of theology and philosophy.[76] Bureaucracy represents the wedding of legal and monetary gradation. As suggested in

---

[74] For a clear description of this earlier arithmetical ideal of measure and order, see Edouard-Henri Wéber, "Commensuratio de l'agir par l'objet d'activité et par le sujet agent chez Albert le Grand, Thomas d'Aquin, et Maître Eckhart," in *Mensura: Mass, Zahl, Zahlensymbolik im Mittelalter* (Berlin, 1983), 43–64.   [75] Murdoch, "Medieval Language of Proportions," 270.
[76] For the overlapping of law, theology, and philosophy at Oxford, especially from the late thirteenth century, see Courtenay, *Schools and Scholars*, 37–41. See also John Murdoch's connection of law and philosophical measurement in "Unitary Character," 272 n. 2, discussed in the introduction to this book.

chapter 6, the Calculators' preoccupation with the question of *limits*, and their creation of limit languages – *incipit et desinit, de primo et ultimo instanti, de maximo et minimo* – can be viewed within the context of their considerable experience as administrators searching for similar rules governing limits in practical affairs.[77] The practical management of the agricultural estates supporting Merton College required the application of detailed measurement and calculation to a wide range of subjects: weights, areas, seed-to-yield ratios, profits, taxes, salaries, etc.[78] Those involved with managing the estates might also have been familiar with the numerous treatises devoted to field measurement and the estimation and maximization of agricultural yields that were circulating in the fourteenth century.[79]

The influence of physical instruments of measurement and relation in daily use should be added to that of abstract measuring systems found in administration and law. Due to its constant utilization in the everyday affairs of market exchange, the mechanical scale attained an ideational weight which permitted it to become the visual symbol of equalization for both commerce and law.[80] Perhaps less visible, but no less important as a measuring tool, and, therefore, no less available as a practical model for conceptual calculation, was the merchant's abacus.[81]

Whatever the influence of these non-philosophical measuring systems on philosophical thought, they lacked a number of crucial components

[77] For a characterization of these "limit languages," see Murdoch, "Unitary Character," 280–85, and chapter 6, 77–80. For the bureaucratic activity of William Heytesbury whose *Rules for Solving Sophisms* involves all of these limit languages, see chapter 1.

[78] Most relevant here are the voluminous records surviving from the manor of Cuxham, Oxfordshire – a manor owned by Merton College and managed by the fellows of Merton in the thirteenth and fourteenth centuries, both in person and through appointed officers. See P. D. A. Harvey (ed.), *Manorial Records of Cuxham, Oxfordshire, c. 1200–1359* (London, 1976). Harvey writes in the introduction (4): "we see the officers [of Merton College] in their rôle as landlords, keeping watch on their local reeves and bailiffs, supervising harvest and threshing, estimating stocks of corn, attending manorial courts, auditing accounts, paying out tips and bribes to their own workers, to neighbouring landlords and their officers, and to royal officials. In an age when . . . the College was likely to find its Fellows' agricultural experience of as much value as their theological skill, this picture of their activities may claim to contribute to the history of both the medieval college and the medieval university."

[79] For a collection of thirteenth-century treatises on this subject, see Walter of Henley, *Walter of Henley and Other Treatises on Estate Management and Accounting*, Dorothea Oschinsky (ed.) (Oxford, 1971). Walter's well-known discussion of the relative costs and benefits of oxen vs. horses (319) illustrates the degree to which agricultural management involved measurement and calculation. See also the emphasis on measurement, record-keeping, and calculation in the rules governing estate management written by the philosopher and mathematician Robert Grosseteste for the countess of Lincoln (388–99).

[80] See chapter 4, 106, for Henry of Ghent's use of the mechanical scale to define his ideal model of economic equalization.

[81] Discussed in relation to the "practical" use of the conceptual *latitudo qualitatum* in chapter 6. The influence on proto-scientific speculation of the proliferation of mechanical instruments such as mechanical mills and the mechanical clock is considered later in this chapter.

that the model of money and market exchange possessed. They lacked the idea, necessary to the development of mechanistic thinking, that systems were capable of functioning within themselves, without the constant intervention of mind and an ordering intelligence. The equalization of both judge and bureaucrat is conscious and arithmetical. As Aristotle made clear in his discussion of justice in the *Ethics*, such equalization involves deliberate bisection, addition, and subtraction to and from the line of loss and gain.[82] It is governed by mind in conformance to humanly definable ideals of order and equality. The systems of law and bureaucracy, then, as important as they were to the creation of the Calculators' sense of a gradable and measurable world, lacked the sense of dynamic equalization central to the most forward-looking developments in fourteenth-century scientific thought.

As is clear from his position on incommensurability, no one was more willing to accept nature on its own terms than Oresme – even when those terms proved upsetting to long-held human definitions of symmetry and regularity. In an important way, scientific investigation only began to make sense of the natural world when it abandoned the idea that the natural world made sense, i.e., conformed to human ideals of form and perfection. Important steps were taken in this direction within the philosophical tradition. From the Condemnations of 1277 at Paris to the philosophical systems of Duns Scotus and William of Ockham, the direction within philosophical and theological thought was to free God's power to act in the world from the limitations of human (Aristotelian in particular) definitions of possibility and order.

The intellectual tradition influencing Oresme's acceptance of natural order on its own terms should be weighed against the influences provided by experience in the marketplace of the fourteenth century. The marketplace provided the first and clearest example of a system in which order or equality was derived, *and was understood to derive*, from definitional disorder – the desire of each individual exchanger for unequal profit and gain.[83] The acceptance of the theologically and philosophically untenable position that objective inequality in the individual case could be converted into equality within a dynamic system was pioneered in the scholastic *economic* debate as a way to comprehend the workings of market exchange.[84]

Oresme framed his acceptance of celestial incommensurability in terms of the *natural* production of order out of seeming disorder within the

---

[82] Aristotle, *Ethics* v [1131b25–1132b10].
[83] Olivi, *Tractatus*, 51; Buridan, *Ethics* v.10; Buridan, *Politics* i.15; ii.2; Oresme, *Ethics* viii, ch. 18.
[84] See John Noonan's decisive comment on this development in chapter 5, 157.

system of the heavens. In my view, he was able to so thoroughly redefine the functional possibilities of the system of the heavens because his experience and understanding of monetized exchange had redefined for him the possibilities of an equalizing system. In the *De commensurabilitate vel incommensurabilitate motuum celi*, after discussing the "startling" implications of incommensurability for the possibilities of heavenly conjunctions over time, he wrote:

In the description of these angles and from the multiplication of such points, a diligent mind can consider the wonderful way in which some things arise from the incommensurability and irregularity of motions, so that I could utter such [expressions] as "rational irrationality," "regular non-uniformity," "uniform disparity," "harmonious discord." *Thus, by means of the greatest inequality, which departs from every equality, the most just and established order is preserved* [emphasis added].[85]

The acceptance of such a possibility for equalization was central to Oresme's understanding of nature, and yet he recognized the novelty of this concept in the realm of natural philosophy.[86] He did not arrive at his position through choosing and then combining various insights of his contemporaries or his philosophical predecessors. Although a new concept within Oresme's natural philosophy, the recognition of a system that produced order out of individual discord already had more than a half-century's history in scholastic economic-ethical thought. As the regularizing and equalizing system of exchange transformed the ethically disturbing desire for unequal gain into equality, so, in Oresme's scheme, the system of the heavens, the *machina mundi*, was capable of transforming incommensurable motions and times repugnant to human sensibility, into "the most just and established order." No better terms could be found to express the scholastic conception of market equalization in the fourteenth century than Oresme's "rational irrationality," "regular non-uniformity," and "harmonious discord."

---

[85] Oresme, *De commensurabilitate*, 256–57: "In quorum angulorum descriptione et talium punctorum multiplicatione diligens theoricus spectare potest modum mirabilem quo ex incommensurabilitate et regularitate motuum oritur quedam ut ita dicam rationalis irrationalitas, regularis difformitas, uniformis disparitas, concors discordia. Et cum summa inequalitate que ab omni equalitate degenerat equissima atque ratissima ordinatio perseverat." See Grant's commentary to this passage in the same volume, 46.

[86] Oresme, *De commensurabilitate*, from the Prologue, 174–75: "But if another has set out the more fundamental principles [or elements found in this book], I have yet to see them. For this reason I did not release this little book without [first] submitting it for correction to the Fellows and Masters of the most sacred University of Paris [Sed principaliores si alius tradidit nundum vidi. Non ergo dimisi quin hoc opusculum committerem sociis et magistris huius sacratissime universitatis Parisiensis sub eorum correctione]."

Historians of science have long noted early evidence of mechanistic thinking and imagery in fourteenth-century natural philosophy.[87] Oresme is often credited with one of the clearest examples of such imagery. In his commentary on Aristotle's *De caelo* he suggested that, at creation, God may have implanted in the heavens motive powers by means of which the heavens were moved continually without violence. "Thus," he wrote,

violence excepted, the situation is much like that of a man making a clock and letting it run and continue its own motion by itself. In this manner did God allow the heavens to be moved continually according to the proportions of the motive powers to the resistances and according to the established order [of regularity].[88]

Marshall Clagett has warned against ascribing to Oresme a mechanical conception of the universe on the basis of his precocious use of the metaphor of the clock. He noted that Oresme's insistence on the difference between the mechanics governing terrestrial motion and celestial motion, and his failure to entirely abandon the idea of the intelligences as moving powers, precludes his thought from being categorized as "proto-mechanical."[89] It is possible, however, to avoid the anachronistic ascription of a mechanistic worldview to Oresme, and yet still see the very use of the image of the clock and God as clockmaker as indicative of a markedly new way of perceiving the system of the heavens. While

---

[87] They have noted as well that this early mechanistic thinking developed within the philosophical and theological context (associated with William of Ockham) of a contingent, only conditionally predictable, causal order. For a discussion of this seeming paradox, see Grant, "Buridan and Oresme," 109–10; Ernest Moody, *Studies in Medieval Philosophy, Science, and Logic: Collected Papers 1933–1969* (Berkeley, 1975), 157. On mechanistic thinking in the thought of Ockham, see Moody, *Studies*, 428–29; Carlo Giacon, "Una 'Nota Magistri Fratris Occam De quantitate," in *La filosofia della natura nel medioevo: atti del 3. Congresso internazionale de filosofia medioevale* (Milan, 1966), 625–33.

[88] Oresme, *Livre du ciel* II.2, 288–89: "Et selon verité, nulle intelligence n'est simplement immobile et . . . posé que les cielz soient meuz par intelligences, car par aventure, quant Dieu les crea, Il mist en eulz qualitéz et vertus motivez aussi comme Il mist pesanteur es choses terrestes, et mist en eulz resistences contre ces vertus motivez. Et sont ces vertus et ces resistences d'autre nature et d'autre matiere que quelconque chose sensible ou qualité qui soit ici-bas. Et sont ces vertus contre ces resistences telement moderees, attrempees et acordees que les mouvemens sont faiz sanz violence; et excepté la violence, c'est aucunement semblable quant un honme a fait un horloge et il le lesse aler et estre meu par soy. Ainsi lessa Dieu les cielz estre meuz continuelment selon les proporcions que les vertus motivez ont aus resistences et selon l'ordenance establie."

[89] Clagett, "Oresme," esp. 300–02. Edward Grant discusses the profound differences between a mechanical conception of the heavens and the generally accepted Aristotelian conception in his article on "Cosmology," in Lindberg, *Science in the Middle Ages*, 265–302, 286. He notes how poorly the mechanical model of the clock fits the accepted conception (including Oresme's conception) of the form and composition of the heavenly spheres as "perfectly round and smoothly polished." For Oresme's statements on this point in the *Livre du ciel*, see II.2; II.5 (316–17); II.9 (386–87); II.15 (438–39). On the other hand, Oresme talks at times as if he has a mechanical model of the universe in mind and as if the heavens are moved as part of a great "machine corporelle." See *Livre du ciel* I.29 (200–04); II.2 (288–89); II.22 (508–09); II.25 (520–21).

maintaining the existence of celestial intelligences, Oresme greatly attenuated their actual role as moving and ordering powers within his schema. Similarly, while maintaining the Aristotelian distinction between terrestrial mechanics and celestial motion, Oresme nevertheless used arguments based on functioning mechanical systems such as mill wheels, automata, and clocks to understand and illustrate the working of the heavens.[90]

If Oresme's image of the clock only implied the absence of the celestial intelligences as movers, his fellow natural philosopher (and economic commentator) at the University of Paris, Jean Buridan, actually put forward such a suggestion:

And thus one could imagine that it is unnecessary to posit intelligences as the movers of celestial bodies since the Holy Scriptures do not inform us that intelligences must be posited. For it could be said that when God created the celestial spheres, He began to move each of them as He wished, and they are still moved by the impetus which He gave to them because, there being no resistance, the impetus is neither corrupted nor diminished.[91]

As radical as this speculation might appear, it fits in quite well with other elements of Buridan's thought. On many levels and in many areas, one can see what one of Buridan's editors has called his "constant preference for mechanical rather than metaphysical explanations of the dynamic order of the universe."[92] Rather than illustrate Buridan's mechanical preference through his well-studied speculation on impetus, I have chosen the lesser-known but no less striking example of his mechanical model of geological change, developed over a series of questions in his commentary on Aristotle's *De caelo*.

Thinkers of the fourteenth century inherited from Aristotle a model of the cosmos that contained strong elements of systematic regularity and

---

[90] See Oresme's mathematical demonstration, based on his concept of celestial incommensurability, of the possibility (*contra* Aristotle) that a motion could have a beginning in time and yet still last forever. This proof establishes a scientific basis for the Christian assertion (again *contra* Aristotle) that the earth was created in time. To demonstrate this, Oresme uses a mechanical system based on the wheels of a clock, where motion is controlled by weights and counterweights and results explicitly from mechanical contact: *Livre du ciel* I.29, 200–04. See Grant, "Buridan and Oresme," 113–14; Grant in his introduction to Oresme, *De commensurabilitate*, 38–39.

[91] *Iohannis Buridani: Quaestiones super libris quattuor de caelo et mundo*, Ernest A. Moody (ed.) (Cambridge, Mass., 1942), II.12, 180–81: "Et sic aliquis posset imaginari, quod non oporteat ponere intelligentias moventes corpora caelestia, quia nec habemus ex scriptura sacra quod debeant poni. Posset enim dici quod quando deus creavit sphaeras caelestes, ipse incepit movere unamquamque earum sicut voluit; et tunc ab impetu quem dedit eis, moventur adhuc, quia ille impetus non corrumpitur nec diminuitur, cum non habeant resistentiam" (translated by Clagett in *Science of Mechanics*, 561). See also Clagett, "Oresme," 300; Clagett, *Science of Mechanics*, 522–25, and 532–38 which includes a translation of the relevant argument from Buridan's *Questiones super octo phisicorum libros Aristotelis* VIII.12.

[92] E. A. Moody in his introduction to Buridan's *Quaestiones de caelo*, xviii.

causal necessity. In the case of Buridan's geological speculation, perhaps the most important of these Aristotelian influences come from his *De generatione et corruptione*. In Book II, Aristotle states the general proposition that things which come to be by natural processes all exhibit either an absolute or, at the least, a highly regular uniformity in their coming to be.[93] Aristotle applied the idea of causal regularity in nature to the process of elemental alteration. The result was a model which involved a strong sense of conservation: where the coming to be of one thing is always a passing away of another through its contrary,[94] where the elements generate and are generated, act and suffer action, reciprocally,[95] and where the elements of air, water, earth, and fire are constantly transforming and being transformed through their reciprocal action.[96]

Buridan accepted this physical model of reciprocal action from Aristotle and then applied it to specific geological questions with a systematic rigor that far surpassed Aristotle's own treatment of the same questions in the *De caelo* or in the *Meteorologica*.[97] In the course of answering the question "Whether the whole world is habitable" (Commentary on the *De caelo* II.7), Buridan was led to inquire why, given the fact of the observable erosion of land by wind, rain, and flood, and given the age of the earth, erosion had not succeeded in wearing down and washing all the high places into the sea.[98]

Buridan virtually dismissed Aristotle's argument from final causes that the earth was ordered toward the existence of animals and plants and therefore must remain habitable.[99] He formulated instead a mechanical argument, centering on the mixture of earth dissolved in water and water in earth, to explain why, despite the fact that erosion was always carrying dry land into the sea, the proportion of dry land to water remained, and would *always* remain, constant. In his illustrations of the working of this model of mechanical equalization, parts of earth that were above the water are taken into the water by erosion where they begin to build up

[93] Aristotle, *De generatione et corruptione* II.6 [333b 5–9].
[94] Aristotle, *De generatione et corruptione* I.3 [318b12–30]; [319a20–23].
[95] Aristotle, *De generatione et corruptione* I.10 [327a27–28].
[96] Aristotle, *De generatione et corruptione* II.2 [329b23–24]. Book II, 4–5, considers particular cases of elemental transformation.
[97] Ernest Moody characterizes Buridan's discussion here as "a strictly mechanical explanation of a geological problem," and he briefly discusses the differences between Buridan's treatment of these questions and earlier treatments, particularly Aquinas' in his *Expositio in meteorologicam*. See Ernest A. Moody, "John Buridan on the Habitability of the Earth," *Speculum* 16 (1941), 415–25, esp. 420. See also Duhem's treatment of this question in *Système*, vol. IX, and his labeling of Buridan's solution as a "théorie mécanique," 171, 186–202.
[98] Buridan, *Quaestiones de caelo* II.7, 159: "Sed tunc cum quaeris ultra, cum partes illius terrae elevatae fluant cum fluviis ad aliam partem ad fundum maris, quo modo potest salvari ista elevatio."
[99] *Ibid.* See Duhem, *Système*, vol. IX, 176, for the acceptance of final cause as an explanation in the early part of the century, notably by Walter Burley.

until eventually they become dry land again. Given the fullness of time, Buridan believed that this gradualist mechanical model could explain even the generation of the highest mountains.[100]

Thus, in Buridan's vision, the whole earth was being continually transformed over time, with parts that were deep within the center gradually moving to the habitable surface, as parts from the surface were carried to the depths. And throughout this ceaseless process of transformation, the proportion of land to sea remained a dynamic constant.[101] Moreover, in a second question within his commentary on the *De caelo*, "Whether the Earth is always at rest in the center of the universe," Buridan tied the constant shifting of the earth's masses due to erosion to the rectilinear motion of the earth around the center point of the universe. He considered such motion a logical (mechanical) necessity given the Aristotelian assumption that the center of the earth's weight always coincided with the center point of the universe.[102] He wrote:

from the higher [part of] the earth many parts of the earth continually flow along with the rivers to the bottom of the sea, and thus the earth is augmented in the covered part and is diminished in the uncovered part. Consequently, the center of gravity does not remain the same as it was before. Now therefore with the center of gravity changed, that which has newly become the center of gravity is moved so that it will coincide with the center of the universe, and that point which was the center of gravity before ascends and recedes, and thus the whole earth is elevated toward the uncovered part.[103]

[100] Buridan, *Quaestiones de caelo* II.7, 160: "Sic etiam salvatur generatio altissimorum montium; quia intra terram sunt partes terrae bene dissimiles, prout experiuntur fodientes – aliquae enim sunt lapidosae et durae, aliae sunt magis tenerae et citius divisibiles. Cum ergo illae partes interiores terrae elevantur modo praedicto ad superficiem terrae, illae quae sunt tenerae et divisibiles per ventos et pluvias et fluvios, iterum moventur ad profundum maris; aliae autem magis durae et lapidosae non possunt sic dividi et fluere, ideo manent et continue per longissima tempora elevantur per totalem terrae elevationem; et ita possunt fieri montes altissimi."

[101] Buridan, *Quaestiones de caelo* II.7, 160: "Ideo oportet quod totalis terra moveatur et elevetur versus plagam discoopertam; et tunc ex hoc sequitur ultra, per processum temporis, quod partes quae sunt in centro terrae tandem venient ad superficiem terrae habitabilis, propter hoc quod continue removentur de ista terra partes quae fluunt ad partem oppositam; et sic semper salvatur elevatio terrae."

[102] Since Buridan's solution to this question turned on the concept of the relativity of motion, it will be considered more fully below. A partial translation of this important question "Utrum terra semper quiescat in medio mundi (II.22)" appears in Clagett, *Science of Mechanics*, 594–98.

[103] From Clagett's translation, *Science of Mechanics*, 597–98; Buridan, *Quaestiones de caelo* II.22, 231–32: "Et per hoc solvitur alia dubitatio, scilicet utrum terra aliquando moveatur secundum se totam motu recto. Et possumus dicere quod sic, quia continue de ista terra altiori cum fluviis fluunt multae partes terrae ad profundum maris, et sic augetur terre ad profundum maris, et sic augetur terra in parte cooperta et diminuitur in parte discooperta; et per consequens non remanet idem medium gravitatis sicut ante fuit. Modo ergo, mutato medio gravitatis, illud quod de novo factum est medium gravitatis movetur ut sit medium mundi, et illud quod ante erat medium gravitatis ascendit et recedit; et sic elevatur tota terra versus partem discoopertam, ut semper medium gravitatis fiat medium mundi."

Within Buridan's model of dynamic equalization, the logic of mechanical effect had achieved sufficient force to move the mass of the earth against the great weight of religious and philosophical tradition.[104]

Where does such mechanical thinking come from? Of course there were the practical examples of functioning mechanical devices in common use. The development of the mechanical clock and other highly complex automata in the first half of the fourteenth century indicates the high level of sophistication that mechanical instruments and techniques had attained by Buridan's time. On a slightly lower level, there was the ubiquitous presence of water, wind, and hand mills even within the confines of the great cities and university towns – anchored to the very piers of the bridges of Paris.[105] Buridan has left us ample evidence that he paid close attention to the practical mechanical devices in place in his society. His critique of Aristotelian explanations of motion as well as his important formulation of the idea of impressed *impetus* rested on his actual observations and "experience" of mechanical devices, particularly observations of the properties of the mill wheel.[106] But in addition to these physical instruments, there were immensely important mechanisms of equalization functioning within society itself. The introduction of the hour-striking clock in the middle of the fourteenth century encouraged ever greater and more widely experienced refinements in the measurement of time.[107] Jacques Le Goff has suggested that the medieval town in itself should be seen as a system of acquisition and distribution, the very scale of which required new attention to number and quantitative measurement, and within which evolved a new sense of rationalized space and time.[108]

While all of these factors should be considered in relation to develop-

---

[104] In his discussion of Buridan's solution to this question, Grant notes that the projected mechanical motion of the earth violated a "fundamental tenet of Aristotle's cosmic picture." See Edward Grant, "Medieval Departures from Aristotelian Natural Philosophy," in Caroti, *Studies in Medieval Natural Philosophy*, 237–56, esp. 253.

[105] See the miniatures depicting these mills at work in the fourteenth century in Egbert, *On the Bridges of Medieval Paris*.

[106] See especially Buridan's commentary on VIII.12, of the *Physics*, translated in Clagett, *Science of Mechanics*, 532–38.

[107] For a discussion of the social repercussions of mechanical development, particularly that connected to the mechanical clock, see Jacques Le Goff, "Labor Time in the 'Crisis' of the Fourteenth Century" (43–52), and "Merchant's Time and Church's Time in the Middle Ages" (29–42), both in Le Goff, *Time, Work, and Culture in the Middle Ages*, Arthur Goldhammer (tr.) (Chicago, 1980); Gerhard dohrn-van Rossum, *History of the Hour: Clocks and Modern Temporal Orders*, Thomas Dunlap (tr.) (Chicago, 1996), 108–17, 197–215, 232–82, and on the relationship between time measurement and science, 282–87. Note that mechanical time measurement (and thus its effects on habits and thought) was still in its infancy in the fourteenth century.

[108] Jacques Le Goff, "The Town as an Agent of Civilization 1200–1500," in Carlo Cipolla (ed.), *The Fontana Economic History of Europe: The Middle Ages* (London, 1972), 71–95.

ments in proto-mechanical thinking in the fourteenth century, none of them had for Oresme, Buridan, or for any of the schoolmen what money and the marketplace had – not only an extremely powerful practical presence as a self-equalizing system, but a strong tradition of legal and philosophical commentary on its functioning principles. The recognition that the system of economic exchange was characterized by "harmonious discord" stretched back both to the Roman legal doctrine of permissible deception in buying and selling and to Aristotle's equalizing *figura* of exchange. It was in the marketplace that the conception first arose of a geometrical system capable of creating equality as a product of willed inequalities. As open as commercial activity was to censure, it was also seen as a model of equality, with the merchant serving as a kind of animate instrument of equalization. "Merchants carry that which is plentiful in one place to another place where the same thing is scarce," wrote Thomas of Chobham in the early thirteenth century.[109] One can see the dim outlines of Buridan's geological model of equalization in this influential and often-repeated economic observation.

Our judgment of the steps taken by Oresme and Buridan in the direction of proto-mechanistic models of natural activity should be seen in the context of their perception of the mechanics of the marketplace. Even their momentous steps toward removing the theologically sanctioned and subjectively conceived "intelligences" as the active movers of the heavenly spheres had their parallel and were prefigured in scholastic economic thought. The equalizing system of exchange was the first system considered extensively within scholastic thought to be conceptually freed of the need for an external, subjective orderer as its principle of order. In both practical law and in scholastic speculation from the early fourteenth century the intervening intelligence or judge (Aristotle's *iustitia animata*) was progressively de-valued and finally eliminated from the normal process of economic exchange. In theory and in practice the subjective orderer as guarantor of equality in the marketplace was replaced by the concept of a dynamic, self-equalizing system constructed around the *instrumentum equivalens* of money.[110]

COMMON VALUATION IN EXCHANGE

Central to every scholastic discussion of economic exchange from the time of Albertus Magnus was the identification of market price with common estimation (*aestimatio communis*) based on an aggregate common

---

[109] Thomas of Chobham, *Summa* VI.4.10, cited in Langholm, *Economics in the Schools*, 55.
[110] This process is discussed in detail in chapter 5, 129–33.

need (*indigentia communis*).[111] The association of market price with just price in both Roman law and scholastic thought was based on the recognition that common estimation, by eliminating the idiosyncrasies of individual needs and choices, represented a more trustworthy guide to fair economic value than did personal estimation. Many aspects of scholastic economic analysis demonstrate a similar choice of common over private valuation. In Book v of the *Ethics* Aristotle links the success of money to its function as a public instrument, whose value was instituted by law (*nomos* hence *nomisma*) and guaranteed by the ruler's stamp.[112] In chapter 6 of the *De moneta* entitled "Who owns the money?," Oresme argues that money as common measure and common sign is an instrument invented by the community to facilitate the exchange of its natural wealth.[113] The central point of his hortatory treatise is that monetary policy is properly ordered toward the public good of the community rather than toward the private needs of the ruler.

The championing of the *bonum commune* over the *bonum privatum* formed one of the richest intellectual themes of the thirteenth and fourteenth centuries. Aristotle's influence on the development of this theme in both political and philosophical speculation was immense, but so too was the influence of the practical, visible success of thousands of self-governing social entities, from the powerful independent urban communes, to guilds and corporations of all stripes, to the self-governing university itself. The practice of election common to many of these entities (and particularly to the university) reinforced the economic recognition that an aggregate common valuation was in many situations more to be trusted than individual choice. In both cases, aggregation was understood to smooth out and correct for personal particularities. The confluence of intellectual and practical influences from many spheres on this understanding renders it impossible to separate the validation of common valuation found in economic thought and experience from other sources of similar validation. It is just this confluence that sensitized philosophers to the implications of common valuation in their speculation.

The privileging of the common can be seen reflected in every category of proto-scientific innovation so far considered in this chapter. Oresme's great claims for his measuring system of configurations rested in large part

---

[111] *Digest*, 9.2.33: "pretia rerum non ex affectione nec utilitate singulorum, sed communiter funguntur." [112] Aristotle, *Ethics* v [1133a30–32].

[113] Oresme, *De moneta*, ch. 6, "Cuius sit ipsa moneta," 10: "Quamvis pro utilitate communi princeps habeat signare nummisma, non tamen ipse est dominus seu proprietarius monete currentis in suo principatu. Moneta siquidem est instrumentum equivalens permutandi divicias naturales, ut patet ex primo capitulo. Ipsa igitur est eorum possessio, quorum sunt huiusmodi divicie."

on their function as a common *medium* of representation, tied to the senses and common experience, and capable of serving as a common measuring referent for all continuous qualities and motions. Similarly, when Geometry bragged to Arithmetic about the capabilities of her lines, she flaunted their ability to serve as a common measure for a greater set of entities (irrational as well as rational ratios) than could number. As she put it: "We have what number has and much more." *Geometria* blunted traditional fears of irrational ratios by bringing them under the common measure of the geometrical line and by integrating them as common elements in the rational system of the heavens.

The notion of common measure, common referent, common system of representation, and common valuation was closely wedded to the scholastic preoccupation with determining ratios and the relations between things. From mathematics to philosophy to economics, the relation of incommensurable entities of different species was understood to depend upon the existence of a common third term or measuring continuum against which they could be measured and compared. Euclid's proposition that "magnitudes commensurable with the same magnitude are commensurable with each other" had enormous range and power within the new conception of nature in the fourteenth century as a "world of lines."[114]

Perhaps of greatest implication for the future of scientific thought was the place of common valuation in the conceptualization of mechanical systems. The recognition that market price was determined by *aestimatio communis in foro* provided the necessary base for the later, more difficult conception of the marketplace as a self-regulating system of equalization. Working against the great weight of philosophical and theological tradition, thinkers moved the location of value from its center in the individual to its identification as an ever-changing aggregate product, shifting in relation to scarcity and need and time and place. Only after the depersonalization of Aristotle's *figura* of exchange and its replacement with a supra-personal system constructed around common estimation could the marketplace serve as a model for the imagining of self-regulating systems within proto-scientific thought.

RELATION AND THE RELATIVITY OF VALUE IN EXCHANGE

The new measuring languages of the Calculators were essentially formulated around questions of relation, whether of the number and order of parts within continua, or the relation of two or more qualitative

---

[114] Euclid, *Elements* x, proposition 12.

intensities within the same species (i.e., comparing two or more latitudes of whiteness or heat or charity), or the relation of latitudes of velocity with latitudes of proportion in the investigation of motion, or the relation of finite species to the infinite creator.[115] The conceptual development of the *latitudo qualitatum* followed on the Calculators' desire to maximize its potential as an instrument of relation. The representation of all qualitative intensities by the common continuum of the geometric line, and the decision to standardize the measurement of intensities from the shared point of *non gradus* or zero degree, greatly expanded the latitude's relational capacities.

A striking application of the latitude as an instrument of relation appears in the work of several natural philosophers associated with the University of Paris. Jean de Ripa, Peter Ceffons, John of Mirecourt, and Jacob of Naples used the measuring *latitudo* to conceptually link and grade the range of natural species leading from the lowest creature up to God.[116] In giving a new logical and geometrical form to the ancient conception of a "great chain of being," they went against accepted philosophical and mathematical principles. According to the medieval interpretation of Euclid (*Elements* v.3) mathematical ratio could be found only between things of the same kind (*eiusdem generis*).[117] The direct mathematical relation of diverse species was consequently held to be impossible: *Sola enim univoca comparabilia sunt.*[118] Moreover, Campanus de Novarra (d. 1297), the influential thirteenth-century commentator on the *Elements*, never realized the relational or representational possibilities of comparison through a common third continuum.[119] This was an insight first realized within the measuring "mania" of fourteenth-century natural philosophy and then applied in many ways by the Oxford Calculators through the use of the measuring latitude. A great conceptual distance

---

[115] Murdoch, "Unitary Character," 283, 300–02. Tachau, *Vision and Certitude*, considers the subject of relation within fourteenth-century metaphysics in depth. See her treatment of relation in reference to cognition in the thought of Peter Aureol (91–104), William of Ockham (141–47), and Adam Wodeham (301–09).   [116] Kaye, "Impact," 262–63.

[117] Clagett (in his introduction to Oresme, *De configurationibus*, 52) cites Campanus of Novarra's influential thirteenth-century commentary on *Elements* v.3, which stressed this opinion: "ratio is the mutual relationship of two magnitudes such that one of them is greater than, less than, or equal to the other. Accordingly it is evidently necessary that they be of the same kind (*eiusdem generis*), e.g., as two numbers, two lines, two surfaces, two bodies, two places, or two times; for a line cannot be said to be greater than or less than a surface or a body, or a time than a place, but rather a line than a line and a surface than a surface."

[118] Campanus of Novarra cited in Clagett, introduction to Oresme, *De configurationibus*, 53: "For univocal things alone are comparable [by ratio]." Also, Murdoch, "*Subtilitates Anglicanae*," 59–60; Murdoch, "*Mathesis*," 239–44; Coleman, "De Ripa," 156–57.

[119] Clagett notes the crucial distinction between Campanus' actual position here, and Oresme's use of Campanus as an authority for his relation of continuous magnitudes through common geometric figures. See Clagett, introduction to Oresme, *De configurationibus*, 52–53; Oresme, *Quaestiones super geometriam Euclidis*, q. 11 (Clagett edn.), 536–37.

separated the fourteenth century from the thirteenth in this area. This distance is clearly apparent in the attempt by Jean de Ripa in the middle of the fourteenth century to formulate general rules governing the relation (ratio) of all created natural species. To accomplish this he conceptualized a single measuring scale along which separate formal perfections of each created species could be ranged.[120] Each perfection standing higher on this unitary measuring scale or "latitude of being" (*latitudo entium*) would, in his view, contain within it all perfections beneath it. De Ripa's *latitudo entium* thus, in his mind, satisfied the logical requirements of the continuum and could be used to relate species in the same way that points are related through their position on the line.[121]

Jacob of Naples and John of Mirecourt proceeded in similar fashion, despite their awareness of philosophical and mathematical prohibitions against doing so. After first restating the general proposition that ratios can be found only between things of the same kind, Jacob nevertheless concluded that diverse species can indeed be compared if they are done so *relative to some third thing they possess in common*. Thus a man and an ass, incomparable in themselves, can still be compared in relation to their common possession of different degrees of "being."[122] Thinking along similar lines, John of Mirecourt sought to create a conceptual latitude of being, stretching from God to the least of his creations. Species located upon this latitude could be compared and measured against each other in relation to their position on it.[123]

Such schemata of relation along a measuring latitude, although new in the context of the philosophical question to which they were applied, reflect with great clarity the technological model of money as an instrument of relation.[124] The central tenet of the scholastic understanding of money as measure is that money cannot measure things in themselves, nor can it commensurate and relate things in their proper natures. From Augustine to Aquinas to Oresme, medieval thinkers hammered at the point that money acts as an instrument of relation by measuring and

---

[120] Murdoch, "*Mathesis*," 241. Murdoch cites Jean de Ripa's *Conclusiones*, André Combes (ed.) (Paris, 1957), 70–71 and 75–76.

[121] Coleman, "De Ripa," 156. Perhaps the best description of this in relation to diverse species is *Conclusiones* II.2, art. 2 (273): "Secunda ymaginatio, quod data tota latitudine essendi simpliciter communicabili, illa tota continetur in omni superiori, puta vita et hujusmodi, ita quod data tota perfectione lapidis, quicumque gradus vite detur, sibi correspondet gradus entis simpliciter qui est perfectior quam totus gradus lapidis et totius latitudinis essendi citra."

[122] This example is taken from Murdoch, "*Mathesis*," 244, and n. 102. Compare it to Albertus Magnus' comment in *Ethics* V (Kübel edn., 344b), where he states that in the common measurement of lengths of different materials, the common measurement is of a shared continuous quantity, rather than of the essences of the things measured. And Albert uses this explanation of measurement of a common third term specifically to explain how things are measured by money. See chapter 3, 67–68.

[123] Murdoch, "*Subtilitates Anglicanae*," 60; Kaye "Impact," 263.   [124] See 170–75.

giving a numerical value to an accidental continuous quality (need or *indigentia*) possessed in common by all goods in exchange.[125] As *esse* functioned in Jacob's scheme in the relation of man and ass, so *indigentia* functioned as the necessary common quality or "third term" (in Jacob's words, the *tertium recipitur pro genere*) in the measurement and relation of commodities in the marketplace. And as money was the continuum recognized to measure *indigentia* in exchange thus permitting commensuration, so the *latitudo entium* was the continuum that certain philosophers devised to measure degrees of being, enabling formerly incommensurable species to be brought into relation.

John Dumbleton, as well as most of the Oxford Calculators, argued against the possibility of constructing a "latitude of being." In Dumbleton's mind, the *latitudo* could only be used to compare intensities within any given species. There was a latitude of whiteness to measure varying intensities of whiteness, a latitude of heat to measure heat, but there was no latitude that could relate whiteness to heat or any other species in their proper natures.[126] He could not accept the use of a *latitudo entium* because it violated the principles of a true continuum. He could not see how human nature could be compared along a continuum with that of asses or of angels because in his view every essence was indivisible in its nature.[127] Perfections, essences, natures cannot form a common continuum because they are not continuous. True continua cannot be composed of indivisibles, nor can natural species be divided infinitely as they must be if they form part of a true continuum.[128]

Although Dumbleton often repeats his theoretical objections to measuring diverse species along a single continuum, there are numerous conditions under which he willingly does so. The clearest example of this is his relation of qualitative and quantitative distance through the *latitudo qualitatum*, even though he explicitly notes that they are "of different genera."[129] Dumbleton denies the possibility of comparing species *in their*

---

[125] For Aquinas' position, see chapter 3. For Oresme's position, see chapter 5.

[126] Dumbleton, *Summa logicae*, in Sylla, *Oxford Calculators*, 571: "omnes qualitates in eadem specie eandem latitudinem habentes equales existere intensive."

[127] Sylla, *Oxford Calculators*, 380; Dumbleton, *Summa logicae*, in Sylla, *Oxford Calculators*, 568.

[128] Dumbleton, *Summa logicae*, in Sylla, *Oxford Calculators*, 568: "Nam si esset aliqua latitudo in perfectione essentie specifice, ista latitudo posset deperdi partibiliter secundum se totum. Ergo in infinitum modice perfectionis posset esse individuum illius speciei quod est impossibili." Yet more arguments by Dumbleton against the notion of a *latitudo perfectionis* are presented by Sylla, *Oxford Calculators*, 380–82.

[129] Dumbleton, *Summa logicae*, cited by Sylla, *Oxford Calculators*, 397: "Dum diversi generis sunt inter se distantia qualitativa et quantitativa, nec unum continuum ex hiis fieri potest, quare propter extensionem intensive nihil accrescit qualitati, sicut nec quantitati accidit maioritas propter qualitatem."

natures along a single continuum, but he tacitly accepts the comparison of diverse species through their relation to a *common third continuum*. Thus, quantity and quality can be linked through their common possession of distance.[130] Dumbleton expanded the possibilities of such relation beyond that of his predecessors. Where earlier Calculators had used different latitudes to describe different kinds of motion – uniform motion, uniformly difform motion, motion of intention, and motion of remission – Dumbleton believed that by focusing on the common factor of intensity shared by all, a single latitude could be used to measure all.[131] The "mean speed theorem" equating uniform and difform motions, invented by the Merton Calculators and put into geometrical form by Oresme (and Galileo), is a clear example of the willingness to relate diverse species of motion through the use of common measuring continua.[132] It is also an excellent example of the proto-scientific insight into the study of motion such relation permits.[133]

Of all scholastic writers, Jean Buridan and Nicole Oresme demonstrated the fullest comprehension of the place of relation and relativity in the economic order. Similarly, in their mathematical and scientific speculation, the concepts of relation and relativity, though partly inherited from the Calculators, were applied with unparalleled comprehension and thoroughness.[134] At times Oresme appears almost intoxicated by the possibilities of relation and ratio. Every aspect of his physical thought demonstrates this concern. His mathematical works centered on the analysis of the ratio and the exponential relationship of ratios which he called the "proportion of proportions." His fixation on the question of commensurability and incommensurability – a fixation which had no precedent in scholastic thought – again hinged on the permutations and

---

[130] Dumbleton, *Summa logicae*, in Sylla, *Oxford Calculators*, 571: "in omni distantia qualitativa gradus non est aliud nisi distantia illa ex partibus suis composita."

[131] Sylla, *Oxford Calculators*, 399–400.

[132] See Buridan's comprehensive treatment of this question in relation to Aristotle's acceptance of the measurement of motion by time, in *Physics* IV.14, 80ra–81va. One striking example is his discussion of how a clock can be said to measure time (80va): "unde sic etiam per horologium mensuramus motum solis diurnum sciendo quot gradus in horlogio signati correspondent uni hore et quot nocti et quot diei sicut enim per gradus pertransitos in horlogio scimus quot sit hora diei vel noctis . . . licet nullus gradus signatus in horologio sit equalis parti motus solis sibi correspondentis nec est cura que proportio sit huius ad illum secundum magnitudinem ita enim per parvum horologium hoc fieret sicut per magnum licet secundum magnitudinem non esset proportio magni et parvi ad celum ad motum solis."

[133] Clagett discusses the development of this theorem at Merton College in *Science of Mechanics*, 255–69. Examples of its formulation are given for Heytesbury (260–89), Richard Swineshead (290–304), Dumbleton (305–25), Oresme (347–81), and Galileo (409–16).

[134] For the central place of relational thinking in the physical thought of William of Ockham, see Goddu, *Physics of William of Ockham*, 188–208; Herman Shapiro, *Motion, Time, and Place According to William of Ockham* (St. Bonaventure, N. Y., 1957); Tachau, *Vision and Certitude*, esp. 141–47.

possibilities of relation.[135] His remarkable graphing system of *configurationes* was based on the use of common geometric forms, derived from the geometric line, to represent and thereby facilitate relation between diverse and even numerically incommensurable qualities and motions. In his *Questiones super geometriam Euclidis*, which preceded the *De configurationibus*, Oresme put forward a novel suggestion concerning the possible relation and commensuration of diverse species.[136] Using his general principle, "qualities are similar in uniformity or difformity which can be imagined by similar surfaces," he argued:

Any such quality is in some way proportional to any other, as frigidity to whiteness, and thus for others. This is proved: Let there be a hot body A and a white body B, and let them be equal [in subject]. Let both of these qualities be difform in the manner of a semi-circle. Then by the previous statements, these are as two equal semi-circles. Therefore, the qualities are equal to each other, and so it ought to be said for others.[137]

Oresme then used the remainder of this question to refute Aristotle's claim in Book VII of the *Physics* that circular and straight motion cannot be compared because they are of different species.[138] His concluding words are: "You say that a straight line and a curve are of different natures. I answer that, in addition, they are, and can be, equal." Again, Oresme has recognized that the problem of incommensurability (whether numerical or generic) can be overcome through the use of a common, continuous, relational *medium*.

No single aspect of the economic order more fascinated and disturbed scholastic thinkers than the fact that economic value was relative value – that a piece of bread, although ontologically lower, was valued higher than a mouse in exchange, or that bread in Paris had a different value than bread in Bruges, or that even within the same market the same commodity might have radically different values from season to season and day to day.[139] Similarly, until the late thirteenth century, relativity was the factor

---

[135] Molland, "Geometrical Background," 118.

[136] Oresme, *Questiones super geometriam Euclidis*, q. 13 (Clagett edn.), 554–55: "qualitates similes in uniformitate aut difformitate sunt que ymaginande sunt per similes superficies."

[137] Oresme, *Questiones super geometriam Euclidis*, q. 14 (Clagett edn.), 566–67: "Ex quo sequitur quod quelibet talis qualitas cuilibet alteri est aliqualiter proportionalis sicud frigiditas albedini et sic de aliis. Hoc probatur: sit *a* calidum et *b* album et sint equalia et sit quelibet istarum qualitatum difformis ad modum semicirculi, tunc per prius dicta iste sunt sicud duo semicirculi equales, ergo inter se sunt equales, et ita dicendum est de aliis."

[138] Aristotle, *Physics* VII.4 [248a10–249b26].

[139] Such examples were often repeated in the scholastic literature. See chapters 2, 3, and 5 under the heading "Relation and the relativity of value in exchange."

that most truly separated the conception of the economic order from the ontologically graded philosophical conception of the natural order.[140] From the late thirteenth century, as I have suggested, in the tension between the comprehension of the two orders, it was the pervasive model of the relativistic economic order that prevailed and came to be projected onto the ontologically ordered plan of nature.

Thirteenth-century intellectual culture had been immensely enriched by absorbing the formulations of relativity within Aristotle's physical and ethical thought. By the middle of the fourteenth century, however, relational thinking had gained such momentum that thinkers like Buridan and Oresme were beginning to criticize and correct Aristotle for his deficiencies in this area.[141] In Book I of the *De caelo* Aristotle spent considerable effort proving that the existence of a plurality of worlds was impossible. His position coincided perfectly with Christian authoritative teaching on this subject. Despite the joint authority of this agreement, Oresme pressed on with his speculation in this area, since, as he said, "it is good to consider the truth of this matter without considering the authority of any human but only of pure reason."[142] Always making clear that his conclusions were purely speculative and subject to correction by Christian authoritative teaching, Oresme arrived at the conclusion that one could *imagine* the existence of a plurality of worlds without violating either logic or physical possibility – *if* one approached the question from a radically relativistic perspective.

He put forward the "imagination" that another similar world existed inside and beneath the circumference of this one.[143] To the argument against this imagination that, if there were another world at the center of ours, it would be too small to be a real world, he wrote:

I point out that *large* and *small* are relative, and not absolute, terms used in comparisons. For each body, however small, is large with respect to the thousandth part of itself . . . Nor does the larger body have more parts than the

---

140 Albert, *Ethics* v (Borgnet edn.), 357b, 360b; Aquinas, *Ethics* v (1969), 294b, 295a.

141 This theme is particularly prominent in Oresme's commentary on Aristotle's *De caelo*, written late in his scholarly career (1372). See for example Oresme, *Livre du ciel* I.4, I.5, II.6, II.7, II.8, II.14, II.24, and II.25.

142 Oresme, *Livre du ciel* I.24, 166–67: "Et est bon de considerer selonc verité ce que l'en puet dire en ceste matiere sanz regarder a auctorité de honme, mais seullement a pure rayson." Duhem considers the history of the question of the plurality of worlds and Oresme's place in that history in his *Système*, vol. IX, 363–430.

143 Oresme, *Livre du ciel* I.24, 166–67: "Une autre ymaginacion puet estre laquelle je weul traitier par esbatement et pour exercitacion de engin, c'est a savoir que en un meisme temps un monde fut dedens un autre monde, si comme se dedens et dessouz cest monde estoit contenu un autre monde semblable et mendre."

smaller, for the parts of each are infinite in number. Also from this it follows that, were the world to be made between now and tomorrow 100 or 1,000 times larger or smaller than it is at present, all its parts being enlarged or diminished proportionally, everything would appear tomorrow exactly as now, just as though nothing had been changed.[144]

Then, based on this remarkable supposition, he concluded that, if at the center of our earth there was a concavity full of air the size of an apple, an entire world modeled on our own could exist within that concavity, without doing violence to the physical basis of our own world, i.e., its position at the center of the universe.[145] Oresme in no way intended this as a proof for the existence of other worlds or for infinite space, only as an assertion of the possibility of rationally conceiving them given the infinite power of God and the relativistic implications of an infinitely divisible continuum.[146]

One is led to wonder where such a highly developed sensitivity to the implications of relativity came from. Clearly Oresme's position as a scholar working within the culture of the university provides a portion of the answer. The concept of relation had received considerable attention within scholastic philosophy from the middle of the thirteenth century.[147] Over the course of the fourteenth century, a number of influential thinkers placed increasing emphasis on relativity in reference to cognition, time, and place.[148] This is particularly clear in the thought of William of Ockham, where extension, number, time, degree, and quantity itself were conceived of as concepts addressing relations between singulars, rather than as absolute properties in themselves.[149] Furthermore, as the citation above demonstrates, Oresme's sense of relativity was bound up with other intellectual concepts – with the logic of the infinitely divisible continuum and with his fascination with propor-

---

[144] Oresme, *Le Livre du ciel* 1.24, 168–69: "je supouse premierement que tout corps est divisible en partie[s] touzjours divisibles sanz fin . . . Item, que *grant* et *petit* sont nons relatis di[s] en comparoyson et non pas absoluement, quar chascune chose, tant soit petite, est grande ou regart de la .m. partie de elle . . . ne la grande n'a pas plus de parties qu'a la petite, quar de chascun corps les parties sont infinies en multitude. Item, par ce s'ensuit que se le monde estoit fait entre cy et demain plus grant ou plus petit .C. foys ou .M. foys que il n'est maintenant, et toutes ses parties estoient creues ou apeticiees proporcionnelment, toutes choses apparoistroient demain tout aussi comme maintenant aussi comme se rien ne fust mué" (corrections by the editors of Oresme).

[145] *Ibid.*: "Et par semblable, se ou centre de la terre estoit une concavité plaine d'air du grant d'une pomme." The discussion of the size of this concavity became the occasion for yet another discussion of relativity.

[146] This point is made forcefully in Jeannine Quillet, "Nicole Oresme et la science nouvelle dans le *Livre du ciel et du monde*," in Asztalos, Murdoch, and Niiniluoto, *Knowledge and the Sciences*, vol. I, 314–21, esp. 316.   [147] See chapter 5, n. 127.   [148] Tachau, *Vision and Certitude*, 141–43, 301.

[149] Funkenstein, *Theology and the Scientific Imagination*, 310; Goddu, *Physics of William of Ockham*, 124–26, 188–89, 205–07.

tionality – concerns inherited, in part, from the Calculators. But explanations of intellectual innovation resting on a trail of insights passed along and amended from thinker to thinker are inadequate if they do not take into account the continual interplay between textual influences and social influences, between the intellectual culture of the schools and the society beyond the schools.

To take one example, Peter John Olivi has been recognized as a precursor to Ockham in the perception, central to the development of kinematics and most especially to the theory of impetus, that motion was not a separate entity, but rather "a mode of relationship."[150] The historian of science, Michael Wolff, has gone so far as to suggest that one of Olivi's most fruitful philosophical innovations touching on the question of relativity was directly influenced by his experience of commercial society. Olivi, as we have seen, was a pioneer in the comprehension of relativity in the economic order as well as in the natural order.[151] Wolff has argued that Olivi's speculation on the dynamic relation between mover and moved object (one of the first formulations of a theory of *impetus*) grew directly out of his observation of the power of money as *capitale* to absorb and hold the *virtus* of the commercial investor.[152]

By the middle of the fourteenth century, Oresme and Buridan had become sufficiently at home with the implications of relativity to be able to *play* with them. In Oresme's relativistic physical world, the earth and all that is within it could be made one hundred or one thousand times larger or smaller while one sleeps, and yet, if the change were proportionally equal in all things, it would be undetectable on awakening. It would maintain its same essential form and its same essential relationships. Relevant to such an "imagination" is the fact that between 1355 and 1360 (when Oresme and, most probably, Buridan were in Paris) there were fifty-one separate legally mandated devaluations and revaluations of the French coinage – seventeen in 1360 alone.[153] In 1359 a successive revaluation and devaluation occurred over a period of four days, and the value of the *gros tournois* at its lowest point in 1360 was 2.3 percent of what it had been in 1336.[154]

Oresme's sensitivity to this phenomenon is witnessed by his writing a treatise on money and minting, the *De moneta*, specifically to counteract

---

[150] On this see Clagett, *Science of Mechanics*, 517, and Anneliese Maier, "The Significance of the Theory of Impetus for Scholastic Natural Philosophy," in *On the Threshold of Exact Science*, 76–102, esp. 83.   [151] For details on this, see chapter 5, 141–42.

[152] Michael Wolff, *Impetustheorie*, 178–88, 245; also Michael Wolff, "Mehrwert," esp. 419–23, and see chapter 5, 123.   [153] Fournial, *Histoire monétaire*, 98; Kaye, "Impact," 252–53.

[154] Fournial, *Histoire monétaire*, 98–100.

this policy of debasement.[155] His physical "imagination" that the thousandfold increase or decrease in the world's scale overnight would go unnoticed depended on the premise that everything was changed proportionally. The economic advisers of the French king preceded Oresme in holding this premise. From the early fourteenth century they realized that the economic dislocations of debasement could be minimized if the proportions between the values of old and new coins, contracts, and debts were fixed by law. At each revaluation of the coinage they issued detailed *Ordonnances* stating the official relationship of the new coins (sometimes a dozen or more) to the old, and proportionally relating the value of old debts to the new coinage. These *Ordonnances*, publicly posted and cried in the marketplaces by law, reached remarkable levels of detail and complication.[156] Here was the practical experience of relativity and proportionality on a grand scale! Whether on the part of peasants and laborers or officials of institutions that depended on the constant receipt and payment of money (like the grand master of the College of Navarre), the strategies derived to survive such sharp monetary changes necessarily involved the ability to think relatively and to manipulate relativized scales of monetary values with facility and precision.

   The most famous application of relativistic thinking in the work of Oresme and Buridan appears in their commentaries on Aristotle's *De caelo* concerning the question of whether the earth is always at rest at the center of the universe.[157] I conclude with this question because it indicates most clearly that the models related to economic experience were critical not only to the development of scholastic natural philosophy but also to the future of scientific thought. Again, Aristotle's position that the earth was at rest at the center of the universe was in complete accord with religious authority and traditional cosmology. Yet once again both Buridan and Oresme brought it into question through their application of a consistent relativism.[158]

---

[155] Oresme, *De moneta*, 30: "Again, it is a great scandal . . . and contemptible in a prince, that the money of his kingdom never remains the same, but changes from day to day, and is sometimes worth on the same day more in one place than in another. Also, as time goes on and changes proceed, it often happens that nobody knows what a particular coin is worth, and money has to be dealt in, bought and sold, or changed from its value, a thing which is against its nature."

[156] For a discussion of these *Ordonnances* and their level of complication, see chapter 1. The *Ordonnance* of June 1313 contained twenty-two paragraphs defining the relation of old to new coinage; that of September 1329 contained twenty-seven. The numbers of such *Ordonnances* increase in the 1340s and 1350s as does their detail.

[157] Buridan, *Quaestiones de caelo* II.22, 226–33; Oresme, *Livre du ciel* II.25, 518–39. The relevant sections from both discussions have been translated in Clagett, *Science of Mechanics*, 594–98, 600–06. Clagett discusses the central role of relativity in these questions in *Science of Mechanics*, 583–89.

[158] For earlier philosophical applications of the concept of relativity to the perception of motion, see Clagett on Ockham in *Science of Mechanics*, 589–90; Goddu, *Physics of William of Ockham*, 188–89.

Not wishing to appear overly exposed in his thinking, Buridan began his commentary with the observation that many before him had held the rotation of the earth to be a possibility. Astronomical appearances such as the moving stars and the rising and setting sun would remain the same if it were the earth that rotated daily while the spheres of the stars and the sun were at rest.[159] In support of this proposition he chose an example from life.[160] An observer in a moving boat, imagining that he is at rest, upon seeing a boat that actually is at rest, will imagine the other boat to be moving.[161] He imagines this, said Buridan, because his eye remains in the same relationship to the other ship regardless of whether his ship or the other ship is moving.[162] This recognition of relative motion is then applied to the motion of the earth:

We imagine we are at rest, just as the man located on the ship which is moving swiftly does not perceive his own motion nor the motion of the ship, then it is certain that the sun would appear to us to rise and then to set, just as it does when it is moved and we are at rest.[163]

Buridan then extended the relational system between perceiving eye and moving body to cover every visible motion in the heavens: "It is undoubtedly true that . . . [if the earth rotated] all the celestial phenomena would appear to us just as they now appear." Two centuries after Buridan, Copernicus used the same example of ship's motion to illustrate this same truth in the *De revolutionibus*.[164]   Oresme was in complete agreement with this position. He began his commentary on this question in a way similar to Buridan's, noting that certain ancient philosophers had held that the earth moves daily in a circuit around its pole. Aristotle, he surmised, did not bother to refute this opinion in the *De caelo* because he

---

[159] Buridan, *Quaestiones de caelo* II.22, 227: "Sciendum est ergo quod multi tenuerunt tanquam probabile, quod non contradicit apparentibus terram moveri circulariter modo praedicto, et ipsam quolibet die naturali perficere unam circulationem de occidente in orientem revertendo iterum ad occidentem – scilicet si aliqua pars terrae signaretur. Et tunc oportet ponere quod sphaera stellata quiesceret, et tunc per talem motum terrae fierent nobis nox et dies ita quod ille motus terrae esset motus diurnus." All English translations from Buridan's question II.22 are by Clagett, *Science of Mechanics*, 594–99.

[160] For earlier uses of this example in the context of the relativity of perception, see Tachau, *Vision and Certitude*, 141–43, 301–09.

[161] Buridan, *Quaestiones de caelo* II.22, 227: "Et potestis de hoc accipere exemplum; quia si aliquis movetur in navi et imaginetur se quiescere, et videat aliam navem quae secundum veritatem quiescit, apparebit sibi quod illa alia navis moveatur."

[162] *Ibid.*: "quia omnino taliter se habebit oculus ad illam aliam navem, si propria navis quiescat et alia moveatur, sicut se haberet si fieret e contrario."

[163] *Ibid.*: "Et ita etiam ponamus quod sphaera solis omnino quiescat, et terra portando nos circumgiretur; cum tamen imaginemur nos quiescere, sicut homo existens in navi velociter mota non percipit motum suum nec motum navis, tunc certum est quod ita sol nobis oriretur et postea nobis occideret sicut modo facit quando ipse movetur et nos quiescimus."

[164] Copernicus, *De revolutionibus* I.8.

considered it of such slight probability.[165] But Oresme disagrees. Subject, of course, to correction by ecclesiastical authority, he finds the opinion concerning a rotating earth to be as probable as its contrary.[166] He assumes that an observer on earth would naturally imagine that he was at rest while the heavens moved. But he assumes as well that a man "in the heavens" looking down on the earth would imagine that it was the earth that moved while the heavens were at rest. He argues that it is impossible to determine from any experience whatever, or from any rational argument whatever, whether the perceived motion of the heavens is caused by the moving heavens or a moving earth.[167] At the center of his position on this question, Oresme places his recognition of relative motion: "Now, I take as a fact that local motion can be perceived only if we can see that one body assumes a different position relative to another body."[168] He then demonstrates the truth of this proposition by recourse to Buridan's example of the two ships, although Oresme's example is more detailed and better designed to illustrate the point of relativity than Buridan's.

To the charge that the motion of the earth would make false all the previous determinations of astronomers and astrologers, Oresme replies:

all heavenly aspects, conjunctions, oppositions, constellations, figures, and influences would be exactly as they are in every respect [if the earth rotated] . . . and the astronomical tables of the heavenly motions and all other books would remain as true as they are at present, save that, with respect to diurnal motion, one would say that it is *apparently* in the heavens, but *actually* in the earth; no other effect would follow or result from one theory more than from the other.[169]

---

[165] Oresme, *Livre du ciel* II.25, 520–21: "Et Aristote ne reprouve pas ici ces opinions pour ce, par aventure, que il li sembloit que elles ont petite apparence." For positions on the rotation of the earth prior to Oresme and Buridan, see Grant McColley, "The Theory of the Diurnal Rotation of the Earth," *Isis* 26 (1937), 392–402.

[166] Oresme, *Livre du ciel* II.25, 520–21: "Mes, souz toute correction, il me samble qu'l'en pourroit bien soustenir et coulourer la desrenier opinion, ce est a savoir que la terre est meue de mouvement journal et le ciel non."

[167] *Ibid.*: "Et premierement, je veul declarer que l'en ne pourroit monstrer le contraire par quelcunque experience; secondement, que ne par raisons;" 522–23: "Et samblablement, se un home estoit ou ciel . . . il lui sambleroit que la terre [fust] meue de mouvement journal, aussi comme il samble du ciel a nous qui sommes en terre."

[168] Oresme, *Livre du ciel*, 522–23: "Item, je suppose que mouvement locale ne peut estre sensiblement apparceu fors en tant comme l'en apparçoit un corps soy avoir autrement ou resgart d'autre corps."

[169] Oresme, *Livre du ciel*, 530–31: "Je di que non, car tous resgars, toutes conjunctions, opposicions, constellacions, figures et influences du ciel seroient aussi comme il sont du tout en tout . . . et les tables de mouvemens et tous autres livres aussi vrays comme il son fors tant seulement que du mouvement journal l'on diroit que il es ou ciel selon apparence et en terre selon verité, et ne s'ensuit autre effit de l'un plus que de l'autre."

The physical case presenting the greatest difficulty to the imagination of a rotating earth was that of an arrow or other object projected straight upward from a moving ship. If the ship were moving to the east, the combined motion of the ship and the rotating earth should cause the arrow to fall way behind toward the west. Experience shows, however, that the arrow comes down in the same place from which it is projected. Both Buridan and Oresme considered this problem, but Oresme's response was the more successful. Not only the earth moves, he argued, but the air and water surrounding it share its motion. The arrow shot upward from a boat is moved toward the east along with air and the "whole mass of the lower portion of the world."[170] On the basis of his vision of earth, air, and water as an integrated system, he asserts that every conceivable motion on earth will appear the same to the viewer within this system, whether the earth is moving or still.[171] With his solution to this question, Oresme demonstrates the vital connections between relativistic thinking and mechanical thinking, between his placement of the viewer's eye within a self-regulating relational system and his placement of the earth, air, and water within a self-contained mechanical system.[172]

Oresme took his position on relativity very seriously. The arguments he proposed in favor of a rotating earth were consistently stronger than those opposed to it.[173] But stronger than any argument either pro or contra was the argument for relativity itself. At the question's end, Oresme announces that since conclusive proof for either position is impossible, he will hold to the common, religiously sanctioned position of a fixed earth.[174] His proof of reason's inescapable uncertainty led him back to belief in the articles of faith.[175] Given his introductory statements and his concluding position (assuming its sincerity) it seems probable that Oresme's primary purpose in constructing this elaborate question was not

---

[170] Oresme, *Livre du ciel*, 524–25: "la terre seulement ne est pas ainsi meue, mes aveques ce l'eaue et l'aer . . . l'en diroit que la seëtte traite en haut, aveques ce trait est meue vers orient tres isnelement aveques l'aer parmi leguel ell passe, et aveques toute la masse de la basse partie du monde."

[171] *Ibid.*: "touz les mouvemens de cibas sembleroient estre comme se la terre reposast."

[172] Clagett, *Science of Mechanics*, 587: "Buridan hinted at, and Oresme rather specifically outlined, the concept of a closed mechanical system, wherein, due to the relativity of the perception of motion, the observer describes all movements as if they were part of his system only."

[173] Strongest of all was the argument based on the so-called "principle of economy": *Livre du ciel* II.25, 534–35. Oresme posits that it would be much simpler (and thus more rational and more probable) for all the motions of the heavens to be caused by the simple daily rotation of the earth rather than by the motion of each of the many heavenly bodies, which, given their great distance from the earth, would have to move at "marvelously swift and excessively great" speeds, "far beyond belief and estimation."

[174] Oresme, *Livre du ciel*, 536–37: "et nientmoins touz tiennent et je cuide que il est ainsi meu et la terre non: Deus enim firmavit orbem terre, qui non commovebitur, nonobstans les raisons au contraire, car ce sont persuasions qui ne concludent pas evidanment."

[175] Oresme, *Livre du ciel*, 538–39: "Et ainsi ce que je ay dit par esbatement en cest maniere peu aler valoir a confuter et reprendre ceulz que voudroient nostre foy par raysons impugner."

to prove the possible rotation of the earth.[176] Rather, his purpose was to prove the impossibility of possessing certain knowledge (through experience or reason) within a relational system – a relational system now expanded to include the whole of nature and the heavens.[177] Both in the general and particular case, the implications are quite unsettling. We can never know with certainty whether the earth we inhabit is moving. Our knowledge is, at best, approximate.[178] Once again, we see Oresme acknowledging and even flaunting an image of the world disturbing in its fluidity and open-endedness. And once again, as with his notion of "rational irrationality," it is an image rooted in the relativistic economic order of the marketplace.

The inescapable if uncomfortable fact that exchange values are relative values led economic thinkers to focus on the problem of determining equivalences within a purely relational system, generations before this problem appeared in physical speculation. From the late thirteenth century, thinkers like Godfrey of Fontaines, Peter John Olivi, and Duns Scotus pioneered the conceptual integration of relativity and rationality. But, in order to do so, they were forced to introduce and legitimize concepts of estimation, approximation, and probability. Through observation and experience of the system of market exchange, they arrived at the understanding that, in establishing equivalences between relative values, free agreement is rational agreement, even though it is necessarily based on estimation rather than on certain knowledge. Furthermore, they realized that exchange equality based on rational agreement, though essential to the life of the *civitas*, can at best be based on probability and approximation. These were difficult truths, but they were necessary to the comprehension of the monetized marketplace. The same acceptance and legitimization of estimation, approximation, probability, and relativity informed Oresme's most fertile proto-scientific speculation concerning the natural order.

Clearly, there were textual as well as experiential influences on the

---

[176] Pierre Souffrin notes that this question represents the longest gloss in the entire *Livre du ciel* – five folios for one sentence of Aristotle. See his "Oresme, Buridan, et le mouvement de rotation diurne de la terre ou des cieux," in Bernard Ribémont (ed.), *Terres médiévales* (Paris, 1993), 277–99, 279. Souffrin examines the various historical interpretations both of Oresme's purpose in constructing this question and of his ultimate acceptance of the earth's fixity.

[177] This is essentially Souffrin's conclusion as well, "Oresme, Buridan," 283–84, 291.

[178] Edward Grant, "Nicole Oresme on Certitude in Science and Pseudo-Science," in Souffrin and Segonds, *Nicolas Oresme: Tradition et innovation*, 31–43, esp. 38: "The imprecision inherent in irrational mathematical relationships seems to have led Oresme to infer that the physical sciences – at least those that depended on mathematics – must also be uncertain and approximative. Approximation was the best that could be had. For Oresme, it was quite sufficient." Oresme's stance on relative motion involves a similar acceptance of the approximate as a sufficient basis for scientific judgment.

relativity of Buridan and Oresme. Oresme cited the analysis of relative position found in Witelo's *Perspectiva* (also from the late thirteenth century) two times in the course of this question, and his reliance on perspective theory is apparent at many points within his work.[179] But the ability of Buridan and Oresme to situate themselves within a natural system in which all measurement and perception is relative, and their ability to play with the possibilities of relativity, is so far in advance of their predecessors that once again an explanation restricted to the influence of intellectual sources appears inadequate. Both men were at home with concepts that would have caused their predecessors in previous generations the most extreme discomfort, if, that is, their predecessors were capable of either thinking or comprehending them. We have seen such "thinking the unthinkable" before within each of our six categories: the Calculators' quantification of qualities through the measuring *latitudo* against Aristotelian authority; the use of common continua to bring diverse species into relation against traditional objections; Oresme's assertion of the probable incommensurability of celestial motions; his vision of nature as a system characterized by "regular non-uniformity" in which "by means of the greatest inequality, which departs from every equality, the most just and established order is preserved"; Buridan's speculation that the heavens might move in the absence of ordering intelligences; the assertion of the mechanical motion of the earth about the center of the universe as its surface masses shifted over time; the imagination of a plurality of worlds based on the principle that "large and small are relative"; and here, the imagination of a rotating earth.

Each of these speculations grew out of a new conception of nature that took shape within university culture between, roughly, 1250 and 1350. But philosophers do not merely conceptualize nature, they inhabit it. Conceptualization is tied to perception. While the physical basis of reality does not change, the social basis of reality does. Over this period society was transformed through the many-faceted social processes of monetization and market development. Every level of society and every layer of institutional growth was affected. Philosophers, from their earliest days as students, were presented with social and economic experiences, rules of conduct, avenues of advancement, and models of success unknown by previous generations. As social experiences change, so too do perceptions about how the world functions and is ordered.

The rigorous intellectual training of the university and its intense

---

[179] Oresme, *Livre du ciel* II.25, 522–23: "Et il appert ou quart livre de *La Perspective* de Witelo que l'en ne apparçoit mouvement fors telement comme l'en apparçoit .i. corps soy avoir autrement ou resgart d'un autre"; 536–37: "Mais se il est de tel corps ou de tel, ce jugement est fait par les sens de dedens, si comme il met en *Perspective*."

atmosphere of challenge and disputation transformed raw perceptions into insights capable of being elaborated and refined through the technical instruments of mathematics and logic. The scholastic habit of synthesis encouraged the linking of insights and principles between different spheres of thought, between the comprehension of the economic order and the comprehension of the natural order. The result was the creation of a new image of nature based on the experience and observation of monetized society: a dynamic, relativistic, geometric, self-ordering, and self-equalizing world of lines. It was upon this model of nature, first imagined in the fourteenth century, that thinkers from Copernicus to Galileo constructed the "new" science.

BIBLIOGRAPHY

PRIMARY SOURCES

Aegidius Aurelius (Gilles d'Orleans), *Super ethicam*, ms. BN lat. 16089 (195r–237v)

Aegidius Lessinus, *De usuris*, edited as part of Thomas Aquinas, *Opera omnia*, vol. XXVIII, 575–608. Paris, 1889

Aegidius Lessinus (?), *Quaestiones super libros ethicorum*, ms. Vat. lat. 2172 (1r–59r)

Albert of Saxony, *Expositio super libri ethicorum*, ms. Bib. Maz. 3516 (84–156)

Albertus Magnus, *Alberti Magni super ethica commentum et quaestiones*, Wilhelm Kübel (ed.), vol. XIV in *Opera omnia*. Monasterii Westfalorum: Aschendorff, 1968–72 (cited as Kübel edn.)

De mineralibus (*Book of Minerals*), Dorothy Wyckoff (tr.). Oxford: Clarendon Press, 1967

Libri quarti sententiarum Petri Lombardi, A. Borgnet (ed.), vol. XXIX in *Opera omnia*. Paris: Apud L. Vivès, 1894 (cited as *IV Sent.*)

Sententia libri ethicorum, A. Borgnet (ed.), vol. VII in *Opera omnia*. Paris, 1891 (cited as Borgnet edn.)

Alexander Lombard, *Un traité de morale économique au XIVe siècle: le "Tractatus de usuris" de maître Alexandre d'Alexandrie*, A.-M. Hamelin (ed.). Analecta mediaevalia Namurcensia. Louvain: Nauwelaerts, 1962

Alexander of Hales, *Summa theologica*, vol. IV. Quaracchi: Collegii S. Bonaventurae, 1948

Aquinas, Thomas, *Sancti Thomae de Aquino sententia libri ethicorum*, vol. XLVII in *Opera omnia*. Rome: Commissio Leonina, 1969 (cited as Aquinas [1969])

St. Thomas Aquinas: Commentary on the Nicomachean Ethics, C. I. Litzinger (tr.), 2 vols. Chicago: Regnery, 1964 (cited as Litzinger trans.)

St. Thomas Aquinas: The Divisions and Methods of the Sciences – Questions V and VI of His Commentary on the "De trinitate" of Boethius, Armand Maurer (ed. and tr.). Toronto: Pontifical Institute of Mediaeval Studies, 1963

Summa theologiae, 61 vols. New York: Blackfriars, 1964–81 (cited as *ST*)

Aristotle, *Ethica Nicomachea: translatio Roberti Grosseteste Lincolniensis, recensio pura*, R.-A. Gauthier (ed.). Aristoteles Latinus 26, 1–3, Fasc. 3. Leiden: E. J. Brill, 1972 (cited as Arist. Lat. 26 [1972])

Bibliography

Ethica Nicomachea: translatio Roberti Grosseteste Lincolniensis, recensio recognita, R.-A. Gauthier (ed.). Aristoteles Latinus 26, 1–3, Fasc. 4. Leiden: E. J. Brill, 1973 (cited as Arist. Lat. 26 [1973])

The Greek Commentaries on the Nicomachean Ethics of Aristotle in the Latin Translation of Robert Grosseteste, H. P. G. Mercken (ed.). Leiden: E. J. Brill, 1973

Nicomachean Ethics, W. D. Ross (tr.); *Categoriae*, E. M. Edghill (tr.); *Physica*, R. P. Hardie and R. K. Gaye (trs.); *Politica*, Benjamin Jowett (tr.); all in Richard McKeon (ed.), *The Basic Works of Aristotle*. New York: Random House, 1941

Politica (Libri I–II), P. Michaud-Quantin (ed.). Aristoteles Latinus 29, 1. Bruges: De Brouwer, 1961 (cited as Arist. Lat. 29, 1 [1961])

Arnald of Villanova, *Aphorismi de gradibus*, Michael McVaugh (ed.), vol. II in *Arnaldi de Villanova opera medica omnia*. Granada: Seminarium historiae medicae Granatensis, 1975

Aureole, Peter, *Commentarium in primum librum sententiarum*. Rome, 1596

Bradwardine, Thomas, *Thomas of Bradwardine. His "Tractatus de proportionibus": Its Significance for the Development of Mathematical Physics*, H. Lamar Crosby, Jr. (ed. and tr.). University of Wisconsin Publications in Medieval Science, 21. Madison: University of Wisconsin Press, 1961

Buridan, Jean, *In metaphysicen Aristotelis quaestiones*. Paris, 1588; reprint Frankfurt a/M: Minerva, 1964

Iohannis Buridani: Quaestiones super libris quattuor de caelo et mundo, Ernest A. Moody (ed.). Cambridge, Mass.: Medieval Academy of America, 1942

Jean Buridan's Logic: The Treatise on Supposition, The Treatise on Consequences, Peter King (ed. and tr.). Dordrecht: D. Reidel, 1985

Quaestiones in decem libros ethicorum Aristotelis ad Nicomachum. Oxford, 1637

Quaestiones super octo libros politicorum. Oxford, 1640

Quaestiones super octo phisicorum libros Aristotelis diligenter recognite et revise a magistro Johanne Dullaert de Gandavo. Paris, 1509; reprint, Frankurt a/M: Minerva, 1964

Burley, Walter, *De intentione et remissione formarum*. Venice, 1496

Expositio in Aristotelis ethica Nichomachea. Venice, 1500

Expositio super octo libros politicorum, ms. Balliol 95 (161r–232r)

Chartularium Universitatis Parisiensis, H. Denifle and E. Chatelain (eds.), vol. II. Paris: Fratres Delalain, 1891

Chronicon Girardi de Fracheto et anonyma ejusdem operis continuatio, vol. XXI in *Recueil des historiens des Gaules et de la France*. Paris, 1855

Chronique Latine de Guillaume de Nangis de 1113 à 1300, avec les continuations de cette Chronique de 1300 à 1368, Hercule Géraud (ed.), 2 vols. Paris, 1843

Chronique Parisienne anonyme de 1316 à 1339, vol. XI in A. Hellot (ed.), *Mémoires de la Société de l'histoire de Paris et de L'Ile-de-France*. Paris, 1885

Codex Iustinianus, Paul Krueger (ed.), vol. II in *Corpus iuris civilis*. Berlin: Weidmanns, 1954

Decretales Gregorii IX, A. Friedberg (ed.), vol. II in *Corpus iuris canonici*, 2 vols.

Bibliography

Graz: Akademische Druck- u. Verlagsanstalt, 1959

Digesta, Theodore Mommsen (ed.), vol. I in *Corpus iuris civilis*. Berlin: Weidmanns, 1954

Dumbleton, John, *Summa logicae et philosophiae naturalis*, appendix 1, pt. 4, in Sylla, *Oxford Calculators*, 565–625

Duns Scotus, John, *Quaestiones in quattuor libros sententiarum (Opus Oxoniense)*, L. Wadding (ed.), vol. XVIII in *Opera omnia*. Paris: Apud L. Vivès, 1894 (cited as *Sent.* IV)

Fabliaux ou contes, fables, et romans du XIIe et XIIIe siècles, Legrand d'Aussy (ed.), 5 vols. Paris, 1829

Godfrey of Fontaines, *Godefroid de Fontaines: Quodlibet III*, M. De Wulf (ed.), vol. II in *Les philosophes belges*, 15 vols. Louvain: Institut supérieur de philosophie, 1904 (cited as *GF* III)

Godefroid de Fontaines: Quodlibet V, M. De Wulf and J. Hoffmans (eds.), vol. III in *Les philosophes belges*, 15 vols. Louvain: Institut supérieur de philosophie, 1914 (cited as *GF* V)

Godefroid de Fontaines: Quodlibet X, J. Hoffmans (ed.), vol. IV in *Les philosophes belges*, 15 vols. Louvain: Institut supérieur de philosophie, 1924 (cited as *GF* X)

Godefroid de Fontaines: Quodlibet XII, J. Hoffmans (ed.), vol. V in *Les philosophes belges*, 15 vols. Louvain: Institut supérieur de philosophie, 1932 (cited as *GF* XII)

Les Grandes Chroniques de France, Jules Viard (ed.), vols. VIII and IX. Paris: Société de l'histoire de France, 1934

Gratian, *Decretum*, A. Friedberg (ed.), vol. I in *Corpus iuris canonici*, 2 vols. Graz: Akademische Druck- u. Verlagsanstalt, 1959

Henricus de Frimaria, *Sententia totius libri ethicorum*, ms. Vat. lat. 2169 (1–284v)

Henry of Ghent, *Henrici de Gandavo: Quodlibet I*, Raymond Macken (ed.), vol. V in *Opera omnia*. Leuven: University Press, 1979 (cited as *HG* I)

Henrici de Gandavo: Quodlibet II, R. Wielockx (ed.), vol. VI in *Opera omnia*. Leuven: University Press, 1983 (cited as *HG* II)

Henrici de Gandavo: Quodlibet VI, G. A. Wilson (ed.), vol. X in *Opera omnia*. Leuven: University Press, 1987 (cited as *HG* VI)

Heytesbury, William, *William Heytesbury. On "Insoluble" Sentences: Chapter 1 of His Rules for Solving Sophisms*, Paul Vincent Spade (ed. and tr.). Toronto: Pontifical Institute, 1979

William Heytesbury. On Maxima and Minima: Chapter 5 of Rules for Solving Sophismata, with an Anonymous Fourteenth-Century Discussion, John Longeway (ed. and tr.). Dordrecht: D. Reidel, 1984

Hostiensis, *Summa aurea*. Venice, 1574; reprint, Turin: Bottega d'Erasmo, 1963

Hugh of St. Victor, *The "Didascalicon" of Hugh of St. Victor: A Medieval Guide to the Arts*, Jerome Taylor (tr.). New York: Columbia University Press, 1961

Isidore of Seville, *Isidori hispalensis episcopi etymologiarum sive originum*, W. M. Lindsay (ed.). Oxford: Clarendon, 1911

Jacob of San Martino (?), *De latitudinibus formarum*. Vienna, 1515

Bibliography

Jean de Jandun, *Tractatus de laudibus Parisius*, in Le Roux de Lincy and L. M. Tisserand (eds.), *Paris et ses historiens aux XIVe et XVe siècles*, 32–79. Paris, 1867

Jean de Ripa, *Conclusiones*, André Combes (ed.). Paris: J. Vrin, 1957

Kilvington, Richard, *Quaestiones morales super libros ethicorum*, ms. Vat. Ottob. lat. 179 (25r–59r)

> *The Sophismata of Richard Kilvington: Introduction, Translation, and Commentary*, Norman Kretzmann and Barbara Ensign Kretzmann (trs.). Cambridge University Press, 1990

> *The Sophismata of Richard Kilvington: Text Edition*, Norman Kretzmann and Barbara Ensign Kretzmann (eds.). Auctores Britannici medii aevi, 12. Oxford University Press, 1990

Latini, Brunetto, *Li livres dou tresor*, Francis J. Carmody (ed.). University of California Publications in Modern Philology, 22. Berkeley, 1948

Liber procuratorum Nationis Anglicanae (Alemanniae) in Universitate Parisiensi, H. Denifle and E. Chatelain (eds.), in *Auctarium chartularii Universitatis parisiensis*, vol. I, *1333–1406*. 2nd edn., Paris: H. Didier, 1937

Mandements et actes divers de Charles V (1364–1380), L. Delisle (ed.). Paris, 1874

Map, Walter, *De nugis curialium*, M. R. James (ed. and tr.), C. N. L. Brooke and R. A. B Mynors (revised). Oxford: Clarendon Press, 1983

Matthew Paris, *Matthew Paris' English History*, J. A. Giles (ed. and tr.), 3 vols. London, 1852–54

Odonis, Geraldus, *Expositio in Aristotelis ethicam*. Brescia, 1482; Venice, 1500

Olivi, Peter John, *Petrus Johannis Olivi: Quaestiones in secundum librum sententiarum*, Bernhard Jansen (ed.). Bibliotheca franciscana scholastica medii aevi, 4–6. Quaracchi: College of St. Bonaventure, 1922–26

> *Quodlibet I*, quests. 16 and 17, and *De contractibus usurariis*, Amleto Spicciani (ed.), "Gli scritti sul capitale et sull'interesse di Fra Pietro di Giovanni Olivi," *Studi francescani* 73 (1976), 289–325

> *Tractatus de emptionibus et venditionibus, de usuris, de restitutionibus*, in Giacomo Todeschini (ed.), *Un trattato di economia politica francescana: il "De emptionibus et venditionibus, de usuris, de restitutionibus" di Pietro di Giovanni Olivi*. Rome: Istituto storico italiano per il medio evo, studi storici, 1980

Ordonnances des roys de France de la troisième race, vol. I, *1254–1327*, and vol. II, *1327–1355*, E. de Laurière (ed.); vol. III, *1355–1364*, Denis François Secousse (ed.). Paris, 1723, 1729, 1732

Oresme, Nicole, *The "De moneta" of Nicholas Oresme and English Mint Documents*, Charles Johnson (ed. and tr.). London: Thomas Nelson, 1956

> *Maistre Nicole Oresme: Le livre de éthiques d'Aristote*, Albert Douglas Menut (ed.). New York: G. E. Stechert, 1940

> *Maistre Nicole Oresme: Le livre de politiques d'Aristote*, Albert Douglas Menut (ed.). Transactions of the American Philosophical Society, 60, pt. 6. Philadelphia, 1970

> *Nicole Oresme and the Kinematics of Circular Motion: "Tractatus de commensurabilitate vel incommensurabilitate motuum celi"*, Edward Grant (ed. and tr). Madison: University of Wisconsin Press, 1971

Bibliography

Nicole Oresme and the Marvels of Nature: A Study of His "De causis mirabilium" with Critical Edition, Translation, and Commentary, Bert Hansen (ed. and tr.). Toronto: Pontifical Institute, 1985

Nicole Oresme and the Medieval Geometry of Qualities and Motions: A Treatise on the Uniformity and Difformity of Intensities Known as "Tractatus de configurationibus qualitatum et motuum", Marshall Clagett (ed. and tr.). University of Wisconsin Publications in Medieval Science, 12. Madison, 1968

Nicole Oresme: "De proportionibus proportionum" and "Ad pauca respicientes", Edward Grant (ed. and tr). Madison: University of Wisconsin Press, 1966

Nicole Oresme: Le livre du ciel et du monde, Albert D. Menut and Alexander J. Denomy (eds. and trs.). Madison: University of Wisconsin Press, 1968

Questiones super geometriam Euclidis, Marshall Clagett (ed.), appendix I of the *Tractatus de configurationibus qualitatum et motuum*, 526–75 (cited as Clagett edn.)

Quaestiones super geometriam Euclidis, H. L. L. Busard (ed.). Leiden: E. J. Brill, 1961

Sacrae conciones, ms. BN lat. 16893 (1–212) (cited as *SC*)

"Sermo coram Papa Urbano V," in Flaccius Illyricus (ed.), *Tomus secundus scriptorum veterum*. London, 1690

Traictie de la première invention des monnoies de Nicole Oresme, L. Wolowski (ed.). Paris, 1864

Statuti di Verona del 1327, Silvana Anna Bianchi and Rosalba Granuzzo (eds.), 2 vols., vol. II, Corpus statutario delle Venezie. Rome: Jouvance, 1992

Swineshead, Richard, *Subtillissimi Ricardi Suiseth Anglici Calculationes noviter emendate atque revise*. Venice, 1520

Terrenia, Guido, *Super ethicam*, ms. BN lat. 3228 (1r–55v)

Walter of Henley, *Walter of Henley and Other Treatises on Estate Management and Accounting*, Dorothea Oschinsky (ed.). Oxford: Clarendon Press, 1971

SECONDARY WORKS

Ashley, Benedict, "St. Albert and the Nature of Natural Science," in Weisheipl, *Albertus Magnus and the Sciences*, 73–104

Aston, T. H., "The External Administration and Resources of Merton College to c. 1348," in Catto, *Early Oxford Schools*, 311–68

Aston, T. H., and Rosamond Faith, "The Endowments of the University and Colleges to c. 1348," in Catto, *Early Oxford Schools*, 265–310

Asztalos, Monika, John Murdoch, and Ilkka Niiniluoto (eds.), *Knowledge and the Sciences in Medieval Philosophy*, vol. I of *Proceedings of the Eighth International Congress of Medieval Philosophy (SIEPM)*. Acta philosophica Fennica, 48. Helsinki: Yliopistopaino, 1990

Babbit, Susan M., *Oresme's "Livre de Politiques" and the France of Charles V*. Transactions of the American Philosophical Society, 75, pt. I. Philadelphia, 1985

Bibliography

Baldwin, John W., *The Government of Philip Augustus: Foundations of French Royal Power in the Middle Ages.* Berkeley: University of California Press, 1986

    Masters, Princes, and Merchants: The Social Views of Peter the Chanter and His Circle, 2 vols. Princeton University Press, 1970

    The Medieval Theories of the Just Price: Romanists, Canonists, and Theologians in the Twelfth and Thirteenth Centuries. Transactions of the American Philosophical Society, 49, pt. 4. Philadelphia, 1959

Becker, Marvin B., *Medieval Italy: Constraints and Creativity.* Bloomington: Indiana University Press, 1981

Bettini, O., "Olivi da fronte ad Aristotile," *Studi francescani* 55 (1958), 176–97

Bisson, Thomas N., *Conservation of Coinage: Monetary Exploitation and Its Restraint in France, Catalonia, and Aragon (c. AD 1000–c. 1225).* Oxford: Clarendon Press, 1979

Bloch, Marc, "Mutations monétaire dans l'ancienne France," *Annales ESC* 7 (1953), 145–58, 433–56

    "Natural Economy or Money Economy: A Pseudo Dilemma" in J. E. Anderson (tr.), *Land and Work in Mediaeval Europe,* 230–43. New York: Harper and Row, 1967

Borchert, Ernest, *Die Lehrer von der Bewegung bei Nicolaus Oresme.* Beiträge zur Geschichte der Philosophie und Theologie des Mittelalters, 31, pt. 3. Münster: Aschendorff, 1934

Bougerol, J. G., *Lexique saint Bonaventure.* Paris: Editions franciscaines, 1969

Bowler, John F., "Intuitive and Abstractive Cognition," in Kretzmann, Kenny, and Pinborg, *Cambridge History of Later Medieval Philosophy,* 460–78

Braudel, Fernand, *Capitalism and Material Life 1400–1800,* Miriam Kochan (tr.). New York: Harper and Row, 1973

    "Prices in Europe from 1450 to 1750," in *The Cambridge Economic History of Europe,* vol. IV, 374–486. Cambridge University Press, 1967

Bridrey, Emile, *La théorie de la monnaie au XIVe siècle: Nicole Oresme.* Paris: V. Giard, 1906

Britnell, R. H., "Commercialisation and Economic Development in England 1000–1300," in Britnell and Campbell, *A Commercialising Economy,* 7–26

    The Commercialisation of English Society 1000–1500. Cambridge University Press, 1993

    "The Proliferation of Markets in England 1200–1349," *Economic History Review* 34 (1981), 209–21

Britnell, R. H., and B. M. S. Campbell (eds.), *A Commercialising Economy: England 1086–1300.* Manchester University Press, 1995

Burr, David, *Olivi and Franciscan Poverty: The Origins of Usus Pauper Controversy.* Philadelphia: University of Pennsylvania Press, 1989

    The Persecution of Peter Olivi. Transactions of the American Philosophical Society, 66, pt. 5. Philadelphia, 1976

Cahn, Kenneth S., "The Roman and Frankish Roots of the Just Price of Medieval Canon Law," *Studies in Medieval and Renaissance History* 6 (1969), 3–52

Bibliography

Callus, D. A., "The Date of Grosseteste's Translations and Commentaries on the Pseudo-Dionysius and the Nicomachean Ethics," *Recherches de théologie ancienne et médiévale* 14 (1947), 186–210

Campbell, Bruce, "Measuring the Commercialisation of Seigneurial Agriculture c. 1300," in Britnell and Campbell, *A Commercialising Economy*, 132–93

Capitani, Ovidio, "Il 'De peccato usure' di Remigio de' Girolami," *Studi medievali*, ser. 3, 6, Fasc. II (1965), 537–662

"Sulla questione dell'usura nel Medio Evo," in Capitani (ed.), *L'etica economica medievale: testi à cura di Ovidio Capitani*, 23–46. Bologna: Il mulino, 1974

Caroti, Stefano, "La perception du mouvement selon Nicole Oresme (*Questiones super physicam* III.1)," in *Comprendre et maîtriser la nature au moyen âge: mélanges d'histoire de science offerts à Guy Beaujouan*, 83–99. Haute études médiévales et modernes, 73. Geneva: Librairie Droz, 1994

Caroti, Stefano (ed.), *Studies in Medieval Natural Philosophy*. Florence: Olschki, 1989

Catto, J. I., "Citizens, Scholars, and Masters," in Catto, *Early Oxford Schools*, 151–92

"Ideas and Experience in the Political Thought of Aquinas," *Past and Present* 71 (1976), 3–21

Catto, J. I. (ed.), *The Early Oxford Schools*, vol. 1 of *The History of the University of Oxford*. Oxford: Clarendon Press, 1984

Cazelles, Raymond, *Nouvelle histoire de Paris de la fin du règne de Philippe Auguste à la mort de Charles V 1223–1380*. Paris: Hachette, 1972

"Quelques réflexions à propos des mutations de la monnaie royale française (1295–1360)," *Le moyen âge* 72 (1966), 83–105, 251–78

Société politique, noblesse, et couronne sous Jean le Bon et Charles V. Geneva: Droz, 1982

"La stabilization de la monnaie par la création du franc (décembre 1360) – blocage d'une société," *Traditio* 32 (1976), 293–311

Challis, C. E. (ed.), *A New History of the Royal Mint*. Cambridge University Press, 1992

Cipolla, Carlo M., *Before the Industrial Revolution: European Society and Economy*. New York: Norton Press, 1976

Money, Prices, and Civilization in the Mediterranean World. New York: Gordian Press, 1967

Clagett, Marshall, "Nicole Oresme and Medieval Scientific Thought," *Proceedings of the American Philosophical Society* 108 (1964), 298–310

"Richard Swineshead and Late Medieval Physics," *Osiris* 9 (1950), 131–61

The Science of Mechanics in the Middle Ages. University of Wisconsin Publications in Medieval Science, 4. Madison, 1959

"Some Novel Trends in the Science of the Fourteenth Century," in Charles Singleton (ed.), *Art, Science, and History in the Renaissance*, 275–303. Baltimore: Johns Hopkins University Press, 1967

Clanchy, Michael, *From Memory to Written Record: England 1066–1307*. 2nd edn., Oxford: Blackwell, 1993

Bibliography

Cobban, Alan, *The Medieval English Universities: Oxford and Cambridge to c. 1500.*
    Aldershot: Scholar, 1988
Coleman, Janet, "Jean de Ripa, OFM, and the Oxford Calculators," *Mediaeval
    Studies* 37 (1975), 130–89
Cosman, Madeleine Pelner, and Bruce Chandler (eds.), *Machaut's World: Science
    and Art in the Fourteenth Century.* Annals of the New York Academy of the
    Sciences, 314. New York, 1978
Courtenay, William J., "The King and the Leaden Coin: The Economic
    Background of 'Sine qua non' Causality," *Traditio* 28 (1972), 185–210
    "The London *Studia* in the Fourteenth Century," *Medievalia et humanistica* 13
    (1985), 127–41
    "The Registers of the University of Paris and the Statutes Against the *Scientia
    Occamica*," *Vivarium* 29 (1991), 13–49
    "The Role of English Thought in the Transformation of University Educa-
    tion in the Late Middle Ages," in James M. Kittelson and Pamela J. Transue
    (eds.), *Rebirth, Reform, and Resilience: Universities in Transition 1300–1700,*
    103–62. Columbus: Ohio State University Press, 1984
    Schools and Scholars in Fourteenth-Century England. Princeton University Press,
    1987
Crombie, A. C., *Medieval and Early Modern Science,* 2 vols. Garden City, N. Y.:
    Doubleday, 1959
Daly, Lowrie J., "The Conclusions of Walter Burley's Commentary on the
    Politics, Books I to IV," *Manuscripta* 12 (1968), 79–92
    The Medieval University 1200–1400. New York: Sheed and Ward, 1961
de Libera, Alain, "Le développement de nouveaux instruments conceptuels et
    leur utilisation dans la philosophie de la nature au XIVe siècle," in Asztalos,
    Murdoch, and Niiniluoto, *Knowledge and the Sciences,* vol. I, 158–97
De Roover, Raymond A., "The Concept of the Just Price: Theory and Prac-
    tice," *Journal of Economic History* 18 (1958), 418–34
    Money, Banking, and Credit in Mediaeval Bruges. Cambridge, Mass.: Medieval
    Academy of America, 1948
    La pensée économique des scolastiques: doctrines et méthodes. Montreal: Institute
    d'études médiévales, 1971
    *San Bernardino of Siena and Sant' Antonino of Florence: The Two Great Economic
    Thinkers of the Middle Ages.* Kress Library of Business and Economics, 19.
    Boston, 1967
Denholm-Young, N., "Richard de Bury (1287–1345)," in *Collected Papers on
    Mediaeval Subjects,* 1–25. Oxford: Blackwell, 1946
Denholm-Young, N. (ed. and tr.), *Vita Edwardi Secundi: The Life of Edward the
    Second,* Medieval Classics. London: Thomas Nelson, 1957
dohrn-van Rossum, Gerhard, *History of the Hour: Clocks and Modern Temporal
    Orders,* Thomas Dunlap (tr.). University of Chicago Press, 1996
Du Boulay, C. E., *Historia Universitatis Parisiensis,* 6 vols. Paris, 1665–73
Duby, Georges, *The Early Growth of the European Economy: Warriors and Peasants
    from the Seventh to the Twelfth Century,* Howard Clarke (tr.). Ithaca, N. Y.:

Bibliography

Cornell University Press, 1974

Rural Economy and Country Life in the Medieval West, Cynthia Postan (tr.). Columbia: University of South Carolina Press, 1968

Duhem, Pierre, *Etudes sur Léonard de Vinci*, 3 vols. Paris: A. Hermann, 1906–13

Le système du monde: histoire des doctrines cosmologiques de Platon à Copernic, 10 vols. Paris: A. Hermann, 1913–59

Dunbabin, Jean, "Careers and Vocations," in Catto, *Early Oxford Schools*, 565–606

"Meeting the Costs of University Education in Northern France c. 1240–1340," *History of Universities* 10 (1991), 1–27

"The Two Commentaries of Albertus Magnus on the Nicomachean Ethics," *Recherches de théologie ancienne et médiévale* 30 (1963), 232–50

Dyer, Christopher, "The Consumer and the Market in the Later Middle Ages," *Economic History Review* 42 (1989), 305–27

Edgerton, Samuel Y., "From Mental Matrix to *Mappamundi* to Christian Empire: The Heritage of Ptolemaic Cartography in the Renaissance," in David Woodward (ed.), *Art and Cartography: Six Historical Essays*, 10–50. University of Chicago Press, 1987

Egbert, Virginia Wylie, *On the Bridges of Medieval Paris*. Princeton University Press, 1984

Ehrle, Franz, "Petrus Johannis Olivi: sein Leben und seine Schriften," *Archiv für Litteratur und Kirchengeschichte des Mittelalters* 3 (1887), 409–552

Emden, A. B., *A Biographical Register of the University of Oxford to AD 1500*, 3 vols. Oxford: Clarendon, 1957–59

Eschmann, Thomas, "Studies on the Notion of Society in St. Thomas Aquinas," pt. I, *Mediaeval Studies* 8 (1946), 1–42; pt. II, *Mediaeval Studies* 9 (1947), 19–55

Faral, Edmund, "Jean Buridan: maître ès arts de l'université de Paris," in *Histoire littéraire de la France*, vol. xxxviii, 462–605. Paris: Imprimerie nationale, 1949

"Jean Buridan: notes sur les manuscrits, les éditions, et le contenu de ses oeuvres," *Archives d'histoire doctrinale et littéraire du moyen âge* 15 (1946), 1–53

Finley, M. I., *The Ancient Economy*. Berkeley: University of California Press, 1973

"Aristotle and Economic Analysis," Jonathan Barnes et al. (eds.), in *Articles on Aristotle II: Ethics and Politics*, 140–58. London: Duckworth, 1977

Fournial, Etienne. *Histoire monétaire de l'occident médiéval*. Paris: F. Nathan, 1970

Les villes et l'économie d'échange en Forez aux XIIIe et XIVe siècles. Paris: Presses du Palais Royal, 1967

Funkenstein, Amos, *Theology and the Scientific Imagination from the Middle Ages to the Seventeenth Century*. Princeton University Press, 1986

Gabriel, Astrik L., *The College System in the Fourteenth-Century Universities*. Baltimore: n. p., n. d.

Student Life in Ave Maria College, Mediaeval Paris. Notre Dame University Press, 1955

Bibliography

Gauthier, René-Antoine, *Ethica Nicomachea: praefatio*, Aristoteles Latinus 26, Fasc. 1. Leiden: E. J. Brill, 1974

La morale d'Aristote. Initiation philosophique, 34. Paris: Presses universitaires de France, 1958

"Trois commentaires 'averroistes' sur L'Ethique à Nicomaque," *Archives d'histoire doctrinale et littéraire du moyen âge* 23 (1947–48), 187–336

Gauthier, René-Antoine, and Jean-Yves Jolif, *L'Ethique à Nicomaque: introduction, traduction, et commentaire*, 2 vols. Louvain: Publications universitaires, 1970

Géraud, Hercule, *Paris sous Philippe le Bel: le rôle de la taille imposé sur les habitants de Paris en 1292*. Documents inédits sur l'histoire de France, 33. Paris, 1837

Giacon, Carlo, "Una 'Nota Magistri Fratris Occam De quantitate,'" in *La filosofia della natura nel medioevo: atti del 3. Congresso internazionale de filosofia medioevale*, 625–33. Milan: Società editrice vita e pensiero, 1966

Gieben, Servus, "Bibliographia Oliviana," *Collectanea franciscana* 38 (1968), 167–95

Gillard, Lucien, "Nicole Oresme, sujet théorique, objet historique," in Quillet, *Autour de Nicole Oresme: Actes du Colloque Oresme*, 195–233

Glorieux, Palemon, *La littérature quodlibétique*, 2 vols., vol. II. Paris: J. Vrin, 1935

Goclenius, Rudolph, *Lexicon philosophicum*. Frankfurt, 1613

Goddu, André, *The Physics of William of Ockham*. Studien und Texte zur Geistesgeschichte des Mittelalters, 16. Leiden: E. J. Brill, 1984

Gordon, Barry J., "Aristotle and the Development of Value Theory," *Quarterly Journal of Economics* 78 (1964), 115–28

Economic Analysis Before Adam Smith: Hesiod to Lessius. New York: Barnes and Noble, 1975

Grant, Edward, "Cosmology," in Lindberg, *Science in the Middle Ages*, 265–302

"Medieval Departures from Aristotelian Natural Philosophy," in Caroti, *Studies in Medieval Natural Philosophy*, 237–56

"Nicole Oresme on Certitude in Science and Pseudo-Science," in Souffrin and Segonds, *Nicolas Oresme: Tradition et innovation*, 31–43

"Scientific Thought in Fourteenth-Century Paris: Jean Buridan and Nicole Oresme," in Cosman and Chandler, *Machaut's World*, 105–24

A Source Book in Medieval Science. Cambridge, Mass.: Harvard University Press, 1974

Grant, Edward, and John Murdoch (eds.), *Mathematics and Its Applications to Science and Natural Philosophy in the Middle Ages: Essays in Honor of Marshall Clagett*. Cambridge University Press, 1987

Hadden, Richard, *On the Shoulders of Merchants: Exchange and the Mathematical Conception of Nature in Early Modern Europe*, SUNY Series in Science, Technology, and Society. Albany, N. Y., 1994

Hagenauer, Selma, *Das "justum pretium" bei Thomas von Aquino: ein Beitrag zur Geschichte der objecktiven Werttheorie*. Beihefte zur Vierteljahrschrift für Sozial- und Wirtschaftsgeschichte, 24. Stuttgart: W. Kohlhammer, 1931

Hardie, W. F. R., *Aristotle's Ethical Theory*. Oxford: Clarendon Press, 1968

Bibliography

Harris, G. L., *King, Parliament, and Public Finance in Medieval England*. Oxford: Clarendon Press, 1975

Harvey, P. D. A. (ed.), *Manorial Records of Cuxham, Oxfordshire, c. 1200–1359*. Oxfordshire Record Society, 50. London: Historical Manuscripts Commission, 1976

Haskins, Charles H., "Letters of Mediaeval Students," in *Studies in Mediaeval Culture*, 1–35. New York: F. Ungar, 1929

Heath, Thomas, *A History of Greek Mathematics*, 2 vols., vol. 1. Oxford: Clarendon Press, 1921

Heidingsfelder, Georg, *Albert von Sachsen: sein Lebensgang und sein Kommentar zur nikomachischen Ethik des Aristoteles*. Beiträge zur Geschichte der Philosophie und Theologie des Mittelalters, 22, pts. 3–4. Münster: W. Aschendorff, 1927

Henneman, John Bell, *Royal Taxation in Fourteenth-Century France: The Development of War Financing 1322–1356*. Princeton University Press, 1971

Henninger, Mark, *Relations: Medieval Theories 1250–1325*. Oxford: Clarendon Press, 1989

Herlihy, David, "Treasure Hoards in the Italian Economy 960–1139," *Economic History Review* 10 (1957), 1–15

Hollander, Samuel, "On the Interpretation of the Just Price," *Kyklos* 18 (1965), 615–34

Hoskin, M. A., and A. G. Molland, "Swineshead on Falling Bodies: An Example of Fourteenth-Century Physics," *British Journal for the History of Science* 3 (1966), 150–82

Ibanès, Jean, *La doctrine de l'église et les réalités économiques au XIIIe siècle*. Paris: Presses universitaires de France, 1967

Johnson, Harold, "Just Price, Aquinas, and the Labor Theory of Value," in Johnson (ed.), *The Medieval Tradition of Natural Law*, 75–86. Studies in Medieval Culture, 22. Kalamazoo, Mich.: Medieval Institute Publications, 1987

Jordan, Mark D., "Aquinas Reading Aristotle's *Ethics*," in Jordan and Kent Emery, Jr. (eds.), *Ad litteram: Authoritative Texts and Their Medieval Readers*, 229–49. Notre Dame Conferences in Medieval Studies, 3. University of Notre Dame Press, 1992

Kaulla, Rudolph, "Der Lehrer des Oresmius," *Zeitschrift für die gesamte Staatswissenschaft* 60 (1904), 433–62

Kaye, Joel, "The Impact of Money on the Development of Fourteenth-Century Scientific Thought," *Journal of Medieval History* 14 (1988), 251–70

Kelsen, Hans, "Aristotle's Doctrine of Justice," in James J. Walsh and Henry L. Shapiro (eds.), *Aristotle's Ethics: Issues and Interpretations*, 102–19. Belmont, Calif.: Wadsworth, 1967

Kent, Bonnie, *Virtues of the Will: The Transformation of Ethics in the Late Thirteenth Century*. Washington, D. C.: Catholic University of America, 1995

Kirshner, Julius, "The Moral Theology of Public Finance: A Study and Edition of Nicholas de Anglia's *Quaestio disputata* on the Public Debt of Venice," *Archives fratrum praedicatorum* 40 (1970), 47–72

Bibliography

"Storm Over the 'Monte Comune': Genesis of the Moral Controversy over the Public Debt of Florence," *Archivum fratrum praedicatorum* 53 (1983), 219–76

Kirshner, Julius, and Kimberly Lo Prete, "Peter John Olivi's Treatises on Contracts of Sale, Usury, and Restitution: Minorite Economics or Minor Works?," *Quaderni fiorentini* 13 (1984), 233–86

Kittell, Ellen, *From Ad Hoc to Routine: A Case Study in Medieval Bureaucracy*. Philadelphia: University of Pennsylvania Press, 1991

Koyré, Alexandre, *From the Closed World to the Infinite Universe*. Baltimore: Johns Hopkins University Press, 1957

Kretzmann, Norman, Anthony Kenny, and Jan Pinborg (eds.), *The Cambridge History of Later Medieval Philosophy*. Cambridge University Press, 1982

Lagarde, Georges de, "La philosophie sociale d'Henri de Gand et de Godefroid de Fontaines," in *L'organisation corporative du moyen âge à la fin de l'ancien régime*, 57–134. Louvain: Bibliothèque de l'université, 1943

Langholm, Odd, *The Aristotelian Analysis of Usury*. Bergen: Universitetsforlaget, 1984

Economics in the Medieval Schools: Wealth, Exchange, Value, Money, and Usury According to the Paris Theological Tradition 1200–1350. Leiden: E. J. Brill, 1992

Price and Value in the Aristotelian Tradition. Bergen: Universitetsforlaget, 1979

"Scholastic Economics," in Lowry, *Pre-Classical Economic Thought*, 117–34

Wealth and Money in the Aristotelian Tradition. Bergen: Universitetsforlaget, 1983

Langlois, C. V., "Gurial Ot" (Geraldus Odonis), in *Histoire littéraire de la France*, vol. XXXVI, 203–25. Paris: Imprimerie nationale, 1927

Le Bras, Gabriel, "Usure," in *Dictionnaire de théologie catholique*, vol. XV (1950), cols. 2316–90. Paris: Letouzey et Ane, 1923–50

Le Goff, Jacques, "Labor Time in the 'Crisis' of the Fourteenth Century," in Le Goff, *Time, Work, and Culture in the Middle Ages*, Arthur Goldhammer (tr.), 43–52. University of Chicago Press, 1980

"Merchant's Time and Church's Time in the Middle Ages," in Le Goff, *Time, Work, and Culture in the Middle Ages*, Arthur Goldhammer (tr.), 29–42. University of Chicago Press, 1980

"The Town as an Agent of Civiliation 1200–1500," in Carlo Cipolla (ed.), *The Fontana Economic History of Europe: The Middle Ages*, 71–95. London, 1972

"The Usurer and Purgatory," in *The Dawn of Modern Banking*, 25–52. New Haven: Yale University Press, 1979

Your Money or Your Life: Economy and Religion in the Middle Ages. New York: Zone Books, 1990

Lefebvre, Charles, "Hostiensis," in *Dictionnaire de droit canonique*, 7 vols. (1935–65), vol. V (1953), cols. 1211–27. Paris Letouzey et Ane

Leff, Gordon, *Medieval Thought: St. Augustine to Ockham*. Baltimore: Penguin Books, 1962

Lindberg, David, "Roger Bacon and the Origins of *Perspectiva* in the West," in

Bibliography

Grant and Murdoch, *Mathematics and Its Applications*, 249–67

Lindberg, David C. (ed.), *Science in the Middle Ages*. University of Chicago Press, 1978

Little, Lester K., "Evangelical Poverty, the New Money Economy, and Violence," in David Flood (ed.), *Poverty in the Middle Ages*. Werl: D. Coelde, 1975

Religious Poverty and the Profit Economy in Medieval Europe. Ithaca, N. Y.: Cornell University Press, 1978

Livesey, Steven J., "Mathematics *iuxta communem modum loquendi*: Formation and Use of Definitions in Heytesbury's *De motu locali*," *Comitatus* 10 (1979), 9–20

Lopez, Robert S., "An Aristocracy of Money in the Middle Ages," *Speculum* 28 (1953), 1–43

Lopez, Robert S., and Irving Raymond (eds.), *Medieval Trade in the Mediterranean World*. New York: Columbia University Press, 1955

Lottin, Odon, "A propos de la date de certains commentaires sur l'Ethique," *Recherches de théologie ancienne et médiévale* 17 (1950), 127–33

"La connexion des vertus chez Saint Thomas d'Aquin et ses prédécesseurs," in Lottin, *Psychologie et morale aux XIIe et XIIIe siècles*, 6 vols, vol. III, 197–252. Gembloux: J. Duculot, 1957

Lowry, S. Todd, "Aristotle's Mathematical Analysis of Exchange," *History of Political Economy* 1 (1969), 44–66

"The Greek Heritage in Economic Thought," in Lowry, *Pre-Classical Economic Thought*, 7–30

Lowry, S. Todd (ed.), *Pre Classical Economic Thought*. Boston: Kluwer, 1987

Lunt, William E., *Papal Revenues in the Middle Ages*, 2 vols. Columbia University Records of Civilization, Sources, and Studies, 19. New York, 1965

McColley, Grant, "The Theory of the Diurnal Rotation of the Earth," *Isis* 26 (1937), 392–402

McGrade, Arthur Stephen, "Ethics and Politics as Practical Sciences," in Asztalos, Murdoch, and Niiniluoto, *Knowledge and the Sciences*, vol. I, 199–219

McLaughlin, T. P., "The Teachings of the Canonists on Usury," *Mediaeval Studies* 1 (1939), 81–147

McVaugh, Michael, "Arnald of Villanova and Bradwardine's Law," *Isis* 58 (1967), 56–64

Mahoney, Michael S., "Mathematics," in Lindberg, *Science in the Middle Ages*, 145–78

Maier, Anneliese, "La doctrine de Nicolas d'Oresme sur les 'configurationes intensionum,'" in Maier, *Ausgehendes Mittelalter*, 3 vols., vol. I, 335–52. Rome: Edizioni di storia e letteratura, 1964–77

"Der Funktionsbegriff in der Physik des 14. Jahrhunderts," in Maier, *Die Vorläufer Galileis*, 81–110

On the Threshold of Exact Science: Selected Writings of Anneliese Maier on Late Medieval Natural Philosophy, Steven Sargent (ed. and tr.). Philadelphia: University of Pennsylvania Press, 1982

Bibliography

"Das Problem der intensiven Grösse in der Scholastik," in *Zwei Grundprobleme der scholastischen Naturphilosophie: Studien zur Naturphilosophie der Spätscholastik*, 1–79. Rome: Edizioni di storia e letteratura, 1951

Die Vorläufer Galileis im 14. Jahrhundert. Rome: Edizioni di storia e letteratura, 1966

Zwischen Philosophie und Mechanik. Rome: Edizioni di storia e letteratura, 1958

Marcuzzi, Giorgio, "Una soluzione teologico-giuridica al problema dell'usura in una questione *De quolibet* inedita de Guido Terreni (1260–1342)," *Salesianum* 41 (1979), 647–84

Marrone, Stephen P., *Truth and Scientific Knowledge in the Thought of Henry of Ghent*. Cambridge, Mass.: Medieval Academy of America, 1985

Martin, Conor, "Walter Burley," in *Oxford Studies Presented to Daniel Callus*, 194–230. Oxford: Clarendon, 1964

Mate, Mavis, "High Prices in Early Fourteenth-Century England: Causes and Consequences," *Economic History Review* 28 (1975), 1–16

Mayhew, Nicholas, *Coinage in France from the Dark Ages to Napoleon*. London: Seaby, 1988

"Modelling Medieval Monetisation," in Britnell and Campbell, *A Commercialising Economy*, 55–77

"Population, Money Supply, and the Velocity of Circulation in England 1300–1700," *Economic History Review* 48 (1995), 238–58

Menjot, Denis, "La politique monétaire de Nicolas Oresme," in Souffrin and Segonds, *Nicolas Oresme: tradition et innovation*, 179–93

Menut, Albert D., "A Provisional Bibliography of Oresme's Writings," *Mediaeval Studies* 28 (1966), 279–99

Metcalf, D. M., "The Volume of English Currency," in N. J. Mayhew (ed.), *Edwardian Monetary Affairs*. British Archaeological Reports, 36. Oxford, 1977

Meunier, Francis, *Essai sur la vie et les ouvrages de Nicole Oresme*. Paris, 1857

Michael, Bernd, *Johannes Buridan: Studien zu seinem Leben, seinem Werken, und zur Rezeption seiner Theorien im Europa des späten Mittelalters*, 2 vols. Berlin: n. p., 1985

Minio-Paluello, L., "Remigio Girolami's *De bono communi*: Florence at the Time of Dante's Banishment and the Philosopher's Answer to the Crisis," *Italian Studies* 11 (1956), 56–71

Mises, L. von, *The Theory of Money and Credit*, H. E. Bateson (tr.). New Haven: Yale University Press, 1953

Miskimin, Harry A., *The Economy of Early Renaissance Europe 1300–1460*. Englewood Cliffs, N. J.: Prentice-Hall, 1969

Money, Prices, and Foreign Exchange in Fourteenth-Century France. New Haven: Yale University Press, 1963

"L'or, l'argent, la guerre dans la France médiévale," *Annales ESC* 40 (1985), 171–84

Mitchell, Wesley C., "The Role of Money in Economic History," in Frederick C. Lane and Jelle C. Riemersma (eds.), *Enterprise and Secular Change*,

200–04. Homewood, Ill.: R. D. Irwin, 1953

Molland, A. G., "The Geometrical Background to the 'Merton School,'" *British Journal for the History of Science* 4 (1968), 108–25

"Mathematics in the Thought of Albertus Magnus," in Weisheipl, *Albertus Magnus and the Sciences*, 463–78

"The Oresmian Style: Semi-Mathematical, Semi-Holistic," in Souffrin and Segonds, *Nicolas Oresme: Tradition et innovation*, 13–30

Mollat, Guillaume, *Les papes d'Avignon 1305–1378*. Paris: Letouzey et Ane, 1949

Mollat, Michel, and Philippe Wolff, *The Popular Revolutions of the Late Middle Ages*, A. L. Lytton-Sells (tr.). London: Allen and Unwin, 1973

Moody, Ernest A., "Jean Buridan," in *Dictionary of Scientific Biography*, C. C. Gillespie (gen. ed.), 18 vols., vol. II (1970), 603–08. New York: Scribner, 1970–90

"John Buridan on the Habitability of the Earth," *Speculum* 16 (1941), 415–25

Studies in Medieval Philosophy, Science, and Logic: Collected Papers 1933–1969. Berkeley: University of California Press, 1975

Moulin, Léo, *La vie des étudiants au moyen âge*. Paris: A. Michel, 1991

Murdoch, John E., "The Analytic Character of Late Medieval Learning: Natural Philosophy Without Nature," in L. D. Roberts (ed.), *Approaches to Nature in the Middle Ages: Papers of the Tenth Annual Conference of the Center for Medieval and Early Renaissance Studies*, 171–213. Binghamton, N. Y.: Medieval and Renaissance Texts and Studies, 1982

"From Social into Intellectual Factors: An Aspect of the Unitary Character of Late Medieval Learning," in Murdoch and Sylla, *Cultural Context*, 271–348

"The Involvement of Logic in Late Medieval Natural Philosophy," in Caroti, *Studies in Medieval Natural Philosophy*, 3–28

"*Mathesis in philosophiam scholasticam introducta*: The Rise and Development of the Application of Mathematics in Fourteenth-Century Philosophy and Theology," in *Arts libéraux et philosophie au moyen âge: Actes du Quatrième Congrès international de philosophie médiévale*, 215–46. Montreal: Institut d'études médiévales, 1969

"The Medieval Language of Proportions: Elements of the Interaction with Greek Foundations and the Development of New Mathematical Techniques," in A. C. Crombie (ed.), *Scientific Change: Historical Studies in the Intellectual, Social, and Technical Conditions for Scientific Discovery and Technical Invention, from Antiquity to the Present*, 237–71. New York: Basic Books, 1963

"Naissance et développement de l'atomisme au bas moyen âge latin," in *La science de la nature: théories et pratiques*, 11–32. Montreal: Bellarmin, 1974

"*Subtilitates Anglicanae* in Fourteenth-Century Paris: John of Mirecourt and Peter Ceffons," in Cosman and Chandler, *Machaut's World*, 51–86

"Thomas Bradwardine: Mathematics and Continuity in the Fourteenth Century," in Grant and Murdoch, *Mathematics and Its Applications*, 103–37

"William of Ockham and the Logic of Infinity and Continuity," in Norman Kretzmann (ed.), *Infinity and Continuity in Ancient and Medieval Thought*,

165–206. Ithaca, N. Y.: Cornell University Press, 1981

Murdoch, John, and Edith Sylla (eds.), *The Cultural Context of Medieval Learning.* Boston Studies in the Philosophy of Science, 26. Dordrecht and Boston: D. Reidel, 1975

Murdoch, John, and Edward Synan, "Two Questions on the Continuum: Walter Chatton (?), OFM, and Adam Wodeham, OFM," *Franciscan Studies* 26 (1966), 212–88

Murray, Alexander, *Reason and Society in the Middle Ages.* Oxford: Clarendon Press, 1978

Muscatine, Charles, *The Old French Fabliaux.* New Haven: Yale University Press, 1986

Nelson, Benjamin, *The Idea of Usury: From Tribal Brotherhood to Universal Otherhood.* 2nd edn., University of Chicago Press, 1969

Neveux, François, "Nicole Oresme et le clergé normand du XIVe siècle," in Quillet, *Autour de Nicole Oresme: Actes du Colloque Oresme,* 9–36

Noonan, John, *The Scholastic Analysis of Usury.* Cambridge, Mass.: Harvard University Press, 1957

Olivier-Martin, François, "Ordonnonces des rois de France de la troisième race," in *Les travaux de l'Academie des inscriptions et belles-lettres: histoire et inventaire des publications,* 37–44. Paris: Klincksieck, 1947

Omrod, W. M., "The Crown and the English Economy 1290–1348," in Bruce Campbell (ed.), *Before the Black Death: Studies in the "Crisis" of the Early Fourteenth Century,* 149–83. Manchester University Press, 1991

Origo, Iris, *The Merchant of Prato: Francesco di Marco Datini.* London: J. Cape, 1957

Pacetti, D., "Un trattato sulle usure e le restituzioni di Pietro di Giovanni Olivi falsamente attribuito a fr. Gerardo da Siena," *Archivum franciscanum historicum* 46 (1953), 448–57

Pantin, W. A., *The English Church in the Fourteenth Century.* University of Notre Dame Press, 1962

Pardessus, J. M., *Table chronologique des Ordonnances des rois de France de la troisième race.* Paris, 1847

Partner, Nancy F., *Serious Entertainments: The Writing of History in Twelfth-Century England.* University of Chicago Press, 1977

Pelzer, Auguste, "Le cours inédit d'Albert le Grand sur la Morale à Nicomaque, recueilli et rédigé par saint Thomas d'Aquin," *Revue néo-scolastique* 24 (1922), 333–61, 479–520

"Les versions latines des ouvrages de morale conservés sous le nom d'Aristote en usage au XIIIe siècle," *Revue néo-scolastique* 23 (1921), 316–41, 378–412

Perroy, Edouard, "A l'origine d'une economie contractée: les crises du XIVe siècle," *Annales ESC* 4 (1949), 167–82

"Social Mobility Among the French Noblesse in the Later Middle Ages," *Past and Present* (1962), 25–38

Piquet-Marchal, Marie-Odile, "Doctrines monétaires et conjonture aux XIVe et XVe siècles," *Revue internationale d'histoire de la banque* 4 (1971), 327–405

Polanyi, Karl, *Primitive, Archaic, and Modern Economies: Essays of Karl Polanyi,* G.

Bibliography

Dalton (ed.). Garden City, N. Y.: Anchor Books, 1968

Postan, Michael M., *The Medieval Economy and Society: An Economic History of Britain 1100–1500*. Berkeley: University of California Press, 1972

"The Rise of a Money Economy," *Economic History Review* 14 (1944), 123–34

Powicke, F. M., "Robert Grosseteste and the Nicomachean Ethics," *Proceedings of the British Academy* (1930), 85–104

Prestwich, Michael, "Early Fourteenth-Century Exchange Rates," *Economic History Review* 32 (1979), 470–82

Quillet, Jeannine, "Nicole Oresme et la science nouvelle dans le *Livre du ciel et du monde*," in Asztalos, Murdoch, and Niiniluoto, *Knowledge and the Sciences*, vol. I, 314–21

Quillet, Jeannine (ed.), *Autour de Nicole Oresme: Actes du Colloque Oresme organisé à l'Université de Paris XII*. Paris: J. Vrin, 1990

Randall, John Herman, Jr., *Aristotle*. New York: Columbia University Press, 1960

Renouard, Yves, "Les hommes d'affaires italiens et l'avenement de la Renaissance," in *Etudes d'histoire médiévale*, 419–37. Paris: SEVPEN, 1968

"Le rôle des hommes d'affaires italiens dans la Méditerrané au moyen âge," in *Etudes d'histoire médiévale*, 405–18. Paris: SEVPEN, 1968

Ritchie, D. G., "Aristotle's Subdivisions of 'Particular Justice,'" *Classical Review* 8 (1894), 185–92

Rossiaud, Jacques, "The City Dweller," in Jacques Le Goff (ed.), Lydia G. Cochrane (tr.), *Medieval Callings*, 139–79. University of Chicago Press, 1990

Sapori, Armando, "Il giusto prezzo nella dottrina di San Tommaso e nella pratica del suo tempo," *Archivio storico italiano*, ser. 7, 18 (1932), 3–56

Schumpeter, Joseph, *Economic Doctrine and Method*, R. Aris (tr.). New York: Oxford University Press, 1967

History of Economic Analysis. New York: Oxford University Press, 1954

Serene, Eileen, "Demonstrative Science," in Kretzmann, Kenny, and Pinborg, *Cambridge History of Later Medieval Philosophy*, 496–517

Shapiro, Herman, "Motion, Time, and Place According to William Ockham," *Franciscan Studies* 16 (1956), 213–303, 319–72

Motion, Time, and Place According to William of Ockham. St. Bonaventure, N. Y.: Franciscan Institute, 1957

"Walter Burley and the Intension and Remission of Forms," *Speculum* 34 (1959), 413–27

Shell, Marc, *The Economy of Literature*. Baltimore: Johns Hopkins University Press, 1978

Shoaf, R. A., *Dante, Chaucer, and the Currency of the Word: Money, Images, and Reference in Late Medieval Poetry*. Norman, Okla.: Pilgrim Books, 1983

Soudek, Joseph, "Aristotle's Theory of Exchange: An Inquiry into the Origin of Economic Analysis," *Proceedings of the American Philosophical Society* 96 (1952), 45–75

Souffrin, Pierre, "Oresme, Buridan, et le mouvement de rotation diurne de la terre ou des cieux," in Bernard Ribémont (ed.), *Terres médiévales*, 277–99. Paris: Klincksieck, 1993

"La quantification du mouvement chez les scolastiques: la vitesse instantanée chez Nicole Oresme," in Quillet, *Autour de Nicole Oresme: Actes du Colloque Oresme*, 63–83

Souffrin, P., and A. Ph. Segonds (eds.), *Nicolas Oresme: tradition et innovation chez un intellectuel du XIVe siècle*. Paris: Belles Lettres, 1988

Spengler, Joseph J., "Aristotle on Economic Imputation and Related Matters," *Southern Economic Journal* 21 (1955), 371–89

Origins of Economic Thought and Justice. London: Feffer and Simons, 1980

Spicciani, Amleto, *La mercatura e la formazione del prezzo nella riflessione teologica medioevale*. Rome: Accademia nazionale dei Lincei, 1977

Spufford, Peter, "Coinage and Currency," in *The Cambridge Economic History of Europe*, vol. III, 593–95. Cambridge University Press, 1965

Money and Its Use in Medieval Europe. Cambridge University Press, 1988

"Le rôle de la monnaie dans la révolution commerciale du XIIIe siècle," in John Day (ed.), *Etudes d'histoire monétaire*, 355–95. Presses universitaires de Lille, 1984

Stock, Brian, *The Implications of Literacy: Written Language and Models of Interpretation in the Eleventh and Twelfth Centuries*. Princeton University Press, 1983

"Science, Technology, and Economic Progress in the Early Middle Ages," in Lindberg, *Science in the Middle Ages*, 1–51

Strayer, Joseph R., "The Costs and Profits of War," in Harry Miskimin, David Herlihy, and A. L. Udovitch (eds.), *The Medieval City*, 269–91. New Haven: Yale University Press, 1977

The Reign of Philip the Fair. Princeton University Press, 1980

Swanson, R. N., "Learning and Livings: University Study and Clerical Careers in Later Medieval England," *History of Universities* 6 (1987), 81–103

Swetz, Frank J., *Capitalism and Arithmetic: The New Math of the Fifteenth Century*. La Salle, Ill.: Open Court, 1987

Sylla, Edith Dudley, "Autonomous and Handmaiden Science: St. Thomas Aquinas and William of Ockham on the Physics of the Eucharist," in Murdoch and Sylla, *Cultural Context*, 349–91

"Godfrey of Fontaines on Motion with Respect to Quantity of the Eucharist," in A. Maierù (ed.), *Studi sul XIV secolo in memoria di Anneliese Maier*, 105–41. Rome: Edizioni di storia e letteratura, 1981

"Medieval Concepts of the Latitude of Forms: The Oxford Calculators," *Archives d'histoire doctrinale et littéraire du moyen âge* 40 (1973), 223–83

"Medieval Quantifications of Qualities: The 'Merton School,'" *Archive for History of Exact Sciences* 8 (1971), 7–39

"The Oxford Calculators," in Kretzmann, Kenny, and Pinborg, *Cambridge History of Later Medieval Philosophy*, 541–63

The Oxford Calculators and the Mathematics of Motion 1320–1350: Physics and Measurement by Latitudes, Ph.D. dissertation, Harvard University, 1970. Published New York: Garland Press, 1991

Bibliography

"The Oxford Calculators in Context," *Science in Context* 1 (1987), 257–79

Tachau, Katherine, *Vision and Certitude in the Age of Ockham: Optics, Epistemology, and the Foundation of Semantics 1250–1345*. Studien und Texte zur Geistesgeschichte des Mittelalters, 22. Leiden: E. J. Brill, 1988

Thijssen, J. M., "Buridan on Mathematics," *Vivarium* 23 (1985), 55–78

Thomson, S. Harrison, "Walter Burley's Commentary on the *Politics* of Aristotle," in *Mélanges Auguste Pelzer*, 557–78. Recueil de travaux d'histoire et de philologie, ser. 6, 26. Louvain: Bibliothèque de l'Université, 1947

Thorold Rogers, James E., *A History of Agriculture and Prices in England*, 7 vols., vol. II. Oxford, 1886

Thrupp, Sylvia L., *The Merchant Class of Medieval London 1200–1500*. University of Chicago Press, 1948

Todeschini, Giacomo, "Oeconomica Franciscana II: Pietro di Giovanni Olivi come fonte per la storia dell'etica-economica medievale," *Rivista di storia e letteratura religiosa* 13 (1977), 461–94

Unguru, Sabatai, "Mathematics and Experiment in Witelo's *Perspectiva*," in Grant and Murdoch, *Mathematics and Its Applications*, 269–97

Van Egmond, Warren, *The Commercial Revolution and the Beginnings of Western Mathematics in Renaissance Florence, 1300–1500*. Ph.D. dissertation, Indiana University. Ann Arbor: University Microfilm International, 1976

Vance, Eugene, "Chrétien's *Yvain* and the Ideologies of Change and Exchange," *Yale French Studies* 70 (1981), 42–62

Vaughan, Richard, *Matthew Paris*. Cambridge University Press, 1958

Verger, Jacques, "Le coût des grades: droits et frais d'examen dans les universités du Midi de la France au moyen âge," in Astrik L. Gabriel (ed.), *The Economic and Material Frame of the Mediaeval University*, 19–36. Notre Dame University Press, 1977

Walsh, James J., "Buridan and Seneca," *Journal of the History of Ideas* 27 (1966), 23–41

"Is Buridan a Sceptic About Free Will?," *Vivarium* 2 (1964), 50–61

"Some Relationhips Between Gerald Odo's and John Buridan's Commentaries on Aristotle's 'Ethics,'" *Franciscan Studies* 35 (1975), 237–75

Wéber, Edouard-Henri, "Commensuratio de l'agir par l'objet d'activité et par le sujet agent chez Albert le Grand, Thomas d'Aquin, et Maître Eckhart," in *Mensura: Mass, Zahl, Zahlensymbolik im Mittelalter*, 43–64. Miscellanea mediaevalia, 16. Berlin: Walter de Gruyter, 1983

Weijers, Olga, *Terminologie des universités au XIIIe siècle*. Lessico intelettuale europeo, 39. Rome: Aeteneo, 1987

Weisheipl, James A., "Albert's Works on Natural Science in Probable Chronological Order," in Weisheipl, *Albertus Magnus and the Sciences*, Appendix I, 565–77

"Albertus Magnus and the Oxford Platonists," *Proceedings of the American Catholic Philosophical Association* 32 (1958), 124–39

"Developments in the Arts Curriculum at Oxford in the Early Fourteenth Century," *Mediaeval Studies* 28 (1966), 151–75

Bibliography

"Ockham and Some Mertonians," *Mediaeval Studies* 30 (1968), 163–213

"The Place of John Dumbleton in the Merton School," *Isis* 50 (1959), 439–54

"*Repertorium Mertonense*," *Mediaeval Studies* 31 (1969), 174–224

Weisheipl, James A. (ed.), *Albertus Magnus and the Sciences: Commemorative Essays*. Pontifical Institute of Medieval Studies, Studies, and Texts, 49. Toronto, 1980

Wilson, Curtis, *William Heytesbury: Medieval Logic and the Rise of Mathematical Physics*. Madison: University of Wisconsin Press, 1956

Wippel, John F., *The Metaphysical Thought of Godfrey of Fontaines: A Study in Late Thirteenth-Century Philosophy*. Washington, D. C.: Catholic University of America Press, 1981

Wolff, Michael, *Geschichte der Impetustheorie: Untersuchungen zum Ursprung der klassischen Mechanik*. Frankfurt a/M: Suhrkamp, 1978

"Mehrwert und Impetus bei Petrus Johannis Olivi," in Jürgen Miethke and Klaus Schreiner (eds.), *Sozialer Wandel im Mittelalter: Wahrnehmungsformen, Erklärungsmuster, Regelungsmechanismen*, 413–23. Sigmaringen: Jan Thorbecke Verlag, 1994

Wolff, Philippe, *Commerces et marchands de Toulouse (vers 1350–vers 1450)*. Paris: Librairie Plon, 1954

Wulf, Maurice de, *Philosophy and Civilization in the Middle Ages*. Princeton University Press, 1922

Un théologian-philosophe du XIIIe siècle: étude sur la vie, les oeuvres, et l'influence de Godefroid de Fontaines. Brussels: M. Hayez, 1904

Yunck, John A., *The Lineage of Lady Meed: The Development of Mediaeval Venality Satire*. University of Notre Dame Press, 1963

INDEX

abacus (counting table), 190, 221

Accursius, 92, 93, 113

aestimatio communis, in price determination, 77, 92–97, 99, 111–13, 153, 200, 229–31

agreement, as basis for just exchange, 112, 113, 118, 122, 123, 125, 129, 131, 135, 159

Albert of Saxony, 135 n. 74

Albertus Magnus, 10, 13 n. 30, 47, 56, 127
  and geometry of exchange, 71–72, 142, 158 n. 190, 200
  commentary on Aristotle's *Ethics*, Book v, 56–78
  interface between economic and scientific thought, 57–58, 59, 64
  on *aestimatio communis*, 96, 152, 153, 229
  on just price, 65, 96, 153
  on monetized exchange as binder of *civitas*, 71, 72

Alexander III (Pope), 82

Ambrose (Saint), on usury, 81, 82

approximation,
  as basis of just exchange, 99, 100, 102, 114–15, 121, 124, 129, 183, 184, 244
  in natural philosophy, 100, 121, 183, 184, 244

Aquinas, Thomas (Saint), 10, 13 n. 29, 56, 85–101, 104, 110, 127, 152
  acceptance of approximation, 99, 100, 121, 183, 184, 186
  commentary on Aristotle's *Ethics*, Book v, 56–78
  interface between economic and scientific thought, 99–101, 183–84
  linking mind and order in exchange, 106, 114
  linking mind and order in nature, 13, 98, 114
  on equality in exchange, 96–100
  on estimated range of just price, 99, 100, 183
  on geometry of exchange, 66, 76, 97, 200

on just price, 95–99

on money as *instrumentum*, 137

on probability, 85, 86, 119–21

on quantifying qualities, 176, 177, 183, 186

on relativity of price, 70, 97, 233

on usury, 85–87

Aristotle, 10, 49, 116, 143, 149, 156, 164, 183, 186, 225–28, 236, 237, 240, 241
  attitudes toward money, 86, 87
  definition of money's use, 46, 103, 104
  Ethics, Book v, commentaries on, 56–78, 127ff.
  geometric model of exchange, 40–44, 128, 142, 145, 152, 200
  interface between economic and scientific thought, 37, 39, 44, 45
  model of money in exchange, 37–55, 71, 101, 108, 230
  on exchange as binder of *civitas*, 50–52, 72 n. 70, 142, 158
  on the common good, 155, 230
  on the continuum, 175, 176, 179
  on usury, 80, 86, 87
  position against quantifying qualities, 175–77, 186, 191

arithmetical equality in exchange, 60, 108, 113

arithmetical model,
  of equalization in exchange, defined, 94, 95
  replaced by geometrical model in exchange, 117, 121, 122, 132, 136, 144, 153, 159, 200, 201, 211, 212, 219, 220, 222, 229, 244
  replaced by geometrical model in natural philosophy, 131, 137, 144–46, 161, 215–20, 223, 229, 245

arithmetical worldview, in natural philosophy, 213–19

Augustine (Saint), 102, 104, 122, 147, 148
  on relativity of price, 70, 97, 233
  on usury, 81

Index

Index

Cambridge Studies in Medieval Life and Thought
Fourth series

Titles in series

★*Also published as a paperback*

AAO- 9916